142
Advances in Polymer Science

Editorial Board:
A. Abe · A.-C. Albertsson · H.-J. Cantow · K. Dušek
S. Edwards · H. Höcker · J. F. Joanny · H.-H. Kausch
T. Kobayashi · K.-S. Lee · J. E. McGrath
L. Monnerie · S. I. Stupp · U. W. Suter
E. L. Thomas · G. Wegner · R. J. Young

Springer
*Berlin
Heidelberg
New York
Barcelona
Hong Kong
London
Milan
Paris
Singapore
Tokyo*

Branched Polymers I

Volume Editor: J. Roovers

With contributions by
B. Charleux, B. Comanita, R. Faust,
N. Hadjichristidis, H. Iatrou, K. Ito, S. Kawaguchi,
S. Pispas, M. Pitsikalis, J. Roovers, C. Vlahos

 Springer

This series presents critical reviews of the present and future trends in polymer and biopolymer science including chemistry, physical chemistry, physics and materials science. It is addressed to all scientists at universities and in industry who wish to keep abreast of advances in the topics covered.

As a rule, contributions are specially commissioned. The editors and publishers will, however, always be pleased to receive suggestions and supplementary information. Papers are accepted for „Advances in Polymer Science" in English.

In references Advances in Polymer Science is abbreviated Adv. Polym. Sci. and is cited as a journal.

Springer WWW home page: http://www.springer.de

ISSN 0065-3195
ISBN 3-540-64923-9
Springer-Verlag Berlin Heidelberg New York

Library of Congress Catalog Card Number 61642

This work is subject to copyright. All rights are reserved, whether the whole or part of the material is concerned, specifically the rights of translation, reprinting, re-use of illustrations, recitation, broadcasting, reproduction on microfilms or in other ways, and storage in data banks. Duplication of this publication or parts thereof is only permitted under the provisions of the German Copyright Law of September 9, 1965, in its current version, and permission for use must always be obtained from Springer-Verlag. Violations are liable for prosecution under the German Copyright Law.

© Springer-Verlag Berlin Heidelberg 1999
Printed in Germany

The use of registered names, trademarks, etc. in this publication does not imply, even in the absence of a specific statement, that such names are exempt from the relevant protective laws and regulations and therefore free for general use.

Typesetting: Data conversion by MEDIO, Berlin
Cover: E. Kirchner, Heidelberg
SPIN: 10648274   02/3020 - 5 4 3 2 1 0 – Printed on acid-free paper

Volume Editor

Dr. Jacques Roovers
Institute for Chemical Process
and Environmental Technology
National Research Council Canada
Ottawa, Ontario
K1A 0R6
CANADA
E-mail: jacques.roovers@nrc.ca

Editorial Board

Prof. Akihiro Abe
Department of Industrial Chemistry
Tokyo Institute of Polytechnics
1583 Iiyama, Atsugi-shi 243-02, Japan
E-mail: aabe@chem.t-kougei.ac.jp

Prof. Ann-Christine Albertsson
Department of Polymer Technology
The Royal Institute of Technolgy
S-10044 Stockholm, Sweden
E-mail: aila@polymer.kth.se

Prof. Hans-Joachim Cantow
Freiburger Materialforschungszentrum
Stefan Meier-Str. 21
D-79104 Freiburg i. Br., FRG
E-mail: cantow@fmf.uni-freiburg.de

Prof. Karel Dušek
Institute of Macromolecular Chemistry, Czech
Academy of Sciences of the Czech Republic
Heyrovský Sq. 2
16206 Prague 6, Czech Republic
E-mail: office@imc.cas.cz

Prof. Sam Edwards
Department of Physics
Cavendish Laboratory
University of Cambridge
Madingley Road
Cambridge CB3 OHE, UK
E-mail: sfe11@phy.cam.ac.uk

Prof. Hartwig Höcker
Lehrstuhl für Textilchemie
und Makromolekulare Chemie
RWTH Aachen
Veltmanplatz 8
D-52062 Aachen, FRG
E-mail: 100732.1557@compuserve.com

Prof. Jean-François Joanny
Institute Charles Sadron
6, rue Boussingault
F-67083 Strasbourg Cedex, France
E-mail: joanny@europe.u-strasbg.fr

Prof. Hans-Henning Kausch
Laboratoire de Polymères
École Polytechnique Fédérale
de Lausanne, MX-D Ecublens
CH-1015 Lausanne, Switzerland
E-mail: hans-henning.kausch@lp.dmx.epfl.ch

Prof. Takashi Kobayashi
Institute for Chemical Research
Kyoto University
Uji, Kyoto 611, Japan
E-mail: kobayash@eels.kuicr.kyoto-u.ac.jp

Prof. Kwang-Sup Lee
Department of Macromolecular Science
Hannam University
Teajon 300-791, Korea
E-mail: kslee@eve.hannam.ac.kr

Prof. James E. McGrath
Polymer Materials and Interfaces Laboratories
Virginia Polytechnic and State University
2111 Hahn Hall
Blacksbourg
Virginia 24061-0344, USA
E-mail: jmcgrath@chemserver.chem.vt.edu

Prof. Lucien Monnerie
École Supérieure de Physique et de Chimie
Industrielles
Laboratoire de Physico-Chimie
Structurale et Macromoléculaire
10, rue Vauquelin
75231 Paris Cedex 05, France
E-mail: lucien.monnerie@espci.fr

Prof. Samuel I. Stupp
Department of Materials Science
and Engineering
University of Illinois at Urbana-Champaign
1304 West Green Street
Urbana, IL 61801, USA
E-mail: s-stupp@uiuc.edu

Prof. Ulrich W. Suter
Department of Materials
Institute of Polymers
ETZ, CNB E92
CH-8092 Zürich, Switzerland
E-mail: suter@ifp.mat.ethz.ch

Prof. Edwin L. Thomas
Room 13-5094
Materials Science and Engineering
Massachusetts Institute of Technology
Cambridge, MA 02139, USA
E-mail. thomas@uzi.mit.edu

Prof. Gerhard Wegner
Max-Planck-Institut für Polymerforschung
Ackermannweg 10
Postfach 3148
D-55128 Mainz, FRG
E-mail: wegner@mpip-mainz.mpg.de

Prof. Robert J. Young
Manchester Materials Science Centre
University of Manchester and UMIST
Grosvenor Street
Manchester M1 7HS, UK
E-mail: robert.young@umist.ac.uk

Preface

While books have been written on many topics of Polymer Science, no comprehensive treatise on long-chain branching has ever been composed. This series of reviews in Volume 142 and 143 of Advances in Polymer Science tries to fill this gap by highlighting active areas of research on branched polymers.

Long-chain branching is a phenomenon observed in synthetic polymers and in some natural polysaccharides. It has long been recognized as a major molecular parameter of macromolecules. Its presence was first surmised by H. Staudinger and G. V. Schulz (Ber. 68, 2320, 1935). Interestingly, their method of identification by means of the abnormal relation between intrinsic viscosity and molecular weight has survived to this day. Indeed, the most sophisticated method for analysis of long-chain branching uses size exclusion fractionation with the simultaneous recording of mass, molecular weight and intrinsic viscosity of the fractions.

In the 1940s and 1950s, random branching in polymers and its effect on their properties was studied by Stockmayer, Flory, Zimm and many others. Their work remains a milestone on the subject to this day. Flory dedicated several chapters of his "Principles of Polymer Chemistry" to non-linear polymers. Especially important at that time was the view that randomly branched polymers are intermediates to polymeric networks. Further developments in randomly branched polymers came from the introduction of percolation theory. The modern aspects of this topic are elaborated here in the chapter by W. Burchard.

As polymer science developed, greater control over the architecture of polymer molecules was obtained. In polyolefins synthesis, this was due to the introduction of new catalysts. The development of anionic living polymerization with the concomitant formation of narrow molecular weight distribution polymers and an highly reactive functional end group opened the route not only to block copolymers but also to branched polymers with highly controlled architectures such as stars, combs and graft copolymers. The model polymers allowed us to establish relations between the molecular architecture and the physical properties of the branched polymers. This development has been reviewed by e.g. G.S. Grest et al. Adv. Chem. Phys. 94, 65 (1996).

One chapter in this series deals with the newer use of cationic polymerization to form polymers and copolymers with controlled long-chain branched structures. Another chapter deals with the use of anionic polymerization to prepare asymmetric star polymers. The asymmetry is introduced when the arms of the polymer differ in molecular weight, chemical composition or in their topological placement. The synthesis of these polymers has led to new insights in microseparation processes of block copolymers. Anionic and cationic living polymerization has also led to macromonomers. Highlights of recent developments in poly(macromonomers) homo, comb and graft copolymers are reviewed by K. Ito. The poly(macromonomers) with their multiple densely packed small linear subchains often lead to monomolecular micelles.

Very recently, highly regular, highly controlled, dense branching has been developed. The resulting "dendrimers" often have a spherical shape with special interior and surface properties. The synthesis and properties of dendrimers has been reviewed (see e.g. G.R. Newkome et al. "Dendritic Molecules", VCH, 1996). In this series, a chapter deals with the molecular dimensions of dendrimers and with dendrimer-polymer hybrids. One possible development of such materials may be in the fields of biochemistry and biomaterials. The less perfect "hyperbranched polymers" synthesized from A_2B-type monomers offer a real hope for large scale commercialization. A review of the present status of research on hyperbranched polymers is included.

The link between the long-chain branch structure and the properties of the polymer has to be established experimentally by means of model branched polymers. This link can also be derived theoretically or through computer modeling. As a result, a large sub-field of study has emerged. The methods and results of this theoretical work are systematically reviewed by J. Freire. Where available, comparisons with experimental results are made.

The final chapter develops the most modern insights in the relation between the rheological properties and the large scale architecture of polymers. Indeed, the largest effects of branching are encountered in their melt relaxation properties. In the absence of reptation, which dominates relaxation processes in linear polymers, a rich variety of other relaxation processes becomes apparent. The control of the melt properties of polymers by means of their long-chain branch architecture will continue to lead to new industrial applications.

Ottawa, July 1998   J. Roovers

Contents

Synthesis of Branched Polymers by Cationic Polymerization
B. Charleux, R. Faust 1

Asymmetric Star Polymers: Synthesis and Properties
N. Hadjichristidis, S. Pispas, M. Pitsikalis, H. Iatrou, C. Vlahos 71

Poly(macromonomers): Homo- and Copolymerization
K. Ito, S. Kawaguchi 129

Dendrimers and Dendrimer-Polymer Hybrids
J. Roovers, B. Comanita 179

Author Index Volumes 101–142 229

Subject Index 239

Contents of Volume 143

Branched Polymers II

Volume Editor: J. Roovers

Hyperbranched Polymers
A. Hult, M. Johansson, E. Malmström

Conformational Properties of Branched Polymers: Theory and Simulations
J. J. Freire

Solution Properties of Branched Macromolecules
W. Burchard

Entangled Dynamics and Melt Flow of Branched Polymers
T. C. B. McLeish, S. T. Milner

Synthesis of Branched Polymers by Cationic Polymerization

Bernadette Charleux[1], Rudolf Faust[2]

1 Laboratoire de chimie macromoléculaire, Université Pierre et Marie Curie, T44, E1, 4, Place Jussieu F-75252 Paris cedex 05, France E-mail: charleux@ccr.jussieu.fr
2 University of Massachusetts, Lowell Chemistry Department, 1 University Ave. Lowell, MA 01854, USA E-mail: Rudolf_Faust@uml.edu

The synthesis of branched polymers by cationic polymerization of vinyl monomers is reviewed. This includes star, graft, and hyperbranched (co)polymers. The description is essentially focused on the synthetic approach and characterization results are provided as a proof of the structure. When available, specific properties of the materials are also given.

Keywords. Cationic polymerization, Living cationic polymerization, Branched polymers, Star polymers, Graft polymers, Hyperbranched polymers, Microgel core, (Multi)functional initiator, (Multi)functional coupling agent, Grafting from, Grafting onto, Macromonomer

List of Symbols and Abbreviations		3
1	**Introduction**	4
2	**Multi-Arm Star (co)Polymers**	4
2.1	Synthesis Using a Difunctional Monomer as a Linker (Cross-Linked Core)	5
2.1.1	A_n-Type Star Homopolymers	6
2.1.1.1	Poly(vinyl ethers)$_n$	6
2.1.1.2	Poly(alkoxystyrenes)$_n$	9
2.1.1.3	Poly(isobutylene)$_n$	10
2.1.2	$(AB)_n$-Type Star Block Copolymers	13
2.1.2.1	Poly(vinyl ether-b-vinyl ether)$_n$	13
2.1.2.2	Poly(isobutylene-b-styrene)$_n$	14
2.1.3	A_n-Type Star Polymers with a Functionalized Core: Poly(vinyl ether)$_n$	15
2.1.4	A_nB_n-Type Star Copolymers: Poly(vinyl ether)$_n$-Star-Poly(vinyl ether)$_n$	15
2.2	Synthesis Using a Multifunctional Initiator	17
2.2.1	A_n-Type Star Homopolymers	17
2.2.1.1	Poly(vinyl ethers)$_n$	17
2.2.1.2	Poly(p-methoxystyrene)$_n$	19
2.2.1.3	Poly(styrene)$_n$	20
2.2.1.4	Poly(isobutylene)$_n$	21
2.2.2	$(AB)_n$-Type Star Block Copolymers	25
2.2.2.1	Poly(vinyl ether-b-vinyl ether)$_n$	25
2.2.2.2	Poly(isobutylene-b-styrene)$_n$	28

Advances in Polymer Science, Vol.142
© Springer-Verlag Berlin Heidelberg 1999

2.2.2.3	Poly(isobutylene-*b*-*p*-methylstyrene)$_n$	29
2.2.2.4	Poly(isobutylene-*b*-THF)$_n$	29
2.2.2.5	Poly(isobutylene-*b*-methyl methacrylate)$_n$	29
2.3	Synthesis Using a Multifunctional Coupling Agent	30
2.3.1	A$_n$-Type Star Homopolymers	31
2.3.1.1	Poly(vinyl ethers)$_n$	31
2.3.1.2	Poly(isobutylene)$_n$	34
2.3.2	(AB)$_n$-Type Star Block Copolymers	38
2.3.2.1	Poly(vinyl ether-*b*-vinyl ether)$_n$	38
2.3.2.2	Poly(α-methylstyrene-*b*-2-hydroxyethyl vinyl ether)$_n$	38
2.3.3	A$_n$B$_m$-Type Star Copolymers	39
2.3.3.1	Poly(isobutylene)$_2$-Star-Poly(methyl vinyl ether)$_2$	39
2.3.3.2	Poly(isobutylene)-Star-Poly(ethylene oxide)$_m$	40
3	**Graft (co)Polymers**	**41**
3.1	"Grafting From" Technique	41
3.1.1	Synthesis of the Backbone by Cationic Polymerization	41
3.1.1.1	Poly(vinyl ether) Backbone	41
3.1.1.2	Poly(isobutylene) Backbone	41
3.1.2	Synthesis of the Branches by Cationic Polymerization	42
3.1.2.1	Poly(vinyl ether) Branches	42
3.1.2.2	Poly(silyl vinyl ether) Branches	43
3.1.2.3	Poly(isobutylene) Branches	43
3.1.2.4	Poly(styrene) and poly(α-methylstyrene) Branches	44
3.2	"Grafting Onto" Technique	44
3.2.1	Synthesis of the Backbone by Cationic Polymerization	45
3.2.1.1	Poly(vinyl ether) Backbone	45
3.2.1.2	Poly(*p*-bromomethylstyrene-IB-*p*-bromomethylstyrene) Triblock Copolymer Backbone	45
3.2.2	Synthesis of the Branches by Cationic Polymerization: Poly(styrene) Branches	47
3.3	Macromonomers	48
3.3.1	Synthesis of Macromonomers by Living Cationic Polymerization	48
3.3.1.1	Synthesis Using a Functional Initiator	48
3.3.1.2	Synthesis Using a Functional Capping Agent	53
3.3.1.3	Chain End Modification of Poly(isobutylene)	57
3.3.2	Cationic Polymerization of Macromonomers	64
3.3.2.1	Vinyl Ether Polymerizable Group	64
4	**Hyperbranched Polymers**	**65**
5	**Conclusion**	**67**
6	**References**	**67**

List of Symbols and Abbreviations

α-MeS	α-methylstyrene
AcOVE	2-acetoxyethyl vinyl ether
AIBN	azobisisobutyronitrile
ATMS	allyltrimethylsilane
BDTEP	2,2-bis[4-(1-tolylethenyl)phenyl]propane
BMS	bromomethylstyrene
BVE	n-butyl vinyl ether
BzOVE	2-(benzoyloxy)ethyl vinyl ether
CA-PIB	poly(isobutenyl) α-cyanoacrylate
CEVE	2-chloroethyl vinyl ether
CMS	chloromethylstyrene
DIPB	diisopropenylbenzene
DRI	differential refractive index
DP_n	number-average degree of polymerization
DTBP	2,6-di-*tert*-butylpyridine
DVB	divinylbenzene
EO	ethylene oxide
EPDM	ethylene-propylene-diene monomers
EVE	ethyl vinyl ether
f	average number of arms
F_n	number-average end functionality
HEMA	2-hydroxyethyl methacrylate
HOVE	2-hydroxyethyl vinyl ether
IB	isobutylene
IBVE	isobutyl vinyl ether
I_{eff}	initiator efficiency
LCP	living cationic polymerization
MA-PIB	poly(isobutenyl) methacrylate
MeVE	methyl vinyl ether
MMA	methyl methacrylate
M_n	number-average molecular weight
M_v	viscosity-average molecular weight
M_w	weight-average molecular weight
MW	molecular weight
MWD	molecular weight distribution
ODVE	octadecyl vinyl ether
p-MeS	p-methylstyrene
PEO	poly(ethylene oxide)
PIB	poly(isobutylene)
PMMA	poly(methyl methacrylate)
p-MOS	p-methoxystyrene
PS	poly(styrene)
S	styrene
SEC	size exclusion chromatography

SiVE   2-[(*tert*-butyldimethylsilyl)oxy]ethyl vinyl ether
S-PIB   *p*-poly(isobutylene)styrene
*t*BOS *p*-tert-butoxystyrene
THF   tetrahydrofuran
TMPCl   2-chloro-2,4,4-trimethylpentane
VOEM   diethyl 2-(vinyloxy)ethyl malonate
[ ]   molar concentration

1
Introduction

The discovery of living cationic polymerization in the mid-1980s provided a valuable new tool in the synthesis of well-defined macromolecules with controlled molecular weight, narrow molecular weight distribution, and high degree of compositional homogeneity. While linear propagation was the main focus of research in the early years of discovery, recently non-linear polymer architectures such as star, branched, and hyperbranched polymers have gained interest due to their interesting and sometimes unexpected properties opening new areas of applications. This review is intended to cover new developments in branched polymers via cationic polymerization of vinyl monomers. Living cationic ring-opening polymerization (ROP) is outside the scope of this review and therefore only those articles are referred to which make use of cationic vinyl polymerization in addition of ROP. Due to space limitations, a review of monomers that undergo living/controlled cationic polymerization, initiating systems, and general experimental condition is not provided. The reader is referred to two excellent books on the subject matter [1, 2]. This review is intended to be comprehensive, and therefore the literature was thoroughly searched using different key words in September 1997. It is nevertheless conceivable that inadvertently some publications were missed. For this we apologize. Publications that appeared after this date may not have been reviewed.

2
Multi-Arm Star (co)Polymers

Multiarm star (co)polymers can be defined as branched (co)polymers in which three or more either similar or different linear homopolymers or copolymers are linked together to a single core. The nomenclature that will be used follows the usual convention:
- A_n-type star corresponds to a star with n similar homopolymer branches (n>2)
- $(AB)_n$-type star corresponds to a star with n similar AB block copolymer branches
- A_nB_m-type star corresponds to a star with n branches of the homopolymer A and m branches of the homopolymer B

– ABC, ABCD ... etc -type star corresponds to a star with 3, 4 ... etc different branches

Depending on the target structure and on the availability of initiators and linkers, three main methods can be applied for the synthesis: core-first techniques, core-last techniques, and mixed techniques.

In the first case, the arms are grown together from a single core which can be either a microgel with an average number of potentially active sites or a well-defined multifunctional initiator. However, to our knowledge, although there is no specific limitation, cationic polymerization involving a microgel multifunctional initiator has not been reported. Functionalization of the free end of the branches can also be performed by quenching with a functional terminator.

In the second case, first the arms are synthesized separately and then linked together using either a well-defined multifunctional terminator or a difunctional monomer leading to a cross-linked core. The free end of the branches may contain functional groups by using a functional initiator for the preparation of the arms.

Both techniques generally lead to A_n or $(AB)_n$ stars with branches of identical nature and similar composition and length.

Although in anionic polymerization sequential coupling reactions with methyl trichlorosilane or tetrachlorosilane have been used to obtain ABC or ABCD heteroarm stars with three or four different branches respectively, such technique is not available in cationic polymerization due to the lack of suitable coupling agents. To prepare stars with different branches, most methods employ mixed techniques. The first one is derived from the microgel core method applied in three sequential steps: first stage polymerization to give a linear (co)polymer, linking via a divinyl monomer, second stage polymerization initiated by the active sites incorporated in the microgel core. The second method is based on the use of a living coupling agent which is a non-homopolymerizable multivinylic monomer. Upon addition of the living arms to the double bonds, new active species arise that can be used to initiate a second stage polymerization leading to new branches. To date, only one example could be found using living cationic polymerization.

2.1
Synthesis Using a Difunctional Monomer as a Linker (Cross-Linked Core)

In cationic polymerization, this technique has been used only as a core-last technique. It is based on the ability of a linear living polymer chain to act as a macroinitiator for a second monomer. When the second monomer is a divinyl compound, pendant vinyl groups are incorporated in the second block leading to cross-linking reactions which may occur during and after formation of the second block. These reactions provide multi-branched structures where the arms are linked together to a compact microgel core of the divinyl second monomer. This method is particularly suited to prepare stars with many arms. The average number of arms per molecule is a function of several experimental and structural parameters which will be discussed below. With this technique, A_n-, $(AB)_n$-, and A_nB_n-type star polymers could be synthesized.

2.1.1
A_n-Type Star Homopolymers

2.1.1.1
Poly(vinyl ethers)$_n$

The first synthesis of star polymers with a microgel core was reported by Sawamoto et al. for poly(isobutyl vinyl ether) (poly(IBVE)) [3, 4]. In the first step, living cationic polymerization of IBVE was carried out with the HI/ZnI$_2$ initiating system in toluene at −40 °C. Subsequent coupling of the living ends was performed with the various divinyl ethers **1–4**.

$$\text{H}_2\text{C}=\text{CH}-\text{O}-\text{CH}_2\text{CH}_2-\text{O}-\text{C}_6\text{H}_4-\text{C}(\text{CH}_3)_2-\text{C}_6\text{H}_4-\text{O}-\text{CH}_2\text{CH}_2-\text{O}-\text{CH}=\text{CH}_2 \quad (1)$$

$$\text{H}_2\text{C}=\text{CH}-\text{O}-\text{CH}_2\text{CH}_2-\text{O}-\text{C}_6\text{H}_4-\text{O}-\text{CH}_2\text{CH}_2-\text{O}-\text{CH}=\text{CH}_2 \quad (2)$$

$$\text{H}_2\text{C}=\text{CH}-\text{O}-\text{CH}_2\text{CH}_2-\text{O}-\text{CH}_2\text{CH}_2-\text{O}-\text{CH}=\text{CH}_2 \quad (3)$$

$$\text{H}_2\text{C}=\text{CH}-\text{O}-\text{CH}_2\text{CH}_2-\text{O}-\text{CH}=\text{CH}_2 \quad (4)$$

Typically, the coupling reaction was carried out at the end of the first stage polymerization after complete conversion of the monomer, under the same experimental conditions. For example, a living poly(IBVE) with DP_n=38 and narrow MWD was allowed to react with the divinyl ether **1** at r=[**1**]/[living ends]=5 with [living ends]=8.3 mmol l^{-1}. The extent of coupling was followed by SEC of samples withdrawn at various reaction times (Fig. 1) and ^{1}H NMR analysis of the product was used to provide structural information. The coupling agent was progressively consumed and simultaneously the SEC peak of the linear polymer shifted towards slightly lower elution volume (higher MW). This intermediate product strongly absorbed in the UV range at 256 nm, and was ascribed to a block copolymer of IBVE and **1** with only one reacted double bond per divinyl monomer (block copolymer with pendant vinyl functions, see Scheme 1). A still

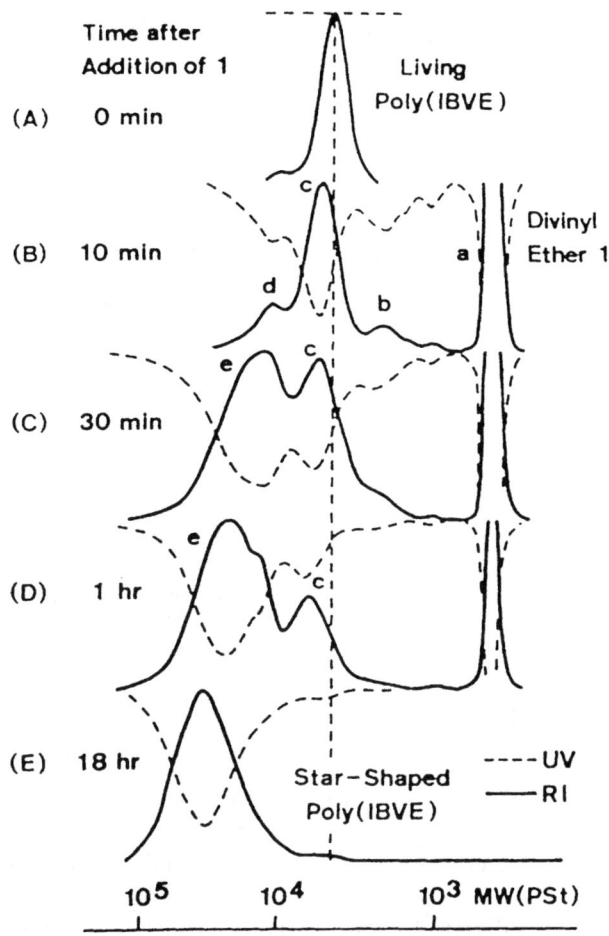

Fig. 1A–E. MWD of the products obtained from the reaction of living poly(IBVE) with divinyl ether **1** in toluene at −40 °C: $DP_{arm}=38$, [living ends]=8.3 mmol l^{-1}, r=5.0: **A** living poly(IBVE): $[IBVE]_0=0.38$ mol l^{-1}, $[HI]_0=10$ mmol l^{-1}, $[ZnI_2]_0=0.2$ mmol l^{-1}, IBVE conversion=100% in 45 min; **B–E** the products recovered after the reaction with **1**. Reaction time after addition of **1**: (B) 10 min, (C) 30 min, (D) 1 h, (E) 18 h [star-shaped poly(IBVE)]. Reprinted with permission from [3]. Copyright 1991 Am. Chem. Soc.

higher MW peak appeared indicating the simultaneous formation of star polymers. Some low MW byproducts, assigned to homopolymer of **1** were also observed. These progressively disappeared from the SEC chromatograms due to their ability to react with the intermediate products of the reaction. As **1** was consumed, the proportion of the intermediate product (block copolymer of IBVE and **1**) slowly decreased while the highest MW peak intensity increased and its position shifted towards higher MW. After 18 h, the coupling agent was

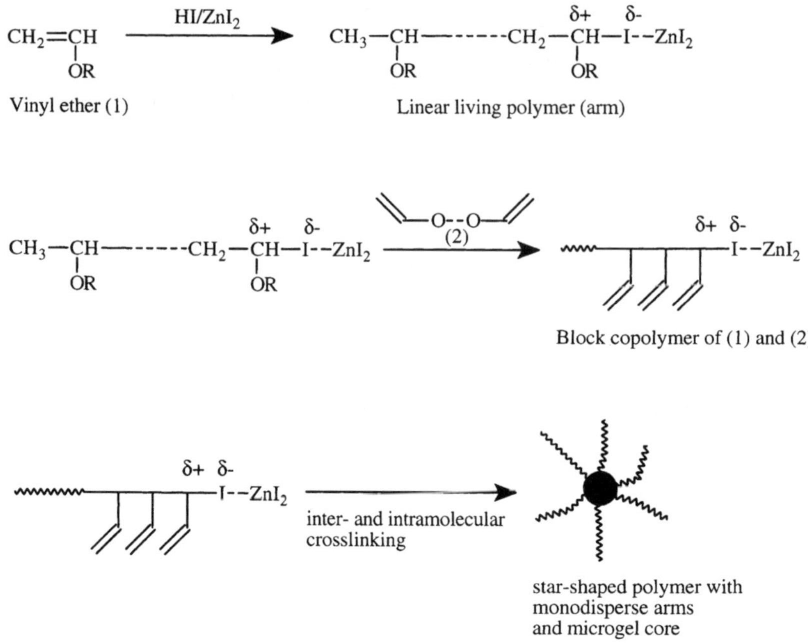

Scheme 1

completely consumed and the SEC showed a main high MW peak of relatively narrow MWD (M_w/M_n=1.35) together with the still remaining lower MW intermediate block copolymer of IBVE and **1**. The yield of the star polymer was not determined.

Based on the ^{1}H NMR spectra, the main product was a star poly(IBVE) where the protons of poly(IBVE) could be recognized together with those of the divinyl monomer in which vinylic protons had completely disappeared. The signals assignable to the aromatic protons of **1** broadened, which indicated more restricted motion supporting the formation of a microgel core. Furthermore, small-angle laser light scattering was used to determine the absolute MWs and allowed one to calculate therefrom the average number of arms. The M_w determined by light scattering was much higher than the corresponding value from SEC, providing additional evidence for the formation of a more compact structure than the linear counterparts. As a conclusion, experimental evidence supported the formation of star poly(IBVE) with monodisperse arms connected to a single cross-linked core. A variety of star polymers were prepared where, depending on experimental conditions, the average number of arms ranged from 3 to 59 and overall M_w from 20,000 to 400,000 g mol^{-1}.

The effect of reaction conditions on the yield, overall molecular weight (MW) and structure of the final polymer was investigated. The studied parameters

were: the length of the arms (DP_n), the initial concentration of the linear precursor [poly(IBVE)], and the value of the molar ratio r=[divinyl compound]/[poly(IBVE)]. The major conclusions are the following:
- when r is increased, the yield of star polymer increases together with its MW and its average arm number; these last two points being correlated with an increase of the weight fraction of the core
- when [poly(IBVE)] is increased, the MW of the final product as well as the average number of arms increases (in the studied series, the star polymer yield was high and independent of [poly(IBVE)] because very favorable conditions were used, i.e., high value of r and short arm)
- when the length of the arms is short, the overall MW is lower but the star polymer yield as well as the number of arms is higher; this indicates that the intermolecular linking reaction of the intermediate block copolymer of IBVE and **1** is sterically less hindered for shorter chains.

In addition to the effect of the experimental conditions, the influence of the nature of the arms and of the divinyl compound was also studied. It was shown that bulkiness of the arms strongly influences the yield of star polymer; for instance, arms of poly(cetyl vinyl ether) were linked in very low yield as compared with poly(IBVE). The influence of the structure of the divinyl ether was investigated and appears to be of great importance. Coupling with **3** and **4** led to low yield of star polymer, while the efficiency of **1** and **2** was much higher. The explanation provided by the authors was that compact and flexible spacers between the two vinyl groups of **3** and **4** could lead to smaller cores where further reaction of incoming chains would be sterically hindered.

2.1.1.2
Poly(alkoxystyrenes)$_n$

Preparation of star polymers of *p*-methoxystyrene (*p*-MOS) and *p-tert*-butoxystyrene (*t*BOS) using two different bifunctional vinyl compounds **1** and **5** was reported by Deng et al. [5].

$$H_2C{=}CH{-}\bigcirc{-}OCH_2CH_2CH_2O{-}\bigcirc{-}CH{=}CH_2 \quad (5)$$

Living cationic polymerization of both styrenic monomers was carried out with the use of the HI/ZnI$_2$ initiating system in CH$_2$Cl$_2$ at –15 °C in the presence of tetra-*n*-butylammonium iodide. The obtained living polymers of *p*-MOS of various lengths were allowed to react with both divinyl monomers **1** and **5** with a ratio r=3 to 7. With **1** the yield of star polymer was very low and a large amount of poly(*p*-MOS) remained unreacted. This was ascribed to the much higher reactivity of the divinyl ether compared with the styrenic monomer. This led to a very fast second stage polymerization and the major part of the linear precursor remained unreacted. In contrast, with the styrenic divinyl compound 5, high yield and quantitative consumption of poly(p-MOS) and 5 were obtained. This result demonstrated that the nature of the divinyl compound is of major importance and that it should have a structure and reactivity similar to those of the living end of the linear polymer chain. Formation of the star polymer (yield >90%) was shown to follow the same pathway as previously described for poly(IBVE) in Scheme 1. NMR and SEC characterization of the final product corroborated the conclusion that star polymers were obtained with monodisperse linear arms linked to a central cross-linked core. The M_w determined by light scattering ranged from 50,000 to 600,000 g mol^{-1} and the average number of arms from 7 to 50 per molecule. The influence of experimental conditions on the stars characteristics were found to be similar to findings with vinyl ether monomers. One unexplained difference however was the near independence of the number of arms on the length of the linear poly(p-MOS) (especially for r=5) whereas for the poly(IBVE), a continuous decrease with increasing DP_n was observed.

Star polymers of poly(t-BOS) were also synthesized in high yield using the divinyl compound 5 indicating that the slight increase in bulkiness of the pendant groups of the linear polymer had little influence.

2.1.1.3
Poly(isobutylene)$_n$

The first synthesis of multiarm star polyisobutylene (PIB), with $DP_{n(arm)}$=116 and the average number of arms=56, was described by Marsalko et al. [6]. The procedure started with the "living" polymerization of IB by the 2-chloro-2,4,4-trimethylpentane (TMPCl)/TiCl$_4$ initiating system in CH$_2$Cl$_2$/hexane (50/50 v/v) at –40 °C in the presence of triethylamine. At ~95% IB conversion, divinylbenzene (DVB, 6, containing 20% ethyl vinylbenzene) was added to effect linking at r=[DVB]/[TMPCl]=10.

$$CH_2=CH-\underset{CH=CH_2}{\bigcirc} \quad (6)$$

The exact time of the addition of the linking agent is important. DVB addition at lower IB conversion led to undesirable ill-defined low MW products, whereas addition of DVB at 100% IB conversion may result in loss of livingness. Linking was relatively slow but efficient, and the final product after 96 h contained less than 4% unlinked PIB arms. Various other reactions such as intramolecular cyclization, star-star linking, etc., were reportedly also involved. The star structure was proven by determining the M_w by light scattering, then selectively destroying the aromatic core by trifluoroperacetic acid, and determining the MW of the surviving PIB arms. The effect of [DVB] was studied in a separate investigation using r=[DVB]/[PIB]=2.5, 5, 7.5, and 10 [7]. The rate of star formation increased with increasing r. Between 48 and 96 h the MW increased dramatically due to intermolecular star-star coupling, the extent of which was proportional to r. Due to star-star linking, the MWD of the final product after 96 h was broad (weight average number of arms=110, M_w/M_n=2.9). For star PIBs with longer arms (M_w=18,700–116,100 g mol^{-1}) the polymerization of IB and linking was performed at −80 °C. It was found that with increasing arm length the rate of star formation rapidly diminished, presumably due to the lower rate of star-star coupling. Based on the observation that star-star coupling is absent when the molecular weight of the arm is higher than M_w=18,700 g mol^{-1}, it was postulated that relatively large arms sterically hinder star-star coupling. This however may not be the only factor determining the presence or absence of star-star coupling since linking of the longer arms was carried out at lower temperature (−80 °C). The effect of the nature of the divinyl monomer was also studied; in contrast to **6**, diisopropenylbenzene (DIPB, **7**) was found to be inefficient.

$$CH_2=\underset{\underset{CH_3}{|}}{C}-\underset{}{\bigcirc}-\underset{\underset{CH_3}{|}}{C}=CH_2 \qquad\qquad (7)$$

The synthesis of multiarm star PIB has also been attempted by the TMP-Cl/TiCl$_4$ initiating system in CH$_3$Cl/hexane (40/60 v/v) at −80 °C in the presence of pyridine using **6** and **7** as the core forming monomers [8]. Similar to findings by Marsalko et al. [7], DIPB was found to be inferior in comparison with DVB, due to slow and incomplete star formation. With DVB the star polymer formed more rapidly and using a [DVB]/[TMPCl]=10 ratio to effect linking, star with $DP_{n(arm)}$=250 and the average number of arms=23 was obtained in 18.5 h with only 4% unlinked PIB arm. Importantly, star-star linking was found to be absent at −80 °C, and thus the product exhibited a relatively narrow MWD. The effect of the PIB arm length on the synthesis of multiarm star PIB was also investigated [9]. Similarly to results reported by Marsalko et al. [7, 10], it was found that with increasing arm MW (from 10,000 to 56,000 g mol^{-1}), dramatically increased linking times (from 24 to 568 h) were necessary to ensure high incorporation of the PIB arms into the star molecule. Simultaneously, the weight average number of arms decreased from 54 to 5 respectively. It was also found that the intrinsic viscosity of the star PIB was much lower than that of a linear analog of an equivalent MW.

Structure-property relationship of multiarm star PIBs has been investigated by a variety of techniques including viscometry, pour points, electron microscopy, and ultrasonic degradation [11]. The intrinsic viscosity of star PIBs changes very little in the 30–100 °C range in contrast to that of linear PIBs of the same MW which increases strongly with temperature. The viscosity of star PIBs was mainly determined by the MW of the arm and was relatively independent of the number of arms. Transmission electron micrograph of a star PIB showed a spherical shape with 55±4 nm in diameter which was in reasonable agreement with the radius of gyration R_g=27 nm determined from light scattering. The pour points of both linear and star PIB oil solutions were found to be similar to a commercial polyisoprene star viscosity improver. Star PIBs are of considerable interest as viscosity modifiers in motor oils, due to their expected shear stability. This was determined using sonic testing which provides qualitative information for mechanical shear degradation. Interestingly, it was found that higher order stars formed by star-star coupling are very sensitive to sonification. These higher order stars strongly increase the kinematic viscosity, although they are unstable under mechanical shear as it was observed that sonification even for 5 min decreases the kinematic viscosity.

Functional star-branched PIBs were prepared in high yield by Wang et al. [12], based on living cationic polymerization via haloboration initiation. First, living PIBs carrying X_2B- head groups (X=Cl or Br) were prepared via haloboration-initiation at -40 °C in CH_2Cl_2 in the presence of 2,6-di-*tert*-butylpyridine (DTBP). For the synthesis of PIBs with very short arms ($DP_n \sim 6$), BBr_3 was used. After 4 h the unreacted monomer was evaporated, the mixture was cooled to -60 °C and BCl_3 was added, followed by the introduction of DVB. After 4 h linking time, the linear PIB arm was completely consumed, and star polymer with a relatively narrow MWD (M_w/M_n=1.4) was obtained. For the synthesis of arm PIBs with DP_n=56, BCl_3 was used. After polymerization the reaction mixtures were warmed to room temperature and the excess BCl_3 and CH_2Cl_2 were evaporated. The solvent mixture CH_3Cl/hexane (40/60 v/v) was added to dissolve the polymer, followed by the addition of $TiCl_4$. The temperature was lowered to -60 °C, and DVB was introduced. $M_{n(star)}$ increased linearly with star polymer yields up to 4 h (86% yield). At longer reaction times intermolecular star-star linking was observed. When the CH_2Cl_2/hexane (40/60 v/v) solvent system was used for the linking reaction the star-star linking was much faster and occurred simultaneously with linking of the PIB arms. This reaction was not observed when BCl_3 was used in the linking reaction. Since intermolecular alkylation is absent in the living polymerization of S with BCl_3, but present with $TiCl_4$, it was suggested that this reaction might be responsible for the star-star linking. The effects of $DP_{n(arm)}$ and the [DVB]/[PIB] mole ratio (r) on the yield, $M_{n(star)}$ and the average number of arms (f) were investigated. As expected, with constant $DP_{n(arm)}$, the yield, $M_{n(star)}$, and f increased with the increase of r. The increase in [DVB]/[PIB] mole ratio led to a parallel increase in $M_{n(core)}$, whereas f increased only modestly. This may suggest that styryl cations add to the double bonds faster than PIB cations. With constant r, higher yields were obtained in 4 h with lower $DP_{n(arm)}$. This is likely due to the higher concentration of the living centers used in these experiments and not the results of lower $DP_{n(arm)}$. The value of f also increased as $DP_{n(arm)}$ decreased. A similar effect was found in the synthesis of star polymers of alkyl vinyl ethers [3] and it was concluded that the intermolecular linking reaction is sterically less hindered for a shorter chain. The results of ^{13}C solid-state NMR spectroscopy were in line with the structure of star-branched PIB consisting of a cross-linked core of poly(DVB) to which almost monodisperse PIB chains are radially attached.

2.1.2
(AB)$_n$-Type Star Block Copolymers

2.1.2.1
Poly(vinyl ether-b-vinyl ether)$_n$

(AB)$_n$-type star polymers of vinyl ethers were prepared by Kanaoka et al. [13] by linking block copolymers of 2-acetoxyethyl vinyl ether (AcOVE) and IBVE with the divinyl compound 1. After hydrolysis of the pendant ester groups, amphiphilic structures were obtained. The arms were prepared by sequential living cationic copolymerization of AcOVE and IBVE using HI/ZnI$_2$ as an initiating system, in toluene at –15 °C (when AcOVE was polymerized first) and in CH$_2$Cl$_2$ at –40 °C (when IBVE was polymerized first). Three series of the linear precursors were prepared: poly(AcOVE-b-IBVE) with DP$_n$=10+30 and DP$_n$=30+10 and poly(IBVE-b-AcOVE) with DP$_n$=10+30. The resulting block copolymers were allowed to react with 1 added in a ratio r=5 after complete conversion of the monomers. For all series the final polymer had much higher MW than the starting arms and the yield of star polymer was claimed to be high although no value was reported. M$_w$ (from about 50,000 to 100,000 g mol^{-1}) was measured by light scattering, from which the average number of arms, ranging from 8 to 16, was calculated. An increase of f was observed when the length of the poly(AcOVE) segment was increased, independently of its position in the copolymer chain. This was attributed to a decrease of steric hindrance. Additional evidence for the star structure was provided by NMR analysis. Depending on which monomer was polymerized first, two types of star-shaped structure could be obtained after complete hydrolysis, i.e., with the hydrophilic segments on the inside or on the outside of the molecule. Their solubility properties were essentially governed by the structure of the outer segments and were clearly different from those of the corresponding linear block copolymers.

Some analogous amphiphilic star block copolymers were prepared by replacing AcOVE by a vinyl ether with a malonate ester pendant group (diethyl 2-(vinyloxy)ethyl malonate; VOEM: CH$_2$=CH-O-CH$_2$CH$_2$CH(COOC$_2$H$_5$)$_2$) [14]. The block copolymers were prepared by sequential living cationic polymerization and were linked together using the difunctional vinyl ether 1 with r=5. With the two following block copolymers, poly(VOEM$_{10}$-b-IBVE$_{30}$) and poly(IBVE$_{30}$-b-VOEM$_{10}$), the average number of arms was six and five respectively and M$_w$, determined by light scattering, was about 40,000 g mol^{-1}. Invariably, a small amount of low MW polymer was recovered which was assigned to the block copolymer with some 1 units as terminal segments. Moreover, due to an increase of steric hindrance in the core, the yield of star polymer was found to be lower when the poly(VOEM) segment was in the inner part. Further alkaline hydrolysis of the esters led to hydrophilic segments with diacid pendant groups. Subsequent decarboxylation led to the monoacid counterparts. As previously, two types of stars were prepared according to the polymerization sequence for the preparation of the arms. The solubility properties of these star block copolymers were essentially determined by the nature of the outer block.

2.1.2.2
Poly(isobutylene-b-styrene)$_n$

In US patent 5,395,885 (1995) Kennedy et al. disclosed and specifically claimed star polymers with PS-PIB block copolymer arms, formed by linking with DVB and related diolefins. Examples however were not provided. Subsequently Storey and Shoemake [15] published the synthesis and characterization of multiarm star polymers based on PS-PIB block copolymer arms using essentially the same method. First S was polymerized with the cumyl chloride/TiCl$_4$ initiating system in the presence of pyridine in hexane/methylchloride (60/40 v/v) at –80 °C. At high (unspecified) S conversion the desired amount of IB was added and polymerized to obtain the PS-PIB block copolymer arm. SEC traces (presumably Differential Refractive Index, DRI) are given to show that the amount of homopolystyrene is small. This is unconvincing, however, in light of the small MW difference between the IB segment (M_n=1900 g mol^{-1}) and PS segment (M_n= 28,500 g mol^{-1}). The SEC UV trace, which would clearly show the extent of blocking, was not shown. The PS-PIB block copolymer arms were next reacted with DVB at [DVB]/[chain end]=10. After 72 h the linking was nearly complete. Further increase in the linking time led to star-star coupling with marginal increase in the incorporation of the arms. To suppress star-star coupling the temperature of the star-forming reaction was increased from –80 °C to room temperature after 0.5 h reaction with DVB at –80 °C. Star formation was more rapid at the higher temperature and reportedly the higher temperature suppressed star-star coupling. This is surprising in view of findings reported by Storey et al. [8] in the synthesis of multiarm star PIBs, namely that star-star coupling can be effectively frozen out by decreasing the temperature from –40 to –80 °C. Properties of the star block copolymers were not reported.

Subsequently, Asthana et al. [16] published the synthesis, characterization, and properties of star polymers with PS-PIB block copolymer arms. In a one pot procedure S was first polymerized with the cumyl chloride/TiCl$_4$ initiating system in the presence of triethylamine in CH$_2$Cl$_2$/hexane (50/50 v/v) at –80 °C. At ~90% S conversion, IB was added and polymerized to ~95% conversion. Then DVB was added to effect linking of the arms. In a two pot procedure the PS-PIB diblock copolymer was isolated and purified. Then it was redissolved in CH$_2$Cl$_2$/hexane (60/40 v/v); triethylamine, TiCl$_4$ and DVB were added and linking was completed in the –56 to –25 °C range. It was reported that linking did not proceed below –56 °C. The SEC DRI trace of the product formed by linking PS-PIB ($M_{n(PS)}$=8900 g mol^{-1}, $M_{n(PIB)}$=30,000 g mol^{-1}) for 72 h showed mainly higher order stars with ~15% unreacted diblock copolymer. Mechanical properties of this star polymer were compared to a linear PS-PIB-PS triblock copolymer of segment MWs of 8900–120,000–8900. It is not clear why the PIB segment was chosen to be twice the desired 60,000 for direct comparison with the star block copolymer. Tensile strength of the star block copolymer (with 15% diblock contamination) was found to be 20.5 MPa, much higher compared to 10 MPa found for the linear triblock (with 7% diblock+homopolymers). However, it is still somewhat lower than the 25–26 MPa reported for well defined linear PS-PIB-PS triblock copolymers. The melt viscosity of star block copolymer was found to be close to an order of magnitude lower than the linear triblock copolymer over a wide range of shear rates. Again this comparison is ambiguous since the PIB middle segment had MW=120,000 g mol^{-1} and not the desired 60,000 g mol^{-1}. It is possible that the melt viscosity of a direct linear analog with 8900–60,000–8900 segment MWs would be more similar.

2.1.3
A_n-Type Star Polymers with a Functionalized Core: Poly(vinyl ether)$_n$

Star copolymers of IBVE and AcOVE were prepared where the second monomer was added together with the cross-linking agent [17]. Typically, IBVE was polymerized first and after complete conversion the resulting living polymer (DP$_n$=30–38) was allowed to react with a mixture of AcOVE and 1 in various proportions. Complete consumption of AcOVE and 1 ensued and soluble high MW star type polymers with 7–10 arms per molecule were obtained, as evidenced by SEC, light scattering, and NMR characterizations. In order to obtain stars with a true functionalized core and not the analogous (AB)$_n$- or A_nB_n-type stars, the second block must be a random copolymer of the monofunctional vinyl ether and of the difunctional one. Although 1 was found slightly more reactive than AcOVE, experimental evidence based on ^{1}H NMR and ^{13}C NMR relaxation time supported the existence of a cross-linked core with incorporated segments of poly(AcOVE). Hydrolysis of the pendant esters of the microgel core was found to yield hydroxyl functions quantitatively and solubility properties of the final products were studied.

2.1.4
A_nB_n-Type Star Copolymers: Poly(vinyl ether)$_n$-Star-Poly(vinyl ether)$_n$

Based on the same technique of core cross-linking, amphiphilic heteroarm star polymers of IBVE and hydrolyzed AcOVE or VOEM were prepared with independent arms of both homopolymers [18]. The first step consisted of the previously described synthesis of a star polymer of IBVE with a microgel core, using 1 as a cross-linker. The final polymer had M$_w$=50,300 g mol^{-1} with an average of ten arms (each of DP$_n$=30) per molecule. This initially formed star polymer still carried living sites within the core which were suitable for initiation of the second monomer, AcOVE or VOEM. Actually, the number of newly growing arms per molecule should be the same as that of the first series since each active site comes from one initial arm. This was verified with a second stage polymerization of IBVE and supported by experimental results (Fig. 2), although the number of living sites in the core after the first stage polymerization could not

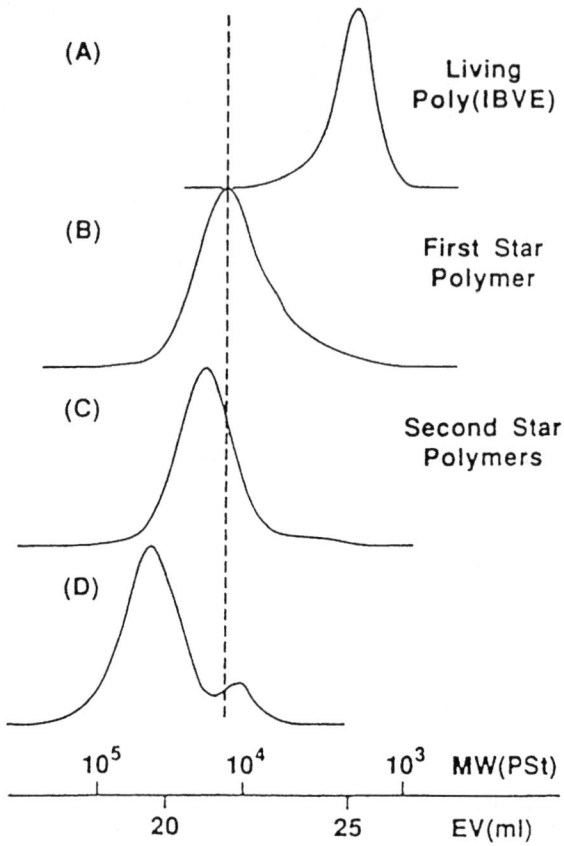

Fig. 2A–D. MWD of star-shaped poly(IBVE) obtained in toluene at − 40 °C: **A** living poly(IBVE): $[IBVE]_0=0.19$ mol l^{-1}, $[HI]_0=10$ mmol l^{-1}, $[ZnI_2]_0=0.2$ mmol l^{-1}, IBVE conversion=100%; **B** first star polymer obtained from the reaction of living poly(IBVE) and divinyl ether **1**: $DP_{arm}=19$, [living ends]=30 mmol l^{-1}, r=3.0; **C,D** the products (second star polymers) obtained by the polymerization of IBVE from the living ends within the core. Molar ratio of the second feed of IBVE to HI (or to the living end): (C) $[IBVE]_{add}/[HI]_0=19$, (D) $[IBVE]_{add}/[HI]_0=76$. Reprinted with permission from [18]. Copyright 1992 ACS

be determined by direct analysis. For the second step with a functional monomer, appropriate conditions to obtain living polymerization were applied and quantitative conversions of the two additional monomers were respectively reached, together with an increase of the average MW as observed by SEC. From the ^{1}H NMR spectra of the final polymer, the degree of polymerization of the second type of arms was in good agreement with the calculated value. Hydrolysis of the pendant ester groups into alcohol or acid led to an amphiphilic material with respectively hydrophobic and hydrophilic homopolymer arms attached to a single microgel core.

Table 1. Multiarm star polymers and copolymers with a microgel core

Monomers	Divinyl monomer	Type of star	Reference
IBVE	1	A_n	[3, 4]
p-MOS, tBOS	5	A_n	[5]
IB	6	A_n	[6–12]
IB	7	A_n	[7, 8]
IBVE/AcOVE	1	$(A)_n$ with functionalized core	[17]
		$(AB)_n$	[13]
		$A_n B_n$	[18]
IBVE/VOEM	1	$(AB)_n$	[14]
		$A_n B_n$	[18]
IB/S	6	$(AB)_n$	[15, 16]

The synthesis of multiarm star polymers and copolymers with a microgel core are listed in Table 1.

2.2
Synthesis Using a Multifunctional Initiator

This technique is based on the use of well-defined soluble multifunctional initiators, which, in contrast to anionic multifunctional initiators, are readily available. From these multiple initiating sites a predetermined number of arms can grow simultaneously when the initiating functions are highly efficient independently of whether the other functions have reacted or not. Under these conditions the number of arms equals the number of initiating functions and living polymerization leads to well defined star polymers with controlled MW and narrow MWD. Subsequent end-functionalization and/or sequential monomer addition can also be performed leading to a variety of end-functionalized A_n or $(AB)_n$ star-shaped structures.

2.2.1
A_n-Type Star Homopolymers

2.2.1.1
Poly(vinyl ethers)$_n$

Three arm star polymers of IBVE were synthesized by living cationic polymerization using trifunctional initiators **8** and **9** with the same trifluoroacetate initiating functions but different cores [19, 20]. The experimental conditions were selected to obtain living polymerization. A series of acetic acid derivatives including trifluoroacetic acid and the IBVE-acid adduct were found to be efficient initiators for the living cationic polymerization of IBVE in conjunction with either $ZnCl_2$, or $EtAlCl_2$ in the presence of a base such as 1,4-dioxane.

(8)

(9)

The polymerizations of IBVE were carried out with the multifunctional initiators **8** and **9** in conjunction with $EtAlCl_2$ and 1,4-dioxane (10 vol.% to the solvent) in *n*-hexane or toluene at 0 °C. To determine their initiating efficiency, the polymerization rates observed with these multifunctional initiators were compared with those observed with their respective monofunctional counterparts at the same concentration of initiating functions. For each system, the polymerization rates were found to be in good agreement indicating that the concentration of growing species was identical, i.e., all functions have initiated. The MWs, determined by SEC with polystyrene calibration, were about three times higher with the trifunctional initiators compared to the monofunctional analog, and the polymers exhibited narrow MWD. For instance, initiator **8** at 3.5 mmol l^{-1} concentration was used to polymerize IBVE at a concentration of 0.76 mol l^{-1}. SEC analysis of the polymer gave apparent M_n=23,300 g mol^{-1} (M_w/M_n=1.08) at complete conversion; using the monofunctional analog at 10 mmol l^{-1} concentration the polymer had M_n=8100 g mol^{-1} (M_w/M_n=1.06). It was also found that additional feeds of monomer after complete conversion of the first monomer increment led to a linear increase of MW in direct proportion with conversion. After quenching with methanol or sodiomalonic ester, the number average end functionality (F_n), calculated using ^{1}H NMR spectroscopy based on integration of characteristic peaks of terminal function and initiator residue, was found to be close to three. Hydrolysis of the ester core of the star polymer obtained with initiator **9** led to a poly(IBVE) with M_n one third of the star itself and the MWD remained narrow. Moreover, ^{1}H NMR spectrum of the isolated arms indicated the expected structure with the hydroxyl terminal function. From this experimental evidence, the authors concluded that well-controlled three arm stars of poly(IBVE) were synthesized for the first time with this monomer. The star polymer from initiator **9** had exactly three arms with uniform and controlled length obtained by a living process and, after quenching, one terminal function per arm. The same conclusion was reached with initiator **8** although no direct experimental evidence of the structure could be given since the arms could not be separated from the core.

To produce four arm star polymers of IBVE the use of a tetrafunctional initiator (**10**) with four trifluoroacetate goups linked to a cyclohexane core was also investigated by the same group [21, 22]. The monomer was polymerized under the same conditions as previously described and the same kinds of analysis were performed. Comparison of rates and MWs with those of the polymerization initiated with the monofunctional analog at a four times higher concentration led to the conclusion that four living arms were growing from the tetrafunctional core. When using a monomer concentration of 0.76 mol l^{-1} and an initiator concentration of 2.5 mmol l^{-1}, SEC measurements based on polystyrene calibration gave an apparent M_n=28,000 g mol^{-1} (M_w/M_n=1.08) whereas 8100 g mol^{-1} (M_w/M_n=1.06) was obtained with the monofunctional initiator at 10 mmol l^{-1}. A value of F_n close to 4 (3.77–3.91) was calculated using ^{1}H NMR spectroscopy after termination with the sodium salt of benzyl malonate.

(10)

2.2.1.2
Poly(p-methoxystyrene)$_n$

Two derivatives of the trifunctional initiators **8** and **9** (respectively, **11**=CH$_3$-C[p-C$_6$H$_4$OCH$_2$CH$_2$OCH(CH$_3$)-I]$_3$ and **12**=C$_6$H$_3$-(1,3,5-)[COOCH$_2$CH$_2$OCH(CH$_3$)-I]$_3$) with an iodine atom at the place of the trifluoroacetate group were used to synthesize three arm star polymers of *p*-MOS using living cationic polymerization with ZnI$_2$ as an activator in toluene at –15 °C [23]. With the typical conditions: [*p*-MOS]$_0$=0.38 mol l^{-1}, [**11**]$_0$=[ZnI$_2$]$_0$=3.3 mmol l^{-1}, living polymerization of *p*-MOS was observed, i.e., a linear increase of MW with conversion and narrow MWD (M_w/M_n<1.1). As determined from SEC analysis using polystyrene calibration, the M_n was in good agreement with the calculated one. However, a small peak with MW about one third of the main peak could be observed and was assigned to species initiated by traces of HI remaining from the initiator synthesis. Linear increase of MW with conversion was also observed when new feeds of *p*-MOS were polymerized after completion of the polymerization of the first monomer increment. Methacrylate-capped three arm poly(p-MOS) was obtained after quantitative end-quenching with 2-hydroxyethyl methacrylate (HEMA). Besides formation of a well-defined trifunctional macromonomer, this reaction could also be used to confirm the structure of the stars using ^{1}H NMR spectroscopy. By integration of the characteristic peaks of the core and of the end group respectively, F_n~3 was found. Using initiator **12** with an ester core, the same star could be prepared. SEC and NMR analysis of the arms after separation from the core by hydrolysis under mild alkaline conditions, confirmed uniformity of the individual arms.

2.2.1.3
Poly(styrene)$_n$

Six arm star polystyrenes were prepared by the core-first method using initiator **13** with six phenylethylchloride-type functions emanating from a central hexasubstituted benzene ring [24].

(13)

Living cationic polymerization of styrene was carried out using $SnCl_4$ and n-$Bu_4N^+Cl^-$ in CH_2Cl_2 at –15 or –30 °C. Polystyrene stars of various MW depending upon the amount of styrene and reaction times were characterized by NMR and SEC equipped with a light scattering detector. M_n values as determined using both techniques were claimed to be in good agreement with each other; moreover, narrow MWDs were found using SEC (M_w/M_n≈1.1). On the basis of these experimental results, the authors concluded that the hexafunctional initiator **13** was efficient to prepare well-controlled six arm star polystyrene up to M_n=90,000 g mol^{-1}. For higher MWs, however, the control was difficult to achieve owing to the possibility of β-proton elimination and subsequent polymerization of the new double bonds. The two-step end-capping of the polystyrene stars with C_{60} was also recently reported [25]. The first step was the introduction of six azido end groups by reaction of the stars with $TiCl_4$ and Me_3SiN_3; the reaction was found to be quantitative according to ^1H NMR analysis. The second step was performed by refluxing the star with an excess of C_{60} in chlorobenzene; ^1H and ^{13}C NMR confirmed quantitative grafting.

2.2.1.4
Poly(isobutylene)$_n$

Three arm star PIBs have been first synthesized by the inifer technique using the tricumyl chloride (TCC, **14**)/BCl_3 initiating system in CH_3Cl at –70 °C [26].

H$_3$C—C(X)—CH$_3$ (attached to benzene ring with two other C(CH$_3$)$_2$X groups)

(14) X=Cl
(15) X=OCH$_3$
(16) X=OCOCH$_3$
(17) X=OH

The inifer technique yields *tert*-chloro telechelic PIBs (Scheme 2) with M_ns determined by the [monomer]/[inifer] ratio. To prepare telechelic products, chain transfer to monomer must be absent, and with BCl_3 as coinitiator this requirement is fulfilled.

Characterization of the three arm star PIB involved a variety of spectroscopic techniques, i.e., ^1H and ^{13}C NMR, IR, and UV, thermal dehydrochlorination, and

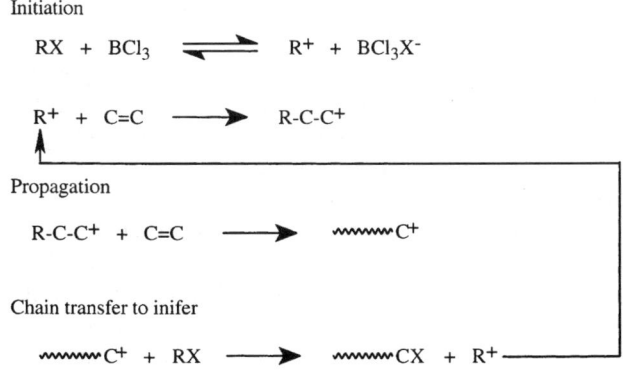

Scheme 2

selective oxidation of the central phenyl ring with CF_3COOH/H_2O_2 followed by M_n determination of the surviving arms. By quantitative dehydrochlorination, three arm star PIB carrying three $-CH_2C(CH_3)=CH_2$ termini could be prepared. This end group in turn could be quantitatively converted to a variety of other valuable functionalities, for instance to primary -OH groups by hydroboration followed by alkaline oxidation. By these functionalization reactions, well documented in [1], star PIBs with different terminal functionality could be obtained.

The conventional batch technique suffers from a number of limitations. The theoretical $M_w/M_n=1.33$ for three arm star polymers can only be obtained at constant [monomer]/[inifer] ratio (low conversion). When the polymerization is carried to high conversion, this ratio changes during the polymerization. Thus, in batch polymerizations, broad or multimodal MWDs have often been reported. In addition, the PIBs carried unfired or once-fired endgroups.

"Unfired" "Once-fired"

While in the presence of these end groups the number average end functionality remained unchanged ($F_n=3$), the reactivity of these end groups might be different from *tert*-chloro terminus of PIB.

Another problem associated with the batch technique is poor reaction control (unsatisfactory stirring, temperature control, etc). To overcome the problems outlined above a semi-continuous polymerization technique has been introduced [27]. In this technique a mixed monomer/inifer feed is added at a sufficiently low constant rate to a well stirred, dilute BCl_3 charge. Due to stationary conditions maintained during the whole polymerization, well-defined telechelic products with symmetrical end groups and theoretical polydispersities could be obtained. The kinetics of the polymerization has been discussed and the DP_n equation has been derived. In contrast to the batch technique, the DP_n for the semi-continuous technique is simply given by the [monomer]/[inifer] ratio. Thus, very reactive or unreactive inifers, unsuitable for batch polymerization, can also be used in the semi-continuous process.

In non-polar solvents BCl_3 is too weak to re-ionize the chloro end of PIB formed in the chain transfer to inifer (or termination) step. However when the polymerization of IB is carried out in polar solvents such as CH_2Cl_2 or CH_3Cl, the chloro end of PIB can be re-ionized by BCl_3. Thus termination is absent and living polymerization is obtained. Living polymerization has also been reported with the tricumyl methyl ether (15)/BCl_3 initiating system, in CH_2Cl_2 or CH_3Cl at $-30\,°C$ [28]. The products, for which the MWs were generally under

15,000 g mol^{-1} due to polymer precipitation, exhibited close to theoretical M_ns and M_w/M_ns in the range 1.3–2.0. The structure of the products has been analyzed by ^{1}H NMR spectroscopy and found to be essentially identical to those obtained by tricumyl chloride. The reaction between tricumyl methyl ether (**15**) and BCl$_3$ was investigated by Zsuga et al. using ^{13}C and ^{11}B NMR spectroscopy in CH$_2$Cl$_2$ at –30 °C [29]. According to the results, tricumyl methyl ether and BCl$_3$ yield tricumyl chloride and BCl$_2$OCH$_3$ in a fast reaction, thus the true initiator may be the chloro derivative. Interestingly the corresponding exchange reaction did not take place with tricumyl acetate (**16**)/BCl$_3$ system which also efficiently initiates the polymerization of IB [30]. The product upon quenching the polymerization however was the chloro functional three arm star PIB. Similarly to tricumyl methyl ether, tricumyl alcohol (**17**), only partially soluble in CH$_3$Cl at –50 °C, was found to be rapidly converted to the soluble choride derivative in a reaction with BCl$_3$. Thus three arm star PIBs have also been obtained by premixing tricumyl alcohol with BCl$_3$ for 10 min followed by the addition of IB [31, 32].

Polar solvents such as CH$_2$Cl$_2$ or CH$_3$Cl are poor solvents for PIB and therefore the MW that can be obtained with BCl$_3$ is limited. In contrast to BCl$_3$, TiCl$_4$ coinitiates the polymerization of IB even in moderately polar solvent mixtures, which dissolve high MW PIB at low temperatures. Organic esters, halides, and ethers can all be used to initiate living polymerization of IB. Ethers are converted to the corresponding chlorides almost instantaneously, while the conversion of esters is somewhat slow. Alcohols are inactive with TiCl$_4$ alone but have been used in conjunction with a mixture of BCl$_3$ and TiCl$_4$; BCl$_3$ converts the alcohols to the active chloride which is activated by TiCl$_4$. Well defined three arm star PIB of controlled MW have been obtained by many groups [32–34] with the **14** or **16**/TiCl$_4$ initiating systems or by using **17** with the combination of BCl$_3$ and TiCl$_4$ under similar conditions, i.e., in CH$_3$Cl or CH$_2$Cl$_2$/hexane (40/60 v/v) solvent mixture at –80 °C in the presence of a Lewis base.

Four arm star PIB has been prepared by living polymerization with the 3,3',5,5'-tetra(2-acetoxy-isopropyl)biphenyl (TCumOAc, **18**)/BCl$_3$ initiating system in dilute solutions in the –35 to –80 °C range [35].

(18)

In CH$_3$Cl/*n*-hexane (40/60 v/v) mixtures, very low conversion and ill-defined products were obtained, presumably due to the very low solubility of the **18**/BCl$_3$ complex. Precipitation was also observed in pure CH$_3$Cl when [IB]>0.129 mol l^{-1}.

At [IB]<0.514 mol l^{-1}, close to the theoretical M_ns ranging from ~3000 to 30,000 g mol^{-1}, and M_w/M_n~2 have been obtained. The products prepared under heterogeneous conditions, i.e., at [IB]>0.129 mol l^{-1} contained appreciable amounts of "once-fired" arms. Under homogeneous conditions, indanyl ring formation, "once-fired" and "non-fired" endgroups were found to be absent and F_n was close to 4.0.

The hexacumyl methyl ether functional initiator **19** was synthesized by Cloutet et al. [36] and used for the living cationic polymerization of IB in conjunction with TiCl$_4$ in CH$_2$Cl$_2$/methylcyclohexane (40/60 v/v) at –80 °C in the presence of a proton trap. The star sample obtained exhibited M_n=13,000 g mol^{-1} and M_w/M_n=1.27.

(19)

The synthesis of eight arm star PIB was recently described by Jacob et al. [37], where eight PIB arms emanate from a calixarene core (multifunctional initiators **20** (*tert*-hydroxy derivative) and **21** (*tert*-methoxy derivative)). The synthetic strategy is shown in Scheme 3.

Model reactions were also carried out using 2-(*p*-methoxyphenyl)-2-methoxypropane, a monofunctional analog of **21**, under conditions employed for the synthesis of eight arm star PIB. IB was polymerized in two stages with BCl$_3$-TiCl$_4$ coinitiators. Stage I was carried out in CH$_3$Cl with a fraction of the required IB plus BCl$_3$ and yielded very low conversions and very low MWs. Stage II was induced by the addition of TiCl$_4$, hexane (to reach CH$_3$Cl/hexane 40/60 v/v) and the balance of IB. In these model experiments, slow initiation was observed (I_{eff}<20%) which was attributed to the formation of resonance stabilized carbocation upon ionization of the initiator. This is questionable, however, in view of possibility of complex formation with BCl_3 via the *p*-methoxy substituent. Since **20** was found to be insufficiently soluble in CH_3Cl at –80 °C, a two-stage process was also used to obtain the eight arm star PIB. The chloride initiator was formed in situ by contacting the alcohol with BCl_3 in the first stage. The product obtained in the second stage exhibited a bimodal MWD. The higher MW product (~70%) was assumed to be the star polymer. Subsequent experiments with **21**, which was found to be soluble in CH_3Cl, produced similar results, i.e., a main product (74%) assumed to be the star PIB and a minor side product (~26%) of lower MW which was UV transparent. It was concluded that this side product was short chain PIB which arose by haloboration initiation. The amount of side product could be decreased to ~10% by decreasing the concentration of BCl_3 and contact time in the first stage. Interestingly, polymerization by **21** and $TiCl_4$ alone produced a gel. A possible route to star-star coupling and cross-linking was suggested to involve proton elimination leading to p-isopropenyl groups, which were subsequently attacked by growing PIB chain ends. Thus, it was concluded that a two stage process using low concentration of BCl_3 is the preferred method. The average number of arms of purified star PIB was determined by core destruction (selective oxidation of the aromatic core) and was found to be 7.6, only slightly lower than the theoretical 8. This is unexpected in light of the low initiator efficiencies obtained with 2-(*p*-methoxyphenyl)-2-methoxypropane and may indicate that the reactivity of the octafunctional *tert*-ether initiator **21** is substantially different, i.e., 2-(*p*-methoxyphenyl)-2-methoxypropane may not be a good model. It is also conceivable that the complexing behavior of the two compounds with BCl_3 might be different due to different steric environment.

2.2.2
(AB)$_n$-Type Star Block Copolymers

2.2.2.1
Poly(vinyl ether-b-vinyl ether)$_n$

Three arm amphiphilic star block copolymers of IBVE and 2-hydroxyethyl vinyl ether (HOVE) were prepared using the trifunctional initiator **8** with sequential cationic polymerization of two hydrophobic monomers, IBVE and AcOVE. Subsequent hydrolysis of the acetates led to the hydrophilic poly(HOVE) segments [38]. Two types of stars were prepared depending on which monomer was polymerized first: three arm star poly(IBVE-*b*-HOVE), with the hydrophobic part inside and three arm star poly(HOVE-*b*-IBVE), with the hydrophobic part outside. When IBVE was polymerized first, the experimental conditions were the same as described in Sect. 2.2.1. After reaching quantitative monomer conversion, AcOVE was added and temperature was raised from 0 to 40 °C to accelerate the reaction since this monomer is less reactive than IBVE. When starting with AcOVE as a first block, both polymerizations were carried out at 40 °C. SEC analysis showed that MWDs were narrow for the two steps whatever the se-

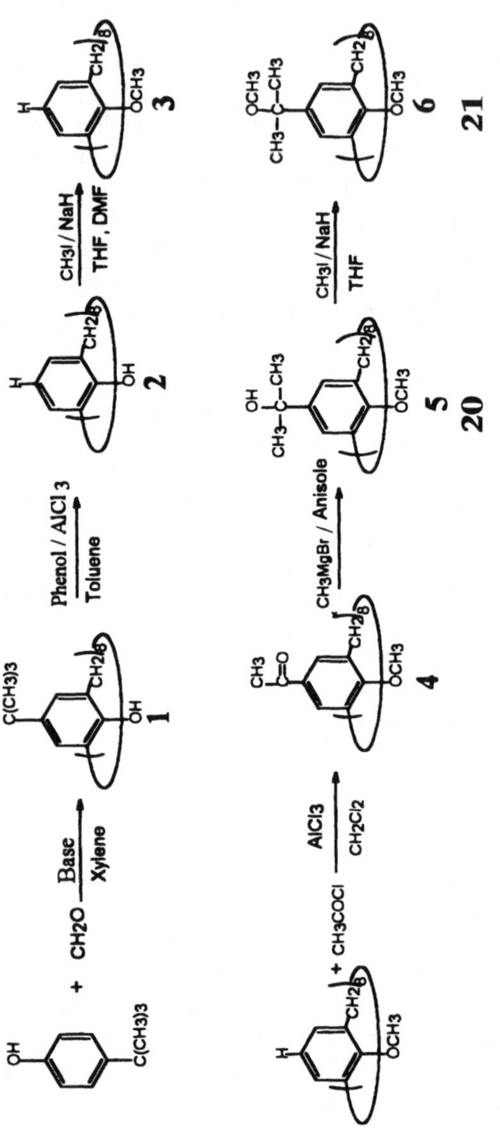

Scheme 3

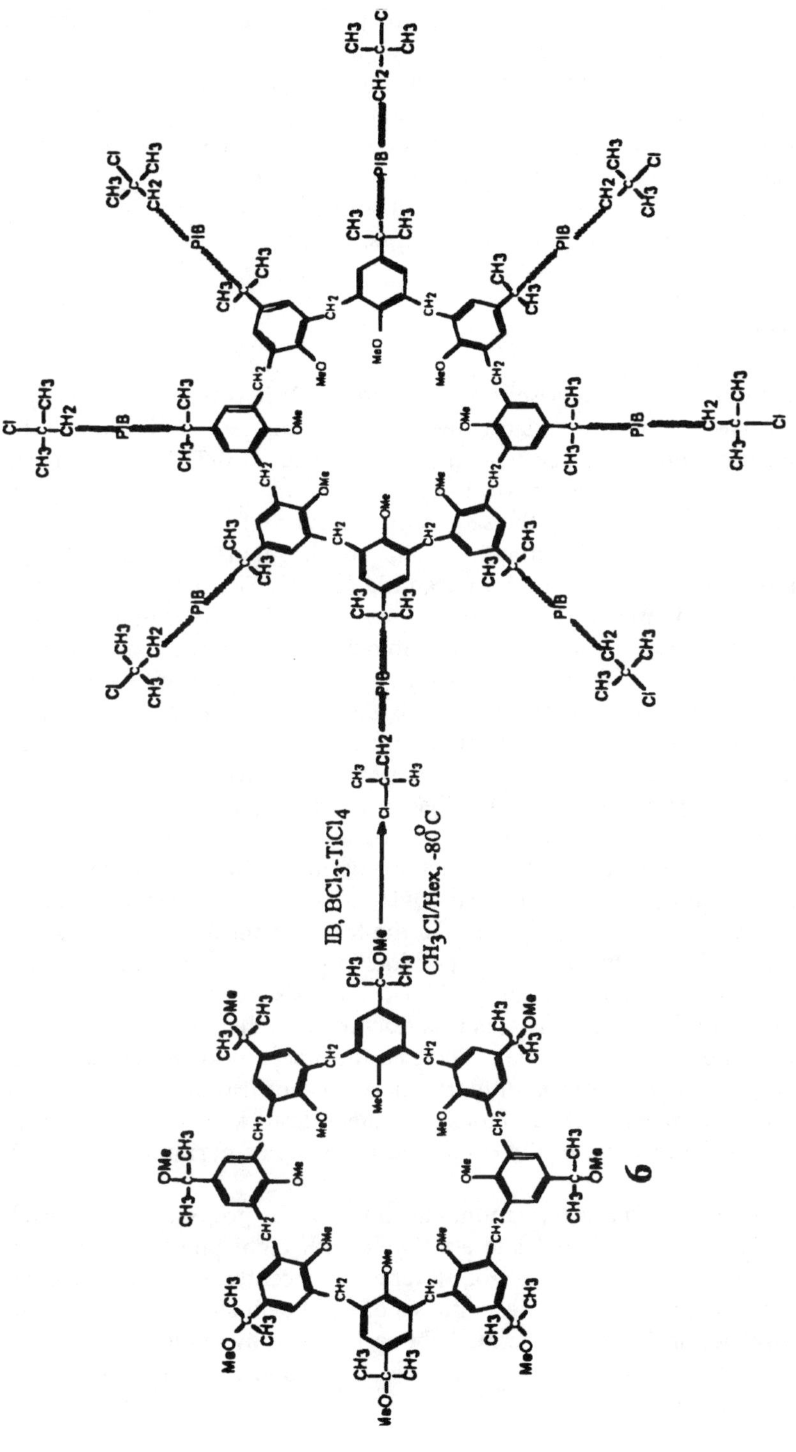

Scheme 3 (continued)

quence order with a complete shift of the peak to higher MW after the second step. The products, obtained after quenching with methanol, were analyzed by ^{1}H NMR spectroscopy to determine DP_n of both segments, which were in good agreement with the calculated values. However, F_n was not given and no experimental evidence of the three arm block copolymer structure was provided. Hydrolysis of the acetate groups was found to be quantitative according to ^{1}H NMR analysis and gave amphiphilic stars with solubility properties essentially determined by the nature of the outer segments.

2.2.2.2
Poly(isobutylene-b-styrene)$_n$

Radial three arm star poly(isobutylene-*b*-styrene)s have been prepared by many groups. The synthesis invariably involved the living polymerization of IB with the tricumyl chloride (**14**) or tricumyl methyl ether (**15**)/$TiCl_4$ initiating system in CH_3Cl/methylcyclohexane (or hexane) (40/60 v/v) in the presence of a Lewis base at −80 °C followed by the sequential addition of S. For instance, tricumyl methyl ether was used as initiator by Kaszas et al. [39] in CH_3Cl/methylcyclohexane in the presence of dimethylacetamide (DMA). The tensile strength of the star block copolymer, which was rather low (13.7 MPa) due to unoptimized conditions, was similar to that of a linear triblock copolymer with comparable composition and MW. For linear triblock copolymers better results were obtained (18.7 MPa) in the combined presence of DMA and DTBP. Star blocks have not been prepared under these conditions, but expectedly they should exhibit similar tensile strength. Storey et al. [40] prepared three arm star block copolymers of poly(isobutylene-*b*-styrene) by slightly modifying the above procedure using tricumyl chloride as initiator in the combined presence of pyridine and DTBP. Interestingly, the three arm star block copolymer exhibited tensile strength of 16 MPa, about twice that of a linear triblock copolymer with similar block segment lengths. This is probably due to the fact that incomplete crossover from PIB to S resulted in the formation of diblock copolymer in the synthesis of linear triblock copolymer, whereas in star block synthesis incomplete crossover would only result in dangling ends. It is well documented that even small amounts of diblock copolymers substantially decrease the mechanical properties of triblock copolymer thermoplastic elastomers. There was no clear difference between the mechanical properties of star block and linear diblock copolymers prepared in CH_3Cl/hexane mixture in the combined presence of pyridine and DTBP at −80 °C.

The synthesis, characterization, and mechanical properties of a novel star block copolymer thermoplastic elastomer with eight poly(isobutylene-b-styrene) arms radiating from a calix[8]arene was recently reported by Jacob et al. [41]. The process involved the synthesis of eight arm star PIB by a method essentially identical to that described above, followed by sequential addition of S after the IB conversion has reached 95%. To minimize alkylation and to obtain high MW PS blocks, moderate $TiCl_4$ concentration (0.059 mol l^{-1}) and a 2- to

2.5-fold excess of S relative to the targeted MW was used. The produced star block copolymer was contaminated by 3–5% homoPS and ~10% linear diblock copolymer. The mechanical properties of selected star blocks have been investigated. All products investigated exhibited excellent tensile strengths up to 26 MPa.

2.2.2.3
Poly(isobutylene-b-p-methylstyrene)$_n$

Well defined three arm star block copolymers were prepared by sequential block copolymerization of IB with p-methylstyrene (*p*-MeS) [42]. First IB was polymerized by the **14**/TiCl$_4$ initiating system in CH$_3$Cl/hexanes (40/60 v/v) at –80 °C in the presence of the proton trap DTBP. When the polymerization was complete the living PIB chain ends were capped with 1,1-diphenylethylene. Subsequently, titanium(IV)isopropoxide was added to decrease the Lewis acidity and *p*-MeS was introduced. The mechanical properties of the star block copolymers were determined and were found to be similar to linear triblocks with the same *p*-MeS segment length and composition. The best star block copolymers exhibited ~22 MPa tensile strength.

2.2.2.4
Poly(isobutylene-b-THF)$_n$

The synthesis of three arm star block copolymers of IB and THF was described by Gadkari and Kennedy [43]. First, three arm star PIB with hydroxyl termini was obtained by dehydrochlorination of three arm star PIB carrying terminal *tert*-chlorine, followed by hydroboration and oxidation. Quantitative conversion of the primary hydroxyl end groups was achieved with triflic acid in the presence of pyridine at 0 °C. The resulting triflate functional PIB was used to induce living polymerization of THF. At room temperature, low initiation rates were observed, which could be increased by increasing the temperature to 60 °C. The star block copolymer which contained considerable amounts of unblocked PIB was purified by column chromatography with hexane/THF mixtures as eluent. The polymer fractions were analyzed and the blocking efficiency was calculated to be >70%. These block copolymers carried an HO- functionality at the polymer end of each arm and thus could be used to prepare polyurethane networks.

2.2.2.5
Poly(isobutylene-b-methyl methacrylate)$_n$

Star block copolymers of IB and methyl methacrylate have been prepared very recently by the combination of living cationic and anionic polymerizations [44]. First, three arm star PIB (M_n=30,000 g mol^{-1}) was prepared by living cationic

Table 2. Multiarm star polymers and copolymers synthesized using a multifunctional initiator

Monomers	Multi-functional initiator	Type of star	Reference
IBVE	8, 9	A_3	[19, 20]
	10	A_4	[21, 22]
p-MOS	11, 12	A_3	[23]
S	13	A_6	[24, 25]
IB	14–17	A_3	[26–34]
	18	A_4	[35]
	19	A_6	[36]
	20, 21	A_8	[37]
IBVE/AcOVE	8	$(AB)_3$	[38]
IB/S	14, 15	$(AB)_3$	[39, 40]
	20, 21	$(AB)_8$	[41]
IB/p-MeS	14	$(AB)_3$	[42]
IB/THF	14	$(AB)_3$	[43]
IB/MMA	14	$(AB)_3$	[44]

polymerization of IB using a trifunctional initiator (tricumyl chloride, **14**), and the living ends were quantitatively capped with 1,1-diphenylethylene. The product obtained upon quenching with methanol was isolated, redissolved in THF, and quantitatively metallated with K/Na alloy. The reaction mixture was filtered and excess LiCl was added to replace K^+ with Li^+, which gives a PIB macroinitiator suitable for anionic polymerization of MMA. The polymerization of MMA was carried out in THF/n-hexane (70/30 v/v) solvent mixture to ensure solubility of PIB at –78 °C. A series of star block copolymers with 27–46% MMA has been prepared with low polydispersity ($M_w/M_n < 1.10$). Physical properties of the star block copolymers have not yet been reported.

The synthesis of star polymer and star block copolymers with multifunctional initiators are detailed in Table 2.

2.3
Synthesis Using a Multifunctional Coupling Agent

Multifunctional coupling agents, bearing several (>2) identical nucleophilic functions sufficiently separated in space to avoid steric hindrance, may be used to link together similar living macromolecular chains. Well defined star structures are obtained when these nucleophilic functions add cleanly and efficiently to the living ends without any side reaction. It is necessary to use strictly stoichiometric concentrations of the chain ends and of the nucleophilic functions to achieve the target structure and to avoid purification.

2.3.1
A_n-Type Star Homopolymers

2.3.1.1
Poly(vinyl ethers)$_n$

Monofunctional malonate ions were shown to terminate quantitatively living cationic chain ends of poly(vinyl ether)s to give stable carbon-carbon bond [45] even when they were used in stoichiometric concentration [46]. The poor solubility of their multifunctional counterparts in organic solvents could be overcome by the use of 18-crown-6 to dissolve them in THF. Coupling reactions of living poly(IBVE), formed by initiation with the HI/ZnI$_2$ system, were performed using the trifunctional coupling agent **22** and the tetrafunctional **23**. With **22**, a three arm polymer was recovered in 56% yield and with **23**, only three out of the four anions reacted to give three arm polymer in 85% yield. Such incomplete reactions were explained by poor solubility as well as steric hindrance at the coupling sites.

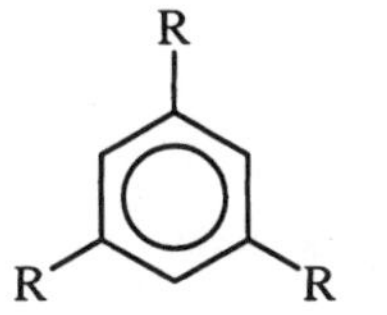

(22)

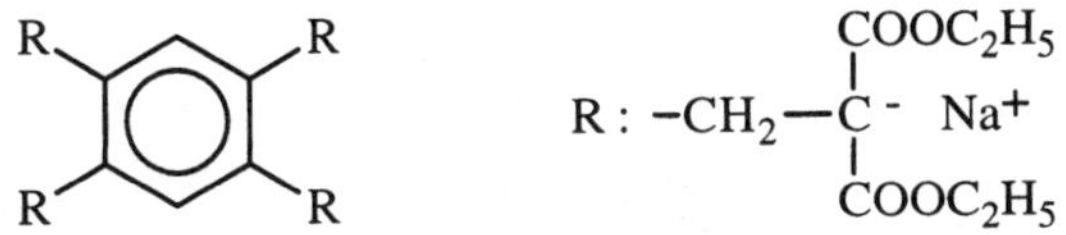

$$R : -CH_2-\underset{COOC_2H_5}{\overset{COOC_2H_5}{C}}{}^- \ Na^+ \quad (23)$$

The same authors chose another very reactive nucleophilic function, the silyl enol ether group, which upon reaction with living cationic chain ends of poly(vinyl ether)s, also leads to a carbon-carbon bond with formation of a ketone (Scheme 4). Model reactions of living poly(IBVE) with various monofunctional silyl enol ethers [47] showed that the α-substituent R should have electron-donating properties in order to increase the electron density on the double bond.

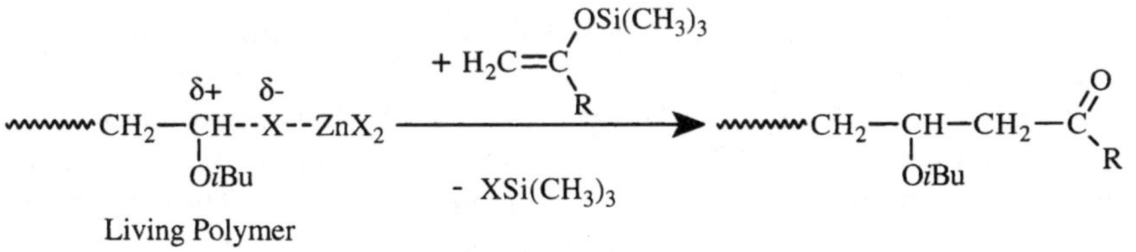

Scheme 4

The coupling efficiency also depended on the length of the polymeric chain, the shorter being the more efficient. Moreover, a chloride counteranion was preferred due to the high affinity of silicon to chlorine.

A tri- and a tetrafunctional coupling agent respectively **24** and **25** [48], both completely soluble in organic solvents, were then designed in order to obtain high yield of coupling of living poly(IBVE). The electron-donating alkoxyphenyl group in the α position enhanced the reactivity of the double bond and the radially shaped structure with rigid phenyl spacers led to well-separated reactive functions suitable for minimizing the steric hindrance previously observed with the malonate derivatives.

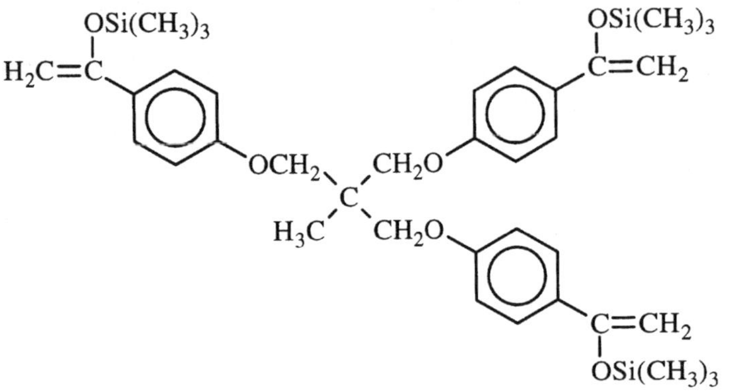

(24)

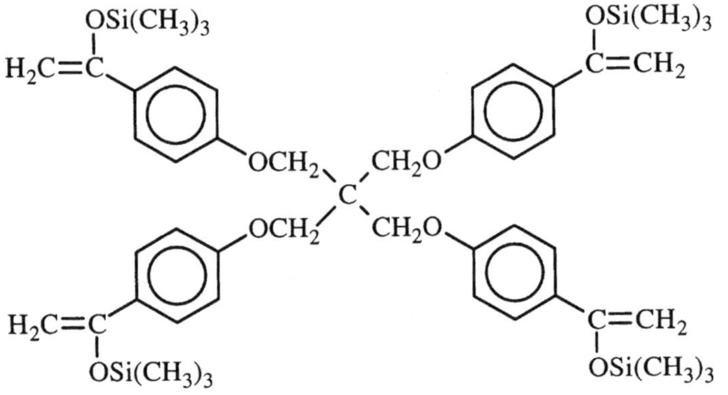

(25)

Short chains (DP$_n$~10) of living poly(IBVE) with Cl$^-$ counter-anion were prepared with the HCl/ZnCl$_2$ initiating system in CH$_2$Cl$_2$ at –15 °C. The coupling reaction with **24** and **25** respectively was carried out by the addition of a solution of the coupling agent in CH$_2$Cl$_2$ at about 80% conversion of IBVE and the reaction mixture was stirred during 24 h at the same temperature. The concentration of the nucleophilic functions was similar to that of the chain ends. In both cases, SEC analysis of the final products showed the complete shift of the low MW peak

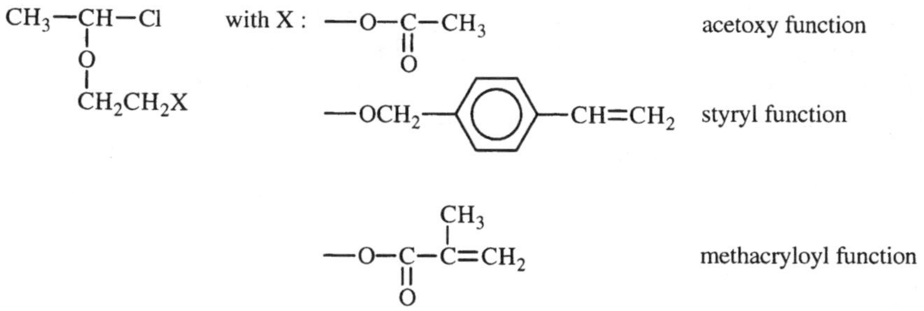

Scheme 5

CH₃—CH—Cl
    |
    O
    |
CH₂CH₂X

with X :

—O—C(=O)—CH₃   acetoxy function

—OCH₂—C₆H₄—CH=CH₂   styryl function

—O—C(=O)—C(CH₃)=CH₂   methacryloyl function

Scheme 5

corresponding to the linear chains towards higher MWs. The higher MW was obtained with the tetrafunctional coupling agent and MWDs remained narrow for both coupled products (M_w/M_n<1.1). Based on these SEC analyses, the overall yields of the coupled products were above 95%. The structure was verified using ^{1}H NMR analysis which evidenced quantitative reaction of each enol ether group for both coupling agents. Moreover, the mole ratio of the aromatic rings in the core to the α-end methyl of the chains was found close to 1 confirming the quantitative coupling. The coupling reaction of similar but longer poly(IBVE) (DP_n~50) was performed in order to study the influence of the chain length. The SEC analysis showed bimodal distributions. The major higher MW peak corresponded to the coupled product and had narrow MWD. The minor lower MW peak corresponded to the unreacted linear precursor. The apparent yield was 85–90% and steric hindrance was assumed to be responsible for incomplete reaction. Nevertheless, it could be concluded that the multifunctional coupling agents based on silyl enol ether functions were superior to those based on malonate ions previously described in the sense that they could lead to three and four arm star poly(IBVE) with short arms in very high yield.

Using the tetrafunctional coupling agent **25**, end-functionalized four arm poly(IBVE)s were synthesized [49]. End-functionalization was performed using functional initiators which were HCl adducts of functionalized vinyl ethers bearing respectively acetoxy, styryl and methacryloyloxy groups (Scheme 5).

Polymerization of IBVE was performed in CH₂Cl₂ at –15 °C using ZnCl₂ as a Lewis acid. The linear polymers quenched with methanol had the expected structure as shown by ^{1}H NMR analysis, with the functional group X at the α-end and an acetal unit at the ω-end. Their MWD was narrow, typically M_w/M_n was lower than 1.1. However, for the initiators with a styryl or a methacryloyloxy group, small amounts of low MW by-products could be seen. The experimental results indicated that poly(IBVE) with a functional α-end group could be synthesized using living cationic polymerization without any significant side reactions affecting the integrity of the functional group. Coupling reaction with **25** was performed at the same conditions as previously described and the same conclusions could be drawn. Based on SEC analysis the initial peak shifted towards higher MWs and the MWD remained narrow. This was especially the case for the coupled products with acetoxy and methacryloyloxy functionality (yield>95%). For the coupled product with the styryl terminal group, the yield was lower (~90%). Structural analysis using ^{1}H NMR spectroscopy was performed after separation of the main product by preparative gel permeation chromatography. F_n was close to the theoretical value 4, indicating that the final product had the expected four arm structure with one functional group at the end of each arm.

2.3.1.2
Poly(isobutylene)$_n$

In view of the excellent shear stability of silicone oils, it was theorized that shear stable multiarm star PIBs could be prepared using cyclosiloxane cores [50]. The synthesis was accomplished in two steps. First, allyl terminated PIB of desired MW was prepared by reacting living PIB with trimethylallylsilane. Linking was effected by hydrosilylation of the allyl-functional PIB with cyclosiloxanes carrying six or eight SiH groups (respectively **26** and **27**) in the presence of H_2PtCl_6 catalyst at 180 °C for an extended period of time. With relatively low MW allyl functional PIB (M_n=5200 g mol^{-1}), after 3 days of linking using **26** at a [C=C]/[Si-H]=1 ratio, six arm star PIB was obtained in ~80% yield. With an arm MW of M_n=12,600 g mol^{-1} however, in addition to the expected star PIB and unreacted PIB arm, the product also contained a much higher MW component. It was theorized that this arose by star-star coupling in the presence of adventitious water. In contrast to allyl-functional PIB, linking of isopropenyl functional PIBs was less successful, as the amount of unreacted PIB arm was ~50%, even with short arms. Experiments with the octafunctional hydrogenoctasilsesquioxane **27** yielded stars with significantly lower than eight arms even with low arm MW. With arm MWs of M_n=12,600–19,200 g mol^{-1}, the number of arms of the primary stars was only ~5. In addition, higher than expected MW stars were also obtained probably by star-star coupling. ^{13}C relaxation NMR studies indicated that the mobility of the arms is not limited by steric compression between them. Apparently, there is enough room around **27** to place eight arms, although access to the unreacted Si-H sites may become limited after five to six arms have been placed.

(26)

```
      R          R
       \        /
       Si―O―Si
       /\    /|\
      O  O  R | O  R
      |   \ /  |/
      |   Si―O―Si
   Si―O―Si    |
  /|\   /| \  |
 R O   O | R O
    \ |   \ |
    Si―O―Si
    /      \
    R       R
```

(27) R=H
(28) R=O-Si(CH$_3$)$_2$-H

The above method appears to have serious limitations. First, the availability of common methylcyclosiloxanes is limited, and second, steric compression prevents quantitative hydrosilylation of neighboring SiH groups.

To eliminate steric congestion, Majoros et al. [51] prepared a new octafunctional siloxane linking agent by moving the SiH group away from the rigid cyclic core skeleton (28). Using H$_2$PtCl$_6$.H$_2$O as catalyst, although star formation was apparent, 60–70% of PIB allyl remained unreacted even after 144 h. Karstedt's catalyst {bis(divinyltetramethyl disiloxane) platinum(0)} was more efficient, although the majority of PIB allyl still remained unreacted. The reaction was further complicated by the formation of higher order stars by star-star coupling, which could be suppressed by increasing the [C=C]/[SiH] ratio from 1.0 to 1.66.

While star-star coupling was considered as a side reaction in the above reports, Omura and Kennedy [52] attempted to exploit core-core coupling of small methylcyclosiloxanes in the presence of moisture under hydrosilylation conditions using PIB allyl to build star PIBs with many arms. The synthetic strategy is shown in Scheme 6.

Kinetic studies of primary and higher order star formation concluded that well-defined first order stars with narrow molecular weigth distribution could be prepared with [SiH]/[C=C]=1.25 at room temperature whereas higher order stars were obtained with [SiH]/[C=C]=4.0 at 120 °C. While primary star formation was very slow and could require up to a week to complete at room temperature, higher order star formation was essentially complete in 24 h. Higher order stars with up to 28 arms have been prepared by this method. Intrinsic viscosities and branching index g' were also studied. The intrinsic viscosities of stars were much lower than those of linear PIBs of the same MW. As expected, it was found that g' values of stars depend on the number of arms and not on the MW of the arms. The stars were found to be resistant to acids and bases suggesting that the PIB corona protects the vulnerable core.

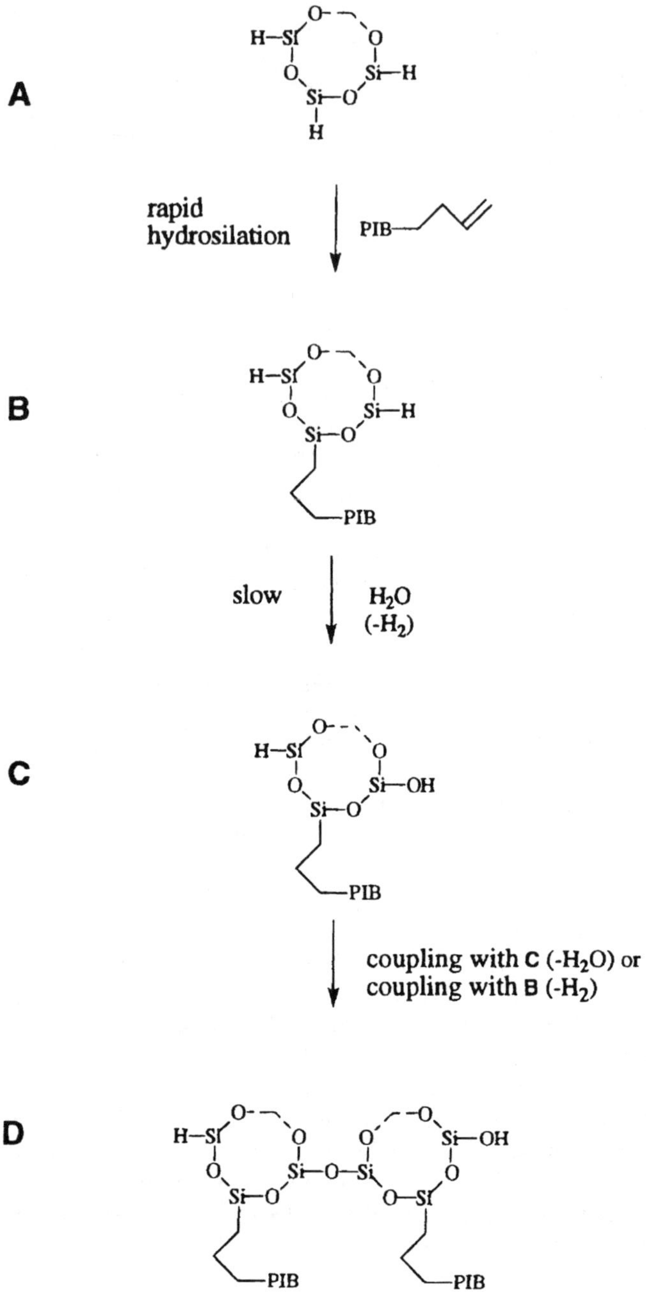
A
rapid
hydrosilation
PIB
B
PIB
slow
H_2O
$(-H_2)$
C
PIB
coupling with C ($-H_2O$) or
coupling with B ($-H_2$)
D
PIB
PIB

Scheme 6

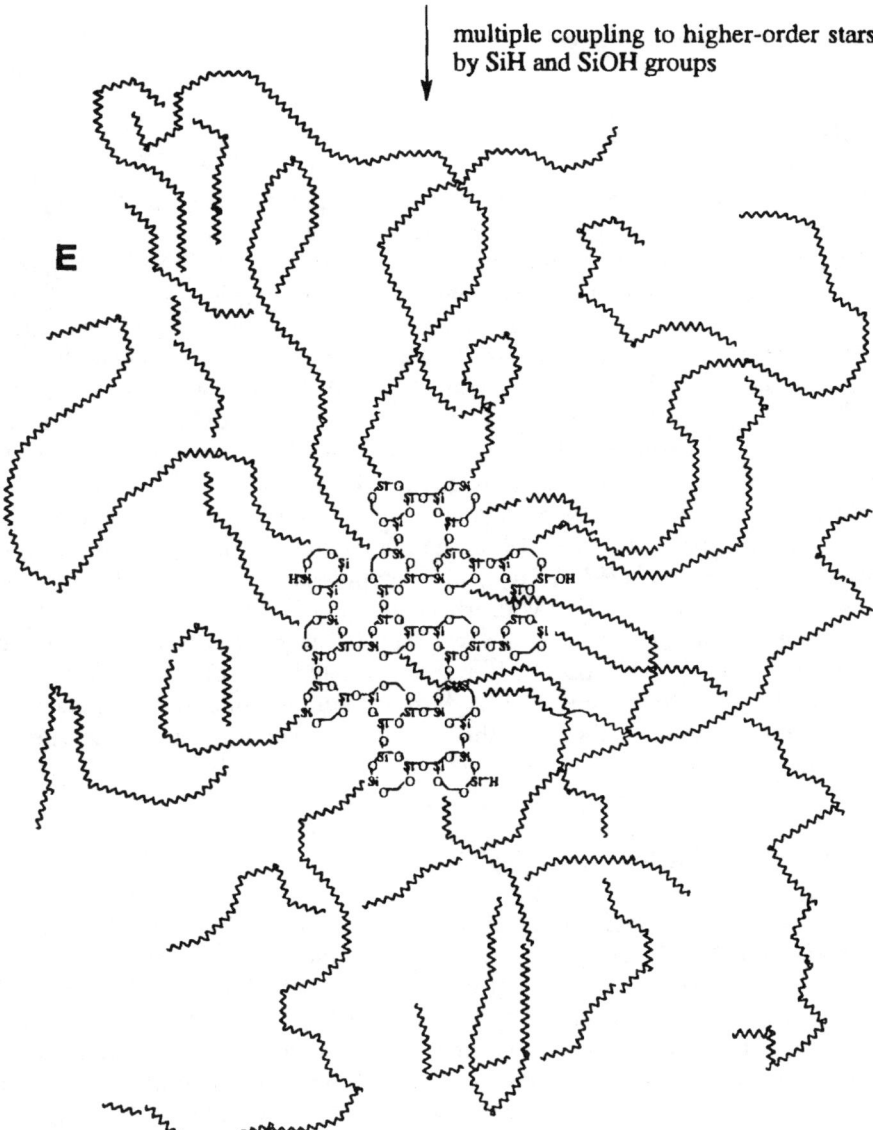

Scheme 6 (continued)

2.3.2
(AB)$_n$-Type Star Block Copolymers

2.3.2.1
Poly(vinyl ether-b-vinyl ether)$_n$

The tetrafunctional coupling agent 25 was used to prepare amphiphilic four arm star poly(vinyl ethers) by linking together hydrophobic AB block copolymers where one block could be further modified chemically to a hydrophilic segment [53]. The block copolymers were prepared by living cationic sequential copolymerization of a hydrophobic vinyl ether (IBVE or 2-chloroethyl vinyl ether, CEVE) and a hydrophobic precursor to a hydrophilic vinyl ether (AcOVE or 2-[(*tert*-butyldimethylsilyl)oxy]ethyl vinyl ether, SiVE). Polymerization was initiated by the HCl/ZnCl$_2$ system in CH$_2$Cl$_2$ at –15 °C and four different copolymers were prepared by changing either the comonomers or the polymerization sequence. Coupling reaction was performed and it was shown that the yield depends on the structure of the copolymer as well as on the monomer sequence in the arms. The yield was higher for monomers with less bulky pendant groups. For instance, poly(CEVE-*b*-AcOVE) block copolymers (where CEVE was polymerized first) were coupled with 93% coupling efficiency whereas only 80% was obtained for the reverse sequence. Even lower yield (73%) was obtained with poly(SiVE-*b*-IBVE). After hydrolysis of the pendant acetoxy or tert-butyldimethylsilyl substituents into hydroxyl groups, amphiphilic four arm star poly(vinyl ethers) were formed and the solubility properties were studied using ^{1}H NMR spectroscopy.

2.3.2.2
Poly(α-methylstyrene-b-2-hydroxyethyl vinyl ether)$_n$

More recently, amphiphilic four arm star block copolymers of α-methylstyrene (α-MeS) and HOVE were prepared using the same coupling agent 25 [54]. In the first step, α-MeS was polymerized using the HCl adduct of CEVE in conjunction with SnBr$_4$ as initiating system in CH$_2$Cl$_2$ at –78 °C. After 95% conversion of this first monomer was reached, SiVE was added and the polymerization was continued to reach 85% conversion. From SEC and NMR analysis of the quenched product, it was shown that a true block copolymer was formed. The coupling reaction was performed in the second step by adding 25 to the living polymerization mixture in the presence of *N*-ethylpiperidine to enhance the coupling efficiency and the reaction mixture was stirred for 24 h at –78 °C. SEC analysis of the isolated product showed that the overall yield was 85%. After separation by preparative gel permeation chromatography, the star structure with an average number of arms of 4.8 per polymer was confirmed by ^{1}H NMR spectroscopy. Like previously, the 2-hydroxyethyl pendant groups were obtained after deprotection using tetra-*n*-butylammonium fluoride at room temperature providing new four arm amphiphilic block copolymer with hard hydrophobic segments of poly(α-MeS) in the outer part and soft hydrophilic segments of poly(HOVE) in the inner part. Solvent effects on ^{1}H NMR spectra were studied and showed the considerable influence of the rigid segments on the properties of the star.

2.3.3
A_nB_m-Type Star Copolymers

2.3.3.1
Poly(isobutylene)$_2$-Star-Poly(methyl vinyl ether)$_2$

Recently, a new concept in cationic polymerization, the concept of living coupling agent was introduced. According to the definition, a living coupling agent must react quantitatively with the living chain ends, the coupled product must retain the living centers stoichiometrically and must be able to reinitiate the second monomer rapidly and stoichiometrically. It was reported that living PIB reacts quantitatively with bis-diphenylethylenes where the two diphenylethylene moieties are separated by an electron-donating spacer group, to yield stoichiometric amounts of bis(diarylalkylcarbenium) ions. Since the resulting diarylalkylcarbenium ions have been successfully employed for the controlled initiation of second monomers such as p-MeS, α-MeS, IBVE, and methyl vinyl ether (MeVE), it was proposed that A_2B_2 star-block copolymers could be synthesized by this method [55]. In the first example, amphiphilic A_2B_2 type star-block copolymers (A=PIB and B=poly(MeVE)) were prepared via the living coupling reaction of living PIB, using 2,2-bis[4-(1-tolylethenyl)phenyl]propane (**29**, BDTEP) as a living coupling agent, followed by initiation of MeVE from the di-cation at the junction of the living coupled PIB [56]. Fractionation of the crude A_2B_2 star-block copolymer was carried out on a silica gel column and the purity of the crude A_2B_2 star-block copolymer was calculated to be=93% based on the weights of fractions. The pure A_2B_2 block copolymer exhibited two T_gs (–60 °C for PIB and –20 °C for poly(MeVE)) indicating the presence of two microphases. Interestingly, an A_2B_2 star-block copolymer with 80 wt% poly(MeVE) composition ((IB_{45})$_2$-star-(MeVE$_{170}$)$_2$) exhibited an order of magnitude higher critical micelle concentration (CMC=4.25×10^{-4} mol l^{-1}) in water, compared to CMCs obtained with linear diblock copolymers with same total M_n and composition (IB$_{90}$-b-MeVE$_{340}$) or with same segmental lengths (IB$_{45}$-b-MeVE$_{170}$). This suggested that block copolymers with star architectures exhibit less tendency to aggregation than their corresponding linear diblock copolymers.

(29)

2.3.3.2
Poly(isobutylene)-Star-Poly(ethylene oxide)$_m$

Amphiphilic multiarm star copolymers of PIB and PEO having one PIB arm and two, three, or four PEO arms with identical length were recently reported by Lemaire et al. [57]. End-chlorinated PIBs with controlled MW (M_n=500 and 1000 g mol^{-1}) and narrow MWD (M_w/M_n=1.1) were prepared by living cationic polymerization and the *tert*-Cl ω-end group was quantitatively converted to anhydride or dianhydride. This species was used as macromolecular coupling agents for α-methoxy-ω-hydroxy PEOs (M_n=750, 2000, 5000 g mol^{-1}) leading to star-shaped polymers upon ester linkage formation. The best coupling efficiency was obtained with *p*-toluenesulfonic acid as a catalyst in mesitylene at 155 °C. The final product, which was characterized by SEC and MALDI-TOF mass spectrometry, was a mixture of the various star-shaped structures together with unreacted PEO and diblock copolymer.

PIB-(di)anhydride

Table 3. Multiarm star polymers and copolymers synthesized using a multifunctional coupling agent

Monomers	Multi-functional coupling agent	Type of star	Reference
IBVE	22	A$_3$(low yield)	[46]
	23	A$_4$(low yield) A$_3$ (major product)	[46]
	24	A$_3$	[48]
	25	A$_4$	[48, 49]
IB	26	A$_6$	[50]
	27	A$_8$ (low yield) A$_5$ (major product)	[50]
	28	A$_8$	[51]
IBVE/SiVE, CEVE/AcOVE	25	(AB)$_4$	[53]
α-MeS/SiVE	25	(AB)$_4$	[54]
IB/MeVE	29	A$_2$B$_2$	[55, 56]
IB/EO	PIB-(di)anhydride	mixture of AB$_2$, AB$_3$, AB$_4$	[57]

The known methods to prepare star polymers and copolymers via living cationic polymerization with multifunctional coupling agents are summarized in Table 3.

3
Graft (co)Polymers

Graft (co)polymers are polymers with a linear backbone to which macromolecular side chains are connected. They can be prepared by three different methods: "grafting from", "grafting onto", and (co)polymerization of macromonomers.

3.1
"Grafting From" Technique

This technique is based on the use of a linear polymer with pendant functional groups that can be activated to initiate the polymerization of a second monomer. Based on this definition, the linear precursor polymer can be considered as a multifunctional macromolecular initiator. The importance of the "grafting from" technique by cationic polymerization of the second monomer increased considerably with the advent of living cationic polymerization. The advantage is the virtual absence of homopolymer formation via chain transfer to monomer.

3.1.1
Synthesis of the Backbone by Cationic Polymerization

3.1.1.1
Poly(vinyl ether) Backbone

Graft copolymers with poly(vinyl ether) backbone and poly(2-ethyloxazoline) branches were reported [58] where the backbone, a random copolymer of poly(EVE) and poly(CEVE), was prepared by conventional cationic polymerization using aluminium hydrogen sulfate as an initiator in pentane at 0 °C. After quenching the copolymer with methanol, quantitative polymerization of 2-ethyloxazoline was performed using the pendant chloroethyl sites as initiators in the presence of sodium iodide in chlorobenzene at 115 °C. The obtained graft copolymer exhibited two glass transition temperatures indicating a phase separated morphology.

3.1.1.2
Poly(isobutylene) Backbone

The simplest method to obtain PIB backbone with pendant functionalities able to initiate polymerization of a second monomer is via copolymerization of IB with a functional monomer such as bromomethylstyrene (BMS) or chlorometh-

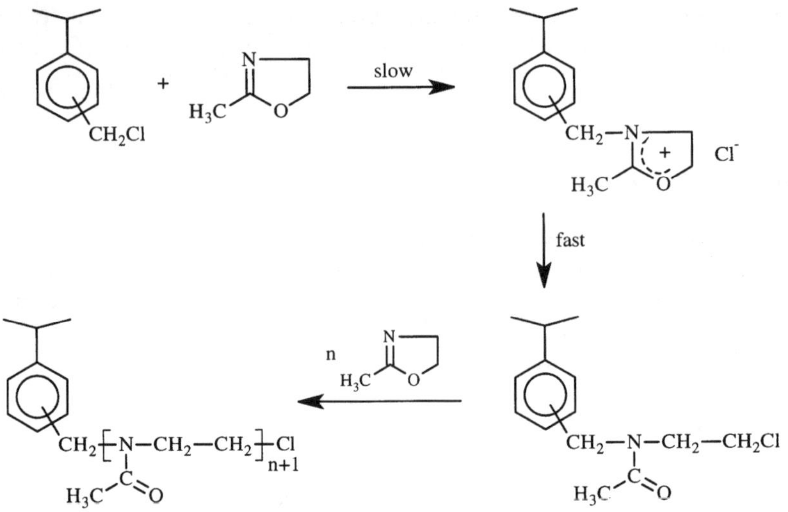

Scheme 7

ylstyrene (CMS). An alternate method to obtain initiating sites along a PIB backbone involves the copolymerization with p-MeS followed by selective bromination [59]. The first method has been used by Nuyken and coworkers [60] to synthesize poly[IB-co-CMS-g-2-methyl-2-oxazoline]. The poly(IB-co-CMS) copolymers were synthesized by cationic copolymerization of IB with CMS in CH_2Cl_2 at –80 °C. The preferred coinitiator was BCl_3 since ionization of the chloromethyl group, which could also act as an initiating site, is negligible in conjunction with BCl_3. The polymerization of 2-methyl-2-oxazoline using poly(IB-co-CMS) was considered to proceed according to Scheme 7.

All products were soluble in water, indicating the formation of the graft copolymer and the absence of ungrafted macroinitiator. Dialysis of the product also revealed the absence of low MW (<6000 g mol^{-1}) homo poly(2-methyl-2-oxazoline). Due to the amphiphilic nature of the graft copolymer, aggregation in water as well as in chloroform was shown by ^{1}H NMR spectroscopy, solution viscosity, and dynamic laser light scattering.

3.1.2
Synthesis of the Branches by Cationic Polymerization

3.1.2.1
Poly(vinyl ether) Branches

Comb-like graft copolymers with polysilane backbone and poly(IBVE) branches were reported by Matyjaszewski and Hrkach [61]. The IBVE monomer was grafted from triflated poly(methylphenyl silylene) at –30 °C using acetone as a promoter (in order to accelerate the initiation step) in the presence of tetrahydrothiophene as a nucleophile. The graft copolymer obtained therefrom had M_n=105,000 g mol^{-1} with broad MWD (M_w/M_n=2.5) and the authors stated that better defined polymers could be prepared by improving the initiating system.

Triflated poly(methyl phenyl silylene)

3.1.2.2
Poly(silyl vinyl ether) Branches

Aldol group transfer polymerization of *tert*-butyldimethylsilyl vinyl ether [62] was initiated by pendant aldehyde functions incorporated along a poly(methyl methacrylate) (PMMA) backbone [63]. This backbone was a random copolymer prepared by group transfer polymerization of methyl methacrylate (MMA) and acetal protected 5-methacryloxy valeraldehyde. After deprotection of the aldehyde initiating group, polymerization proceeded by activation with zinc halide in THF at room temperature. The reaction led to a graft copolymer with PMMA backbone and poly(silyl vinyl ether) or, upon hydrolysis of the *tert*-butyldimethylsilyl groups, poly(vinyl alcohol) branches.

3.1.2.3
Poly(isobutylene) Branches

Grafting IB from PS backbone containing *tert*-benzylic acetate initiating sites was described by Jiang and Fréchet [64]. The backbone was obtained by chemical modification of PS shown in Scheme 8.

Polymerization of IB from the PS macroinitiator was accomplished with BCl$_3$ as coinitiator in CH$_2$Cl$_2$ at –78 °C. Due to the living nature of the polymerization of IB, high grafting efficiencies (~85%) were reported. The resulting ~15% homoPIB was most probably due to initiation from adventitious moisture or direct initiation (haloboration).

A similar multifunctional macroinitiator was obtained by Puskas [65] in a radical copolymerization of S and 4-(1-hydroxy-1-methylethyl)styrene. The macroinitiator was then used to initiate the living cationic polymerization of IB. With relatively short backbone and 8–23 branches with M_n=10,000–20,000 g mol^{-1}, starlike structures, spherical in shape were obtained. The structure was verified by core destruction followed by SEC analysis of the surviving arms.

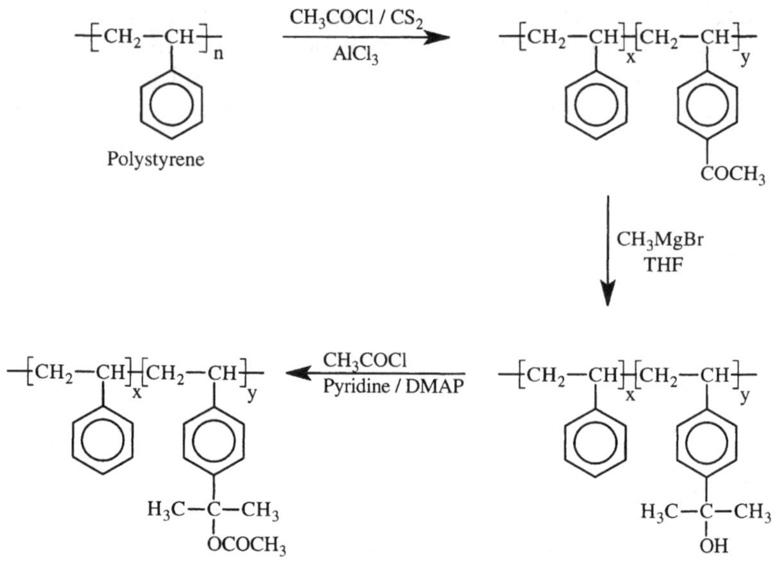

Scheme 8

3.1.2.4
Poly(styrene) and poly(α-methylstyrene) Branches

EPDM graft terpolymers with PS or poly(α-MeS) branches were prepared from chlorinated commercial EPDM polymer (7.7% 5-ethylidene-2-norbornene as diene, M_n=50,000 g mol^{-1}) in conjunction with Et$_2$AlCl, using the "grafting from" technique [66]. Grafting could only be achieved in the presence of 10–20% polar solvent, e.g., CH$_3$Cl or CH$_2$Cl$_2$; in pure heptane no graft copolymer could be isolated. The grafting reaction was carried out at –30 °C, below this temperature the EPDM polymer was not soluble in the mixed heptane/polar solvent used. The grafting efficiencies were determined by selective solvent extraction and found to be quite low, in the range of 10–20%. The tensile strength and ultimate elongation were also determined and found to be quite low when the amount of PS was <30 wt%. The tensile strength increased by increasing the amount of PS or poly(α-MeS) to >50 wt%, although at the expense of the elastomeric properties (increased modulus and decreased ultimate elongation).

3.2
"Grafting Onto" Technique

In the "grafting onto" technique, the macromolecular branches are linked to the main chain by specific reaction between their α or ω end group and reactive functional groups in the backbone.

3.2.1
Synthesis of the Backbone by Cationic Polymerization

3.2.1.1
Poly(vinyl ether) Backbone

Graft copolymers with poly(vinyl ether) backbone and polystyrene or polybutadiene branches were synthesized starting from poly(CEVE) prepared by living cationic polymerization [67]. Typically, the cationic polymerization was performed in toluene at low temperature and was initiated by the HCl adduct of the monomer in the presence of $ZnCl_2$. Polymers with controlled length (DP_n=6–56) and narrow MWD were obtained (M_w/M_n<1.2). The branches were further grafted onto the backbone by reaction of polystyryl- or polybutadienyllithium with the pendant chloroethyl functions (Scheme 9). This technique was based on the recently demonstrated ability of the alkyl chlorides bearing heteroatoms to react efficiently with polystyryl- and polybutadienyllithium [68]. The MWD of the final copolymer was still narrow and the average MW, as determined by light scattering, was in good agreement with the calculated one assuming the complete reaction of the chloroethyl functions. This supports the formation of well-defined graft copolymer with poly(vinyl ether) backbone and polystyrene or polybutadiene branches. Highly branched structures could also be derived from these graft copolymers providing that the anionically prepared polystyrene was initiated by a lithio acetal derivative. Using trimethylsilyl iodide, the acetal function could be transformed into the corresponding α-iodoether which is able to initiate quantitatively (in the presence of $ZnCl_2$) new poly(CEVE) blocks on which PS or polybutadiene branches could be connected again using the same procedure.

3.2.1.2
Poly(p-bromomethylstyrene-IB-p-bromomethylstyrene) Triblock Copolymer Backbone

To replace the laborious synthesis of triblock copolymer thermoplastic elastomers based on PMMA as hard segment and PIB as a soft segment by sequential block copolymerization following cationic to anionic transformation, Gyor et al. proposed an alternative procedure [69]. First poly(p-bromomethylstyrene-IB-p-bromomethylstyrene) triblock copolymer with short poly(p-bromomethylstyrene) segments was synthesized by living cationic sequential copolymerization. In the second step living PMMA anions were connected to both ends of the triblock copolymer by Wurtz-Grignard coupling via the bromomethyl functional groups. When the coupling reaction was carried out in toluene at –78 °C a cross-linked gel formed. Gel formation was absent when the coupling reaction was effected in THF at –78 or at –60 °C. The products were soluble comblike triblock structures. At 5.6 to 23 wt% PMMA, rubbery and non-sticky products were obtained, whereas at 43 wt% PMMA content, the product was a white powder.

Step A

CH₃CHCl + n CH₂=CH-O-CH₂CH₂Cl →(1) ZnCl₂ / 2) MeOH→ CH₃CH(O-CH₂CH₂Cl)[CH₂CH(O-CH₂CH₂Cl)]ₙOCH₃ = ~~~~~~ PCEVE block (1)

Step B

(EtO)₂CH-CH₂CH₂Li + m CH₂=CH-C₆H₅ → (EtO)₂CH-CH₂CH₂[CH₂CH(C₆H₅)]ₘ⁻, Li⁺ = ~~~~~~Li   PS block (2)

Step C

(1) + (2) ⟶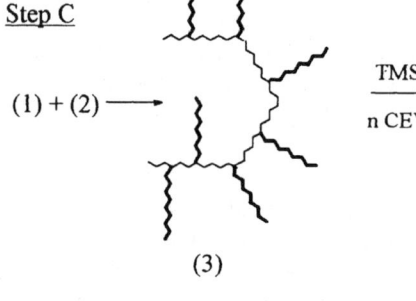

(3)

$\xrightarrow{\text{TMSI} \atop \text{n CEVE}}$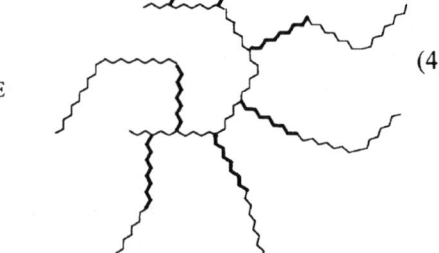

(4)

(4) + (2) ⟶

(5)

(5) + Step C ⟶ Etc.

Scheme 9

3.2.2
Synthesis of the Branches by Cationic Polymerization: Poly(styrene) Branches

Poly(pentadiene-*g*-S) was synthesized by Peng and Dai [70] by grafting PS initiated by adventitious moisture with various Lewis acid coinitiators onto pendant double bonds of polypentadiene in CH_2Cl_2 at 0 °C. After selective solvent extraction, the products were analyzed by IR and ^{1}H NMR spectroscopy. The highest grafting efficiency (73%) and close to complete S conversion were observed with Et_2AlCl. With BCl_3 the grafting efficiency was similar (67%), although S conversion was only ~60%. Cross-linked polymer was not observed which was explained by fast termination of the cation arising after grafting onto the polypentadiene chain. The optimum temperature was ~0 °C; at lower or higher temperatures the grafting efficiencies decreased. Although MWs were not reported, intrinsic viscosity of the graft copolymer was found to be lower than that of polypentadiene, which may indicate chain scission during grafting.

The synthesis of graft copolymers by "grafting from" and "grafting onto" techniques are reported in Table 4.

Table 4. Graft copolymers obtained by "grafting from" and "grafting onto" techniques (names of the polymers obtained by cationic polymerization are italicized)

Nature of the backbone	Nature of the branches	Method used[a]	Reference
Poly(EVE-co-CEVE)	Poly(oxazoline)	f	[58]
Poly(IB-co-CMS)	Poly(oxazoline)	f	[60]
Triflated poly (methyl phenyl silylene)	*Poly(IBVE)*	f	[61]
PMMA	*Poly(silyl vinyl ether)*	f	[63]
PS with *tert*-benzylic acetate	*PIB*	f	[64]
PS-*co*-poly[4-(1-hydroxy-1-methylethyl)styrene]	*PIB*	f	[65]
EPDM	PS *Poly(α-MeS)*	f	[66]
Poly(CEVE)	PS Poly(butadiene)	o	[67]
Poly(BMS-IB-BMS)	PMMA	o	[69]
Poly(pentadiene)	PS	o	[70]

[a] f, Grafting from; o, grafting onto

3.3
Macromonomers

A macromonomer is a macromolecule with a reactive end group that can be homopolymerized or copolymerized with a small monomer by cationic, anionic, free-radical, or coordination polymerization (macromonomers for step-growth polymerization will not be considered here). The resulting species may be a star-like polymer (homopolymerization of the macromonomer), a comb-like polymer (copolymerization with the same monomer), or a graft polymer (copolymerization with a different monomer) in which the branches are the macromonomer chains.

Macromonomers have been synthesized by living cationic polymerization by three different techniques: by the use of a functional initiator, employing functional capping agent or by chain end modification.

3.3.1
Synthesis of Macromonomers by Living Cationic Polymerization

3.3.1.1
Synthesis Using a Functional Initiator

This technique is the simplest as it generally requires only one step since the polymerizable function is incorporated via the initiator fragment. To obtain well-defined macromonomers with one polymerizable end group per chain, controlled length and narrow MWD, the following criteria should be fulfilled:
– initiation only from the initiator (no protic or direct initiation);
– living polymerization conditions (especially no transfer to the monomer);
– during the polymerization the functional group should remain unreacted or it needs to be protected.

3.3.1.1.1
Poly(vinyl ethers)

Most of the reported poly(vinyl ether) macromonomers have been prepared with a methacrylate end group which can be radically polymerized and which is non-reactive under cationic polymerization conditions [71–73]. Generally, the synthesis was based on the use of the functional initiator **30**, which contains a methacrylate ester group and a function able to initiate the cationic polymerization of vinyl ethers. Such initiator can be obtained by the reaction of HI and the corresponding vinyl ether. With initiator **30** the polymerization of ethyl vinyl ether (EVE) was performed using I_2 as an activator in toluene at –40 °C. The MW increased in direct proportion with conversion, and narrow MWD (M_w/M_n = 1.05–1.15) was obtained. The chain length could be controlled by the monomer to initiator feed ratio. Three poly(EVE) macromonomers of different length were prepared by this method: M_n=1200, 5400, and 9700 g mol^{-1}. After complete conversion and quenching with methanol (acetal ω end group), ^{1}H and ^{13}C NMR spectroscopy was used to demonstrate the structural integrity of the macromonomers and to determine the value of F_n for the -methacrylate end group. The latter was always very close to 1, indicating that one methacrylate end group was incorporated per molecule.

Radical homopolymerization and copolymerization with MMA initiated by AIBN in benzene solution or in bulk led to high MW graft (co)polymers.

$$H_2C=C\begin{matrix}CH_3\\ \\C-O-CH_2CH_2-O-\overset{X}{\underset{|}{C}H}-CH_3\\\| \\O\end{matrix} \qquad \begin{matrix}(30) & X=1\\(31) & X=Cl\end{matrix}$$

Macromonomers of the same monomer, EVE, were prepared using initiator 32 bearing an allylic function [73, 74]. This reactive group remained intact during the polymerization and could be further transformed into the corresponding oxirane by peracid oxidation. This epoxy end group can be polymerized by ring-opening polymerization.

$$CH_2=CH-CH_2-O-CH_2CH_2-O-\overset{I}{\underset{|}{C}H}-CH_3 \qquad (32)$$

Other vinyl ethers were also polymerized with initiator 30 under the same experimental conditions [75]. For instance, SiVE and 2-(trimethylsilyloxy)ethyl vinyl ether provided hydrophobic macromonomers which could be desilylated under suitable conditions to obtain the corresponding water-soluble poly(HOVE) without any side reaction of the methacrylate end group.

3.3.1.1.2
Poly(silyl vinyl ether)

Poly(silyl vinyl ether) macromonomers with a styrenic polymerizable function [76, 77] were prepared by the so-called aldol group transfer polymerization of *tert*-butyldimethylsilyl vinyl ether. The polymerization proceeded by activation of the aldehyde in the functionalized initiator *p*-formylstyrene (33) by ZnCl$_2$ or ZnBr$_2$ in THF or CH$_2$Cl$_2$ at 30 °C. Initiation was found to be fast and quantitative. The products, macromonomers of poly(*tert*-butyldimethylsilyl vinyl ether) with an aldehyde ω end group, were analyzed by SEC and ^{1}H NMR spectroscopy. The polymerization was shown to be living resulting in MWs controlled by the monomer to initiator feed ratio and narrow MWD (M_w/M_n<1.3). In the first example [76], the macromonomers were copolymerized with ω-*p*-vinyl-phenyl(polydimethylsiloxane) macromonomers in THF at 60 °C, using AIBN as an initiator. Selective removal of the *tert*-butyldimethylsilyl protective groups led to amphiphilic graft copolymers with hydrophilic poly(vinyl alcohol) branches. In the second example [77], macromonomers with DP_n up to 28 were prepared. After reacting the terminal aldehyde with a silyl ketene acetal (1-methoxy-2-methyl-1-trimethylsilyloxy propene) to provide a more stable ester end group, *tert*-butyldimethylsilyl ether side groups were hydrolyzed to the corresponding alcohol leading to heterotelechelic poly(vinyl alcohol) with a polymerizable styryl α end group and an ester ω end group.

$$H_2C=CH-\underset{}{\bigcirc}-\underset{O}{\overset{}{C}}H \qquad (33)$$

Such hydrophilic macromonomers (DP_n=7–9) were radically homopolymerized and copolymerized with styrene [78] using AIBN as an initiator at 60 °C in deuterated DMSO in order to follow the kinetics directly by ^{1}H NMR analysis. The macromonomer was found to be less reactive than styrene (r_M=0.9 for the macromonomer and r_S=1.3 for styrene). Polymerization led to amphiphilic graft copolymers with a polystyrene backbone and poly(vinyl alcohol) branches. The hydrophilic macromonomer was also used in emulsion polymerization and copolymerized onto seed polystyrene particles in order to incorporate it at the interface.

3.3.1.1.3
Polystyrene and Poly(p-methylstyrene)

Using functional initiator **31**, polystyrene and poly(*p*-MeS) macromonomers bearing a terminal methacrylate group [79] could be prepared by living cationic polymerization in CH_2Cl_2 at –15 °C in the presence of $SnCl_4$ and *n*-Bu_4NCl. To obtain an α end-functionality (F_n) close to 1, mixing of the reagents was carried out at –78 °C. When mixing was performed at –15 °C, the functionality was lower than 1 which was ascribed to a side-reaction, initiation by protons eliminated following intramolecular alkylation. The resulting oligomers could be clearly observed by SEC analysis at low conversion. After complete conversion of the monomers, the polymerization was quenched with methanol and macromonomers with a chloride ω end group were recovered. A polystyrene macromonomer with DP_n=18 (M_w/M_n=1.16 and F_n=0.94) and a poly(*p*-MeS) macromonomer with DP_n=24 (M_w/M_n=1.14 and F_n=0.97) were reported.

3.3.1.1.4
Poly(α-methylstyrene)

A similar procedure was also used for the synthesis of methacrylate functional poly(α-MeS) [80]. Thus, **31** was used in conjunction with $SnBr_4$ in CH_2Cl_2 at –78 °C, to obtain the macromonomer with M_ns substantially (~50%) higher than the theoretical value. This was probably due to the formation of terminated low MW oligomers with indanyl end group structure. The eliminated proton was assumed to remain unreactive and did not initiate new polymer chains, since the functionality of the polymer was found to be close to unity. Allylation of the ω-end of the living polymer was also accomplished by quenching with excess allyltrimethylsilane. In a related development, four arm poly(α-MeS), functionalized with methacrylate end groups, has been synthesized by coupling reaction of α-methacryloyloxy functional living poly(α-MeS), obtained by the procedure given above, with tetrafunctional silyl enol ethers **25** [81].

3.3.1.1.5
Poly(β-pinene)

β-Pinene which is a main component of natural turpentine can be polymerized by living cationic isomerization polymerization [82] (Scheme 10) using TiCl$_3$(OiPr) as a Lewis acid in conjunction with n-Bu$_4$NCl in CH$_2$Cl$_2$ at –40 °C. When initiator **31** was used, polymerization led to a poly(β-pinene) macromonomer with a methacrylate function at the α end and a chlorine atom at the ω chain end [83]. Three macromonomers were prepared with DP$_n$=8, 15, and 25 respectively; they had narrow MWD (M$_w$/M$_n$=1.13–1.22) and the reported functionality was close to 1 (F$_n$=0.90–0.96).

A macromonomer made of a block copolymer of p-MeS and β-pinene was also prepared by sequential living cationic polymerization of both monomers under the same experimental conditions. The first block had 12 p-MeS units and the second had 11 β-pinene units as evaluated by ^{1}H NMR spectroscopy.

The homopolymer and block copolymer macromonomers were copolymerized with MMA by free-radical polymerization in benzene at 60 °C using AIBN as an initiator; typical concentration were [MMA]=1.2 mol l^{-1} and [macromonomer]=0.020 mol l^{-1}. MMA was completely converted in 18 h and the macromonomers conversion reached more than 70% as determined by ^{1}H NMR. Incomplete conversion was explained by steric hindrance. Free-radical copolymerization resulted in high MW graft copolymers with PMMA backbone and relatively rigid, nonpolar poly(β-pinene) branches.

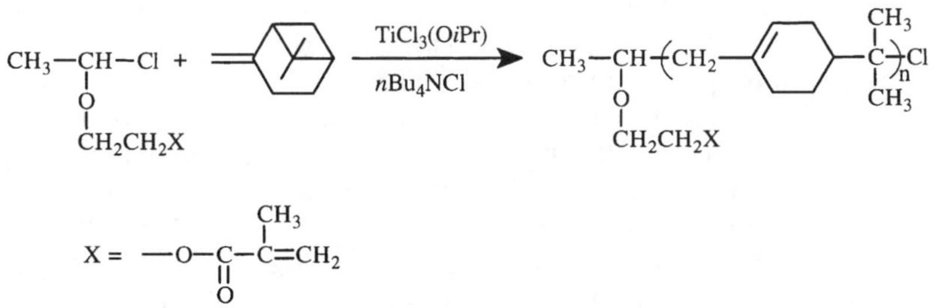

Scheme 10

3.3.1.1.6
Poly(isobutylene)

p-Poly(isobutylene)styrene (S-PIB) was synthesized by polymerizing IB with the *p*-(β-bromoethyl)cumyl chloride (**34**) /(CH$_3$)$_3$Al initiating system in CH$_3$Cl at –55 °C, followed by dehydrobromination with *tert*-BuOK [84]. The polymerization proceeded in the absence of chain transfer to monomer. Although termination was operational with (CH$_3$)$_3$Al, it was apparently slower than propagation since the MWs could be adjusted in the 4000–34,000 g mol^{-1} range by changing the [monomer]/[initiator] ratio. Close to complete monomer conversions could be obtained with [monomer]/[initiator]<100. F_n was determined by UV spectroscopy and found to be close to unity.

$$BrCH_2-CH_2-\underset{}{\underset{}{\bigcirc}}-\underset{CH_3}{\overset{CH_3}{\underset{|}{\overset{|}{C}}}}-Cl \qquad (34)$$

The S-PIB macromonomer was copolymerized by radical copolymerization with MMA and S, and the reactivity ratio of the small comonomer was calculated by a modified copolymer equation [85]. With MMA, r_{MMA}=0.5 was obtained, i.e., close to that reported for conventional S/MMA system. With S however, r_S=2.1 was determined which suggested that the reactivity of S-PIB is lower than that of S, possibly due to steric interference.

Copolymerization of S-PIB with S in dispersion has also been investigated [86]. Conventional emulsion copolymerization with a water soluble initiator resulted in PS homopolymers only, due to the complete insolubility of the S-PIB macromonomer in water. Using AIBN, an oil soluble radical initiator, and very small monomer droplet size (<0.5 µm), initiation and polymerization took place inside the monomer droplets. This method resulted in high yield of poly(S-*g*-IB). The graft copolymer exhibited two T_gs indicating phase separation. However, some phase mixing was suggested by the observed T_g values (–52 and 30 °C), which were considerably different from those of the pure homopolymers. The graft copolymers with relatively long PIB branches (M_n=50,000 g mol^{-1}) exhibited improved impact strength compared to PS.

Asymmetric telechelic α-primary methacrylate ω-*tert*-chloro functional PIB macromonomers (MA-PIB) have been synthesized by living carbocationic polymerization of IB using the 3,3,5-trimethyl-5-chloro-1-hexyl methacrylate (**35**)/TiCl$_4$ initiating system in hexane/CH$_3$Cl (60/40 v/v) [87, 88]. By varying the monomer to initiator ratio, PIBs in the MW range of 2000–40,000 g mol^{-1} were obtained with narrow MWDs. The ester functionality of the polymers (F_n~1) was in good agreement with model reactions indicating that a primary ester functionality was retained after quenching the ester-Lewis acid complexes.

$$CH_2=\overset{CH_3}{\underset{\underset{O}{\|}}{C}}-\overset{}{\underset{}{C}}-O-CH_2-CH_2-\overset{CH_3}{\underset{CH_3}{\underset{|}{C}}}-CH_2-\overset{CH_3}{\underset{CH_3}{\underset{|}{C}}}-Cl \qquad (35)$$

Following quantitative methylation of the ω-*tert*-chloro site by trimethyl aluminum, the methylated polyisobutylene methacrylate macromonomer was co-polymerized with MMA by Group Transfer Polymerization [87]. PMMA-*g*-PIB graft copolymers with controlled MW and composition were obtained. The structure and physical properties were determined by the [MMA]/[MA-PIB] and [MMA]/[Initiator] ratios.

3.3.1.2
Synthesis Using a Functional Capping Agent

In this method, the polymerizable group is incorporated at the ω-end of the macromonomer by a reaction between a functionalized capping agent and the living end of the polymer. To obtain well-defined macromonomers with one polymerizable end group per chain, controlled MW and narrow MWD, the following criteria should be fulfilled:
- living polymerization conditions;
- quantitative coupling of the quencher to the polymer end;
- formation of a stable bond;
- absence of side reaction of the functional group during the quenching process.

3.3.1.2.1
Poly(vinyl ethers)

Sodium salt of malonate carbanions are known to react quantitatively with the living ends of poly(vinyl ether)s to give a stable carbon-carbon bond [45]. This reaction was performed to end-functionalize living poly(vinyl ether)s with a vinyl ether polymerizable end group using the functional malonate ion 36 [73, 89].

$$Na^{+} \ \overset{COOC_2H_5}{\underset{COOC_2H_5}{\underset{|}{\overset{|}{C}}}}-CH_2CH_2-O-CH=CH_2 \qquad (36)$$

IBVE was polymerized with the HI/I$_2$ initiating system in CH$_2$Cl$_2$ at –15 °C. After complete consumption of the monomer, five equivalents of the quenching agent (with respect to the living end) were added. An instantaneous reaction was evidenced by the precipitation of sodium iodide. Poly(IBVE) with controlled MW (by the monomer to HI molar ratio) and narrow MWD was obtained as evidenced by SEC. Moreover, ^{1}H and ^{13}C NMR analyses showed quantitative reaction with the malonate ion and high structural integrity. Particularly, polymerization or side reaction of the vinyl ether function of the quencher could not be detected. The vinylic protons of this vinyl ether end group were recognized together with the other ones of the capping agent. The end functionality was determined using ^{1}H NMR spectroscopy. According to calculation, each polymer chain carried one vinyl ether terminal. The same synthetic method was successfully applied with 2-(benzoyloxy)ethyl vinyl ether (BzOVE) which was polymerized using the HI/I$_2$ initiating system in toluene at –15 °C. Quantitative termination was performed with **36** used in 10–20-fold excess. More recently, the same quencher **36** was used to prepare poly(vinyl ether)s block copolymers with a polymerizable vinyl ether end group [90]. Sequential living cationic polymerization of AcOVE and IBVE was carried out with the HI/ZnI$_2$ system in toluene at –15 °C. After quenching with **36**, well-defined block copolymer macromonomers were obtained: H-(poly(AcOVE))$_m$-(poly(IBVE))$_n$-C(COOC$_2$H$_5$)$_2$-CH$_2$CH$_2$-O-CH=CH$_2$ with m=5/n=15 and m=10/n=30. Further living cationic polymerization of those macromonomers bearing a vinyl ether end group will be described in Sect. 3.3.2.

A hydroxy function is also able to react quantitatively with living end of poly(vinyl ether)s resulting in an acetal end group which, however, has poor stability in acidic media. A proton trap should be added in order to scavenge the protons released during the coupling process. Various end-capping agents with a primary alcohol and a polymerizable double bond were used to produce poly(vinyl ether) macromonomers. The more widely used was 2-hydroxyethyl methacrylate (HEMA, **37**) [91–94] but some alcohols with an allylic or olefinic group were also reported such as allyl alcohol (**38**) [91], 2-[2-(2-allyloxyethoxy)ethoxy] ethanol (**39**) [92] and 10-undecen-1-ol (**40**) [92]. Another capping agent with a methacrylate ester group, 2-(dimethylamino)ethyl methacrylate (**41**) with a tertiary amine as the coupling nucleophilic function [91] was also reported. In that case, capping results in the formation of a quaternary ammonium salt.

$$HO-CH_2CH_2-O-\underset{\underset{O}{\|}}{C}-\underset{\underset{}{|}}{\overset{CH_2}{C}}=CH_3 \qquad (37)$$

$$HO-CH_2-CH=CH_2 \qquad (38)$$

$$HO-(CH_2CH_2O)_3-CH_2-CH=CH_2 \qquad (39)$$

$$HO-(CH_2)_9-CH=CH_2 \qquad (40)$$

$$\underset{H_3C}{\overset{H_3C}{\diagdown}}N-CH_2CH_2-O-\underset{\underset{O}{\|}}{C}-\underset{\underset{}{|}}{\overset{CH_2}{C}}=CH_3 \qquad (41)$$

Well-defined macromonomers of poly(BVE), poly(IBVE), and poly(EVE) with ω-methacrylate end group [91] were prepared by living cationic polymerization of the corresponding monomers initiated by trifluoromethanesulfonic acid in CH_2Cl_2 at -30 °C in the presence of thiolane as a Lewis base. After complete conversion, the polymers were quenched with **37** in the presence of 2,6-lutidine or with **41** to produce macromonomers with M_n up to 10,000 g mol^{-1}, with narrow MWD, bearing one polymerizable methacrylate function per molecule. The same polymers were also quenched with **38** in the presence of 2,6-lutidine to give poly(vinyl ether)s with an allylic terminal group.

A vinyl ether with a mesogenic side group, 3-[4-cyano-4'-biphenyl)oxy]propyl vinyl ether, was polymerized at 0 °C in CH_2Cl_2 with trifluoromethanesulfonic acid as an initiator in the presence of dimethylsulfide [92].

$H_2C=CH-O-CH_2CH_2CH_2-O-\text{C}_6\text{H}_4-\text{C}_6\text{H}_4-CN$

3-[4-cyano-4'-biphenyl)oxy]propyl vinyl ether

Three macromonomers were obtained after quenching with **37**, **39**, and **40** respectively. All had a well defined structure as evidenced by SEC and NMR analyses, with DP_n close to 6 and narrow MWD (M_w/M_n=1.09–1.13) and with F_n very close to 1.

Macromonomers of poly(octadecyl vinyl ether) were prepared by cationic polymerization although this technique is difficult since the corresponding ODVE monomer has a high melting point and poor solubility at low temperature [93]. The polymerization was initiated by the trimethylsilyl iodide/1,1-diethoxyethane system in the presence of ZnI_2 (initiator solution was prepared at -40 °C) and was carried out in toluene at 0 or 10 °C. Linearity of M_n with conversion was observed up to 6000 g mol^{-1}. At higher M_n, deviation was observed and this was assigned to transfer to the monomer. However, although transfer leads to dead chains with a terminal unsaturation, it was shown that upon addition of **37** as a capping agent, all the chains were quenched irrespective of their end group. Actually, the dead chains could also react with the alcoholic quencher after their protonation by HI released in the reaction of the same quencher with the active chains. Thus, although controlled polymerization was not achieved, quantitative end-functionalization of poly(ODVE) with **37** could be obtained. Short macromonomers (M_n<6000 g mol^{-1}) had narrow MWD with M_w/M_n=1.1. Free-radical homopolymerization of these poly(ODVE) macromonomers and copolymerization with acrylates produced highly branched polymers [94] which were soluble in hydrocarbons.

Macromonomers with a well-defined units sequence were reported by Minoda et al. [95–97] (so-called sequence regulated oligomers). They were prepared by sequential introduction of one equivalent of each vinyl ether monomer to the living ends in toluene at -40 °C in the presence of ZnI_2, the first step being the addition of HI to the first monomer. The first example of sequence regulated macromonomer [95, 96] was obtained with the following monomers sequence: n-butyl vinyl ether (BVE), VOEM, BzOVE, and 2-(vinyloxy)ethyl methacrylate, this latter monomer being used to incorporate a terminal methacrylate. The macromonomer was purified from by-products by preparative size exclusion chromatography. The second example was a heterodimer quenched with the malonate anion 36 [97]. It was prepared by a three step technique: quantitative addition of HI to the first vinyl ether, addition of one equivalent of a second monomer (preferably less reactive than the first one) in the presence of ZnI_2, and finally quenching with 36 which allowed introduction of the polymerizable vinyl ether end group. Two macromonomers were synthesized by this method with the respective following sequence: BVE/BzOVE and 2-ethylhexyl vinyl ether/CEVE.

3.3.1.2.2
Poly(p-alkoxystyrenes)

Owing to the lower stability of growing p-alkoxystyrene cations and to the possibility of several side-reactions, some end-capping agents which were successfully used for poly(vinyl ether)s such as sodiomalonic ester and *tert*-butyl alcohol, did not give end-functionalization with poly(p-alkoxystyrene) cations. In contrast, primary and secondary alcohols underwent quantitative reactions to give stable alkoxy functional groups. Thus, 2-hydroxyethyl methacrylate and acrylate were used to introduce a polymerizable group at the ω end [98, 99]. Living cationic polymerizations of p-MOS and *t*BOS were carried out at –15 °C in toluene using HI/ZnI_2 as an initiating system. When the monomer conversion was complete, a large excess of the quencher was added, resulting in a quantitative functionalization. The polymerization was shown to be living and well-defined macromonomers with narrow MWD and one polymerizable acrylate or methacrylate functional group per chain were obtained. Heterotelechelic poly(p-MOS)s were also prepared by the combination of the functional initiator method and the functional end-capping method. This allowed the synthesis of a poly(p-MOS) macromonomer with one malonate diester at the α end and one methacrylate group at the ω end.

3.3.1.2.3
Poly(styrene)

The living cationic polymerization of styrene could be achieved using 1-phenylethyl chloride as an initiator in the presence of $SnCl_4$ and n-Bu_4NCl. However, in contrast to vinyl ethers and p-alkoxystyrenes, quenching with usual bases such as methanol, sodium methoxide, benzylamine, or diethyl sodiomalonate led to the terminal chloride instead of the specific end group. This was explained by the very low concentration of cationic species in comparison with the dormant C-Cl end group and also by the low reactivity of this C-Cl functional group in substitution reactions. This was overcome using organosilicon compounds such as trimethylsilyl methacrylate (**42**) and quantitative functionalization was achieved when the quenching reaction was performed at 0 °C, for 24 h, in the presence of a large excess of the quencher and low concentration of the Lewis acid [100]. The same functionalization reaction could also be performed successfully in two steps, starting from an isolated polystyrene with a C-Cl terminal group.

$$CH_2=\overset{CH_3}{\underset{}{C}}-\underset{\underset{O}{\|}}{C}-O-Si(CH_3)_3 \qquad (42)$$

3.3.1.2.4
Poly(isobutylene)

Allyl terminated linear and three arm star PIBs and epoxy and hydroxy telechelics therefrom have been reported by Ivan and Kennedy [34]. Allyl functional PIBs were obtained in a simple one pot procedure involving living IB polymerization using $TiCl_4$ as coinitiator followed by end-quenching with allyltrimethylsilane (**43**, ATMS). The procedure was based on an earlier report by Wilczek and Kennedy [101] that demonstrated quantitative allylation of PIB-Cl by ATMS in the presence of Et_2AlCl or $TiCl_4$. Structural characterization by ^1H NMR spectroscopy and end group titration by *m*-chloroperbenzoic acid demonstrated quantitative end allylation. Quantitative hydroboration followed by oxidation in alkaline THF at room temperature resulted in -OH functional PIBs, which were used to form PIB-based polyurethanes. Quantitative epoxidation of the double bonds was also achieved with *m*-chloroperbenzoic acid in $CHCl_3$ at room temperature, giving rise to macromonomers able to polymerize by ring-opening polymerization. A three arm star epoxy-telechelic PIB (M_n=4500 g mol^{-1}) with triethyl amine gave a strong rubber exhibiting ca. 300% elongation.

$$CH_2=CH-CH_2-\underset{\underset{CH_3}{|}}{\overset{\overset{CH_3}{|}}{Si}}-CH_3 \qquad (43)$$

3.3.1.3
Chain End Modification of Poly(isobutylene)

The polymerizable function is incorporated by chemical modification of the α or ω end group after isolation of the polymer. Although it is versatile since a wide variety of polymerizable groups can be incorporated, the method generally involves several steps.

The synthesis of polyisobutylene methacrylate (MA-PIB) was first reported by Kennedy and Hiza [102]. The synthesis was accomplished by a multistep process. First IB was polymerized by the cumyl chloride/BCl_3 initiating system.

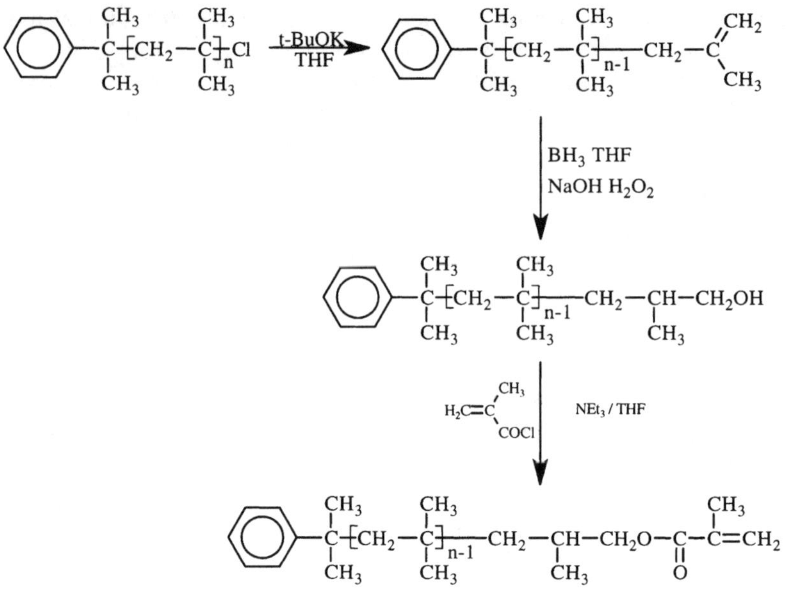

Scheme 11

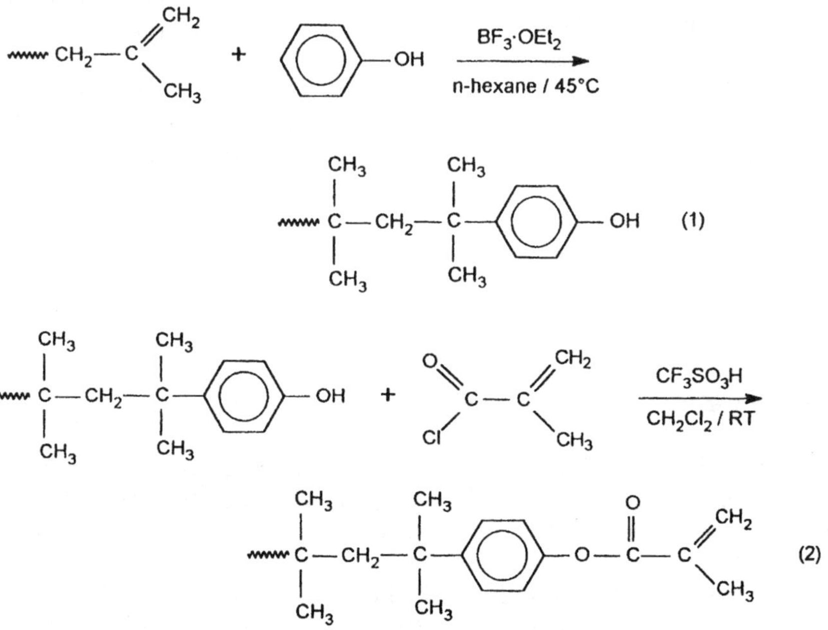

Scheme 12

Dehydrochlorination followed by hydroboration-oxidation resulted in PIB-CH_2OH, which was subsequently esterified with methacryloyl chloride. Scheme 11 helps to visualize the procedure.

The structure and value of F_n were determined by ^1H NMR and IR spectroscopies combined with MW determination by vapor pressure osmometry. According to the results, MA-PIB macromonomers indeed carried close to one methacrylate function per molecule. The homopolymerization of a relatively low MW (M_n=5200 g mol^{-1}) MA-PIB was attempted by radical means. Polymerization did not take place in solution. In bulk however, a portion of the macromonomer polymerized to give a star-like product with high MW (M_v~7.10^5 g mol^{-1}). Free radical nearly ideal copolymerization of MA-PIB with MMA afforded PMMA-g-PIB copolymers which were optically clear but had disappointingly low tensile strength and modulus. All graft copolymers exhibited two T_gs, one at ~–65 °C for PIB and one at ~100 °C for the PMMA component, indicating microphase separated morphology.

Macromonomers with two methacrylate functionalities (MA-PIB-MA) at both ends of the PIB chain have also been synthesized, by a procedure essentially identical to that reported above, but starting with a bifunctional initiator in the polymerization of IB [103]. Free radical copolymerization of the resulting MA-PIB-MA with 2-(dimethylamino)ethyl methacrylate resulted in amphiphilic networks, with a wide range of mechanical and swelling properties.

The acrylate or methacrylate functional PIBs have also been used in UV-curable solventless coatings formulation in the presence of reactive diluents (multifunctional acrylate or methacrylate esters) and a UV-sensitizer [104]. The products were transparent, flexible films, with very little extractables, in which hard polyacrylate or polymethacrylate domains were dispersed in the soft PIB matrix. Tensile strength and ultimate elongation have also been obtained.

The synthesis of MA-PIB macromonomers by three different methods, which were claimed to be less cumbersome than that above, was reported by Maenz and Stadermann [105]. The first procedure, as shown in Scheme 12, involved alkylation of phenol by PIB olefin followed by reaction with methacrylic acid.

The PIB olefins were either commercial products ("Glissopal" by BASF, "HYVIS 5" by BP Chemical Ltd., and "Polybutene" by Amoco Chemicals Co.) or were obtained by selective polymerization of butadiene free C_4-fractions. It should be noted that before esterification non-functional PIBs, present in the commercial products in varying amounts, were removed by column chromatography. The best results were obtained with Glissopal which had the highest double bond functionality, ~0.85. Interestingly, the number average double bonds ($F_{n(DB)}$) determined by ozonolysis or by ^1H NMR spectroscopy differed considerably and relatively good correlation between $F_{n(DB)}$ and F_n of PIB-phenol was obtained only for Glissopal. This suggests that in the other samples a relatively large fraction of the double bonds were not located at the polymer end. The second synthetic route is shown in Scheme 13.

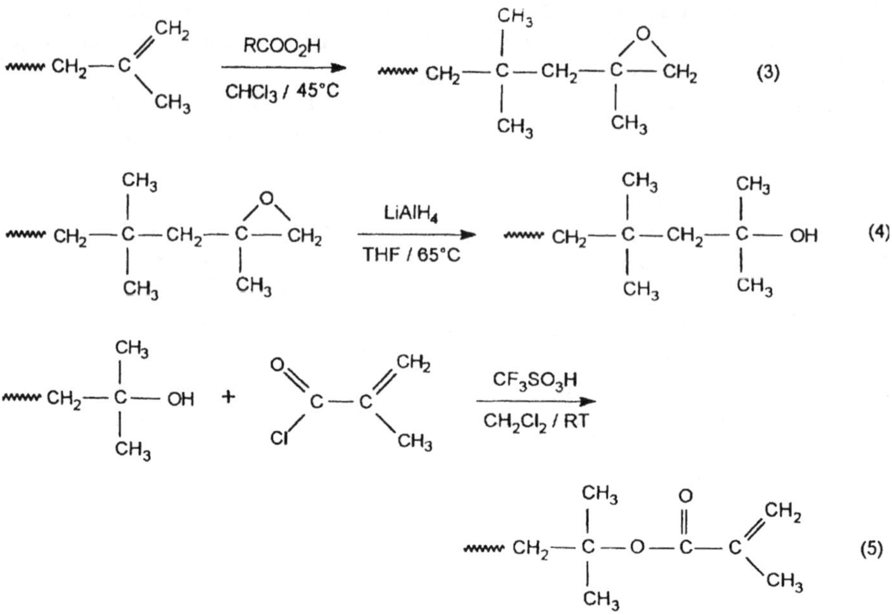

Scheme 13

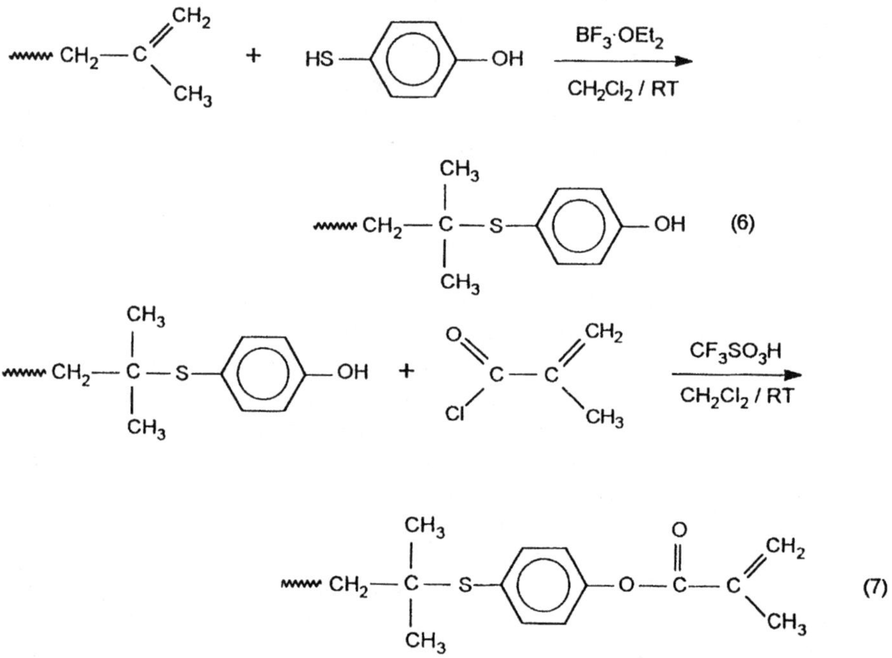

Scheme 14

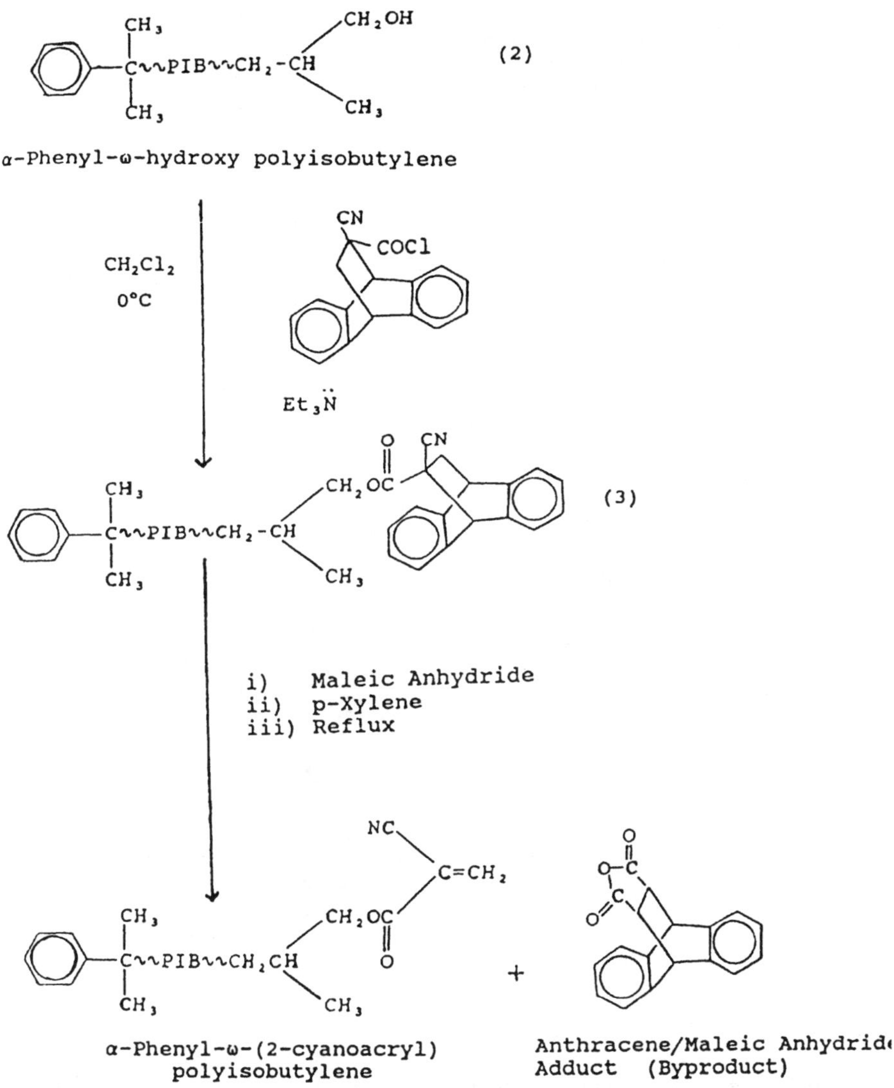

Scheme 15

α-Phenyl-ω-hydroxy polyisobutylene (2)

Reagents: CH$_2$Cl$_2$, 0°C; anthracene-9-carbonitrile-10-carbonyl chloride; Et$_3$N

(3)

i) Maleic Anhydride
ii) p-Xylene
iii) Reflux

α-Phenyl-ω-(2-cyanoacryl) polyisobutylene + Anthracene/Maleic Anhydride Adduct (Byproduct)

In this process, epoxidation of the double bonds was followed by reduction to obtain the *tert*-alcohol which was esterified with methacryloyl chloride in the subsequent step. While epoxidation was found to be close to quantitative based on double bond content, reduction was incomplete and the residual epoxy functional PIB (24–47%) had to be separated by column chromatography before esterification. It should be noted that this macromonomer was a *tert*-ester which might be quite unstable in acidic conditions, and is also more hindered than the

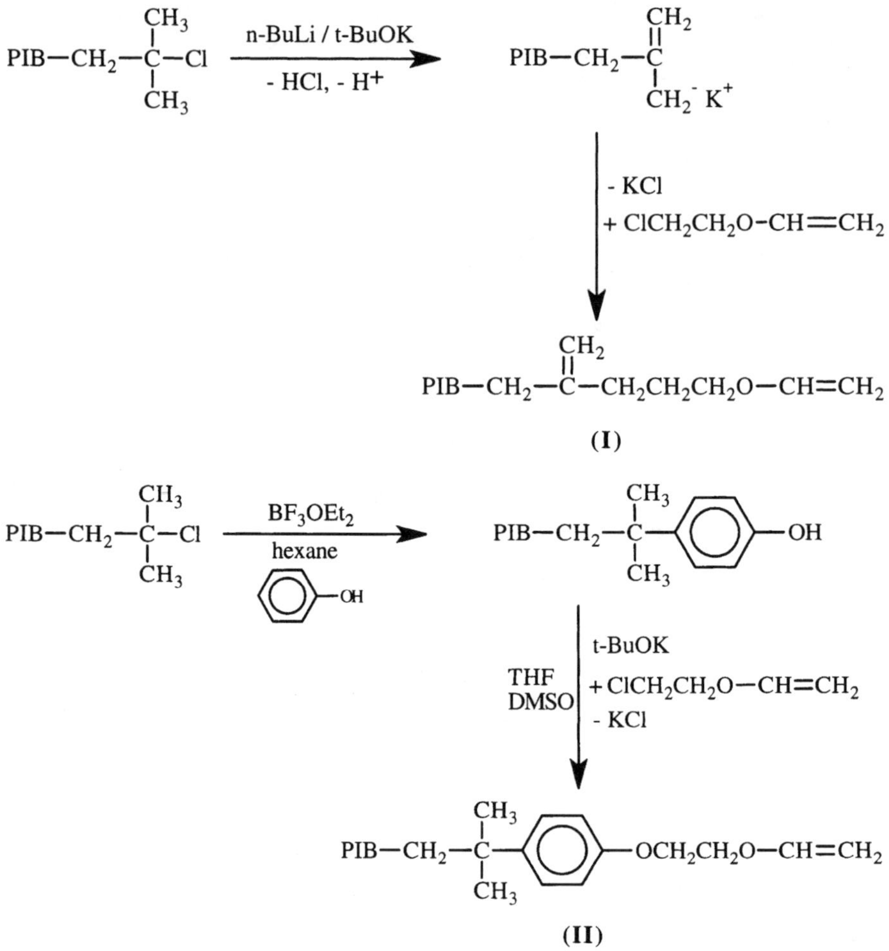

Scheme 16

corresponding primary ester which may affect the copolymerization behavior of the macromonomer.

The third route for the synthesis of PIB macromonomer was based on the addition of *p*-hydroxy-thiophenol onto the PIB double bonds followed by esterification with methacryloyl chloride (Scheme 14).

High conversion of double bonds was found only with Glissopal, but it was necessary, even in this case, to separate from non-functional PIB before esterification. While the feasibility to synthesize methacrylate functional PIBs by all three methods was demonstrated, due to the necessary column chromatography step to obtain high functionality PIB macromonomers, the utility of the method is questionable.

Table 5. Synthesis of macromonomers by cationic polymerization

Nature of the chain	Nature of the polymerizable function	Method used[a]	Reference
Poly(EVE)	Methacrylate	30	[71–73]
Poly(SiVE)	Methacrylate	30	[75]
Poly(EVE)	Allyl, epoxyde	32	[73, 74]
Poly(Silyl vinyl ether) or poly(vinyl alcohol)	Styrene	33	[76–78]
PS Poly(p-MeS)	Methacrylate	31	[79]
Poly(α-MeS)	Methacrylate	31	[80]
Poly(β-pinene) Poly(β-pinene-b-p-MeS)	Methacrylate	31	[83]
PIB	Styrene	34	[84–86]
PIB	Methacrylate	35	[87, 88]
Poly(IBVE)	Vinyl ether	36	[89]
Poly(AcOVE-b-IBVE)	Vinyl ether	36	[90]
Poly(BVE), Poly(IBVE), Poly(EVE)	Methacrylate	37, 41	[91]
Poly(BVE), Poly(IBVE), Poly(EVE)	Allyl	38	[91]
Poly(3-[4-cyano-4'-biphenyl) oxy]propyl vinyl ether)	Methacrylate	37	[92]
	Allyl	39, 40	[92]
Poly(ODVE)	Methacrylate	37	[93, 94]
Sequence regulated oligomers of vinyl ethers	Vinyl ether	36	[97]
Poly(p-MOS)	Methacrylate	37	[98]
Poly(tBOS)	Methacrylate	37	[99]
PS	Methacrylate	42	[100]
PIB	Allyl, epoxyde	43	[34]
PIB	Methacrylate	–	[102–105]
	Cyanoacrylate	–	[106, 107]
	Vinyl ether	–	[108]

[a]Numbers 30–35 = functional initiator; 36–43 = functional terminator; – chain end modification

Cyanoacrylate capped PIB (CA-PIB) has been synthesized by esterification of PIB-CH$_2$OH with the Diels-Alder adduct of 2-cyanoacryloyl chloride and anthracene followed by deprotection (Scheme 15) [106, 107].

The value of F$_n$ was determined by ^{1}H NMR spectroscopy and found to be close to unity. By essentially the same method, bifunctional and trifunctional cyanoacrylate functional PIBs have also been prepared. Anionic polymerization of CA-PIB with N,N-dimethyl-p-toluidine as initiator in solution resulted in high MW product (M$_n$~35,000 g mol^{-1}) [107]. Anionic copolymerization of difunctional and trifunctional PIB yielded clear flexible films with low sol fraction. The di- and trifunctional macromonomers have also been found to undergo chain extension upon contact with proteinaceous materials such as human blood and egg yolk.

Vinyl ether terminated PIBs with different endgroup structures (**I** and **II** in Scheme 16) have been synthesized by Nemes et al. [108]. Scheme 16 summarizes the key transformation steps.

In the first case PIB-Cl was dehydrochlorinated and metallated in a one pot procedure. This was followed by coupling of the resulting PIB anion with CEVE. In the second process, phenol was alkylated with PIB-Cl followed by a reaction with CEVE. The value of F_n determined by ^{1}H NMR spectroscopy indicated close to quantitative functionalization. Copolymerization of the macromonomers has not been reported.

Table 5 summarizes the work on the synthesis of macromonomers by cationic polymerization.

3.3.2
Cationic Polymerization of Macromonomers

Generally, macromonomers are (co)polymerized by free-radical processes owing to convenient experimental conditions, availability of a large number of comonomers, and insensitivity of most chemical functions to the polymerization conditions. Nevertheless, some macromonomers with a suitable end group have been (co)polymerized by cationic polymerization. Provided that living cationic polymerization conditions are applied, well-defined graft homopolymers or copolymers can be prepared with a predetermined and uniform number of branches.

3.3.2.1
Vinyl Ether Polymerizable Group

Macromonomers bearing a vinyl ether end group can be cationically polymerized. This is the case for poly(vinyl ether) macromonomers prepared by living cationic polymerization where the vinyl ether end group was introduced by endcapping with the sodium salt of VOEM (see Sect. 3.3.1.2). For instance poly(IBVE) and poly(BzOVE) macromonomers with homopolymer chain [89] and poly(AcOVE-*b*-IBVE) with block copolymer chain of various length and composition [90] were prepared by this technique. Preliminary studies showed that the first two homopolymer macromonomers underwent quantitative cationic homopolymerization and copolymerization with IBVE using HI/I_2 as an initiating system in CH_2Cl_2 at –15 °C. A more comprehensive study was performed with the block copolymer macromonomer. Living cationic polymerization was carried out using the HI/ZnI_2 initiating system in toluene at –15 °C. The influence of steric effect on conversion was examined by varying the total length of the macromonomers at a constant AcOVE/IBVE molar ratio. It was shown that the shorter the chain, the higher the polymer yield. Influence of the composition was also studied and it appeared that a larger amount of AcOVE was responsible for retardation of the polymerization because the ester group could complex the Lewis acid and reduce its effective concentration. Finally, the best conditions were found for a macromonomer with 5 AcOVE units and 10 IBVE units (M_n= 2600 g mol^{-1} with M_w/M_n=1.13 and F_n=1.10). Homopolymerisation was carried out in toluene at –15 °C and 85% conversion were reached in 3 h to lead to a higher MW polymer with narrow MWD (M_w=15,000 g mol^{-1} as determined by light scattering after fractionation and M_w/M_n=1.16 as determined by SEC). The calculated DP_n was 6.3 which was very close to the theoretical value and indicated that living polymerization conditions were fulfilled leading to a well-defined star-like block copolymer with the predetermined and uniform number of branches. The pendant ester groups of AcOVE were hydrolyzed to their alcoholic counterpart to give the amphiphilic graft copolymer, the solubility properties of which were compared with the corresponding linear and star block copolymers.

Some previously reported sequence-regulated macromonomers [97] were also homopolymerized using non-living (BF_3OEt_2 as an initiator in toluene at –15 °C) and living conditions (HI/ZnI_2 as an initiator in toluene at –15 °C). In both cases, nearly quantitative conversion of the two macromonomers was reached, indicating their ability to undergo cationic polymerization. However, in the first case, a considerable amount of dimer was recovered whereas, under living conditions, the resulting polymer had an average degree of polymerization of 9.4, very close to the theoretical value, with narrow MWD (M_w/M_n<1.1).

4
Hyperbranched Polymers

Highly branched, so called "hyperbranched" macromolecules have recently attracted considerable interest, in the hope that their properties would closely resemble those of dendrimers. Dendrimers have highly regular branched structures, with promising attributes in a variety of applications from catalysis to drug delivery. They are however, only available through laborious multistep procedures. Hyperbranched polymers, formally prepared by polycondensation of AB_2 type monomers, have recently been prepared starting from AB type monomers by a process termed self-condensing vinyl polymerization [109]. In this process, a vinyl monomer with a pendant initiating moieties is used. Addition of this monomer to an active center (radical, cation, or anion) creates two active sites (a propagating one and an initiating one).

Self-condensing vinyl polymerization was first demonstrated with 3-(1-chloroethyl)-ethenylbenzene as an AB type monomer. The cationic polymerization was induced by $SnCl_4$ in CH_2Cl_2 at –15 or –20 °C in the presence of tetrabutylammoniumbromide (Scheme 17).

The MW – time profile closely resembled that of condensation polymerization; a slow initial increase was followed by an exponential growth in MW with time. After 18 h the polymer exhibited M_w~250,000 g mol^{-1}, and M_w/M_n=6. The

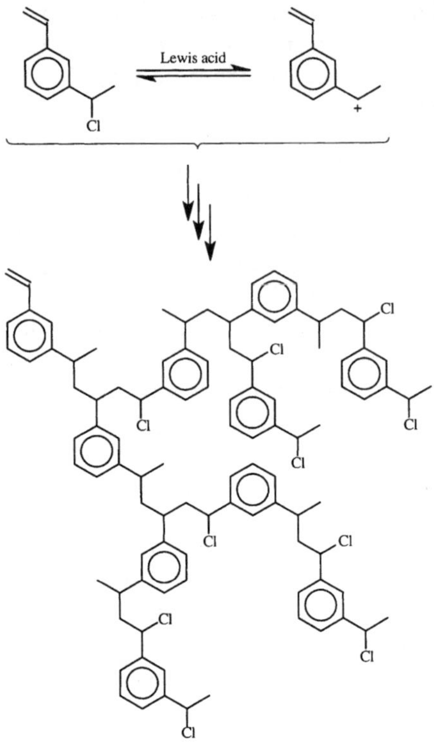

Scheme 17

final product upon quenching with methanol was an irregular branched polymer with numerous chloride functions (it is unlikely that methoxy functions, assumed by the authors are present in measurable amounts).

The effect of reaction parameters such as [monomer]/[SnCl$_4$] ratio, nature of Lewis acid, and quenching agent was also studied [110]. Interestingly, soluble product was only obtained with *m*-substituted styrene. With para-substituted styrene the polymer obtained after precipitation in methanol and drying was insoluble. When 3-(1-chloroethyl)-ethenylbenzene was polymerized under identical conditions with different Lewis acids, the MW and polydispersity of the products increased in the order BCl$_3$<SnCl$_4$<TiCl$_4$. Side reactions such as intra and intermolecular alkylation are also expected to increase in the same order and apparently contribute to the broad MWDs. An attempt to obtain allyl functionality, by adding allyltrimethylsilane after the polymerization resulted in 66% functionalization.

5
Conclusion

This review covered recent developments in the synthesis of branched (star, comb, graft, and hyperbranched) polymers by cationic polymerization. It should be noted that although current examples in some areas may be limited, the general synthetic strategies presented could be extended to other monomers, initiating systems etc. Particularly promising areas to obtain materials formerly unavailable by conventional techniques are heteroarm star-block copolymers and hyperbranched polymers. Even without further examples the number and variety of well-defined branched polymers obtained by cationic polymerization should convince the reader that cationic polymerization has become one of the most important methods in branched polymer synthesis in terms of scope, versatility, and utility.

Acknowledgements. This review could not have been written without support from CNRS (France) and NSF (USA) (INT-Grant No. 9512834) that made collaboration between the two laboratories possible.

6
References

1. Kennedy JP, Ivan B (1992) Designed polymers by carbocationic macromolecular engineering – theory and practice. Hanser Publishers
2. Matyjaszewski K (1996) Cationic polymerizations, mechanisms, synthesis and applications. Marcel Dekker
3. Kanaoka S, Sawamoto M, Higashimura T (1991) Macromolecules 24:2309
4. Sawamoto M, Kanaoka S, Omura T, Higashimura T (1992) ACS Polym Prepr 33(1):148
5. Deng H, Kanaoka S, Sawamoto M, Higashimura T (1996) Macromolecules 29:1772
6. Marsalko TM, Majoros I, Kennedy JP (1993) Polym Bulletin 31:665
7. Marsalko TM, Majoros I, Kennedy JP (1995) Macromol Symp 95:39
8. Storey RF, Shoemake KA, Chisholm BJ (1996) J Polym Sci: Part A: Polym Chem 34:2003
9. Storey RF, Shoemake KA (1996) ACS Polym Prepr 37(1):327
10. Kennedy JP, Marsalko TM, Majoros I (1996) Macromol Symp 107:319
11. Marsalko TM, Majoros I, Kennedy JP (1997) J Macromol Sci-Pure Appl Chem A34:775
12. Wang L, McKenna ST, Faust R (1995) Macromolecules 28:4681
13. Kanaoka S, Sawamoto M, Higashimura T (1991) Macromolecules 24:5741
14. Kanaoka S, Sawamoto M, Higashimura T (1993) Makromol Chem 194:2035
15. Storey RF, Shoemake KA (1996) ACS Polym Prepr 37(2):321
16. Asthana S, Majoros I, Kennedy JP (1997) ACS PMSE Prepr 77:187
17. Kanaoka S, Sawamoto M, Higashimura T (1993) Macromolecules 26:254
18. Kanaoka S, Omura T, Sawamoto M, Higashimura T (1992) Macromolecules 25:6407
19. Shohi H, Sawamoto M, Higashimura T (1991) Macromolecules 24:4926
20. Sawamoto M, Higashimura T (1991) Makromol Chem, Macromol Symp 47:67
21. Shohi H, Sawamoto H, Sawamoto M, Higashimura T (1992) ACS Polym Prepr 33(1):960
22. Sawamoto M, Shohi H, Sawamoto H, Fukui H, Higashimura T (1994) J Macromol Sci Pure Appl Chem A31:1609
23. Shohi H, Sawamoto M, Higashimura T (1992) Makromol Chem 193:2027
24. Cloutet E, Fillaut JL, Gnanou Y, Astruc D (1994) J Chem Soc, Chem Commun 2433

25. Cloutet E, Fillaut JL, Astruc D, Gnanou Y (1997) International Symposium on Ionic Polymerization, Paris (to be published in Macromol Symp) 139
26. Kennedy JP, Ross LR, Lackey JE, Nuyken O (1981) Polym Bull 4:67
27. Faust R, Fehervari A, Kennedy JP (1985) ACS Symp Ser 282:125
28. Mishra MK, Wang B, Kennedy JP (1987) Polym Bull 17:307
29. Zsuga M, Balogh L, Kelen T, Borbély L (1990) Polym Bull 23:335
30. Zsuga M, Kelen T, Borbély J (1991) Polym Bulletin 26:417
31. Chen CC, Kaszas G, Puskas JE, Kennedy JP (1989) Polym Bull 22:463
32. Feldthusen J, Ivan B, Müller AHE (1997) Macromol Rapid Commun 18:417
33. Storey RF, Lee Y (1992) J Macromol Sci Pure Appl Chem A29:1017
34. Ivan B, Kennedy JP (1990) J Polym Sci: Part A: Polym Chem 28:89
35. Huang KJ, Zsuga M, Kennedy JP (1988) Polym Bull 19:43
36. Cloutet E, Fillaut JL, Gnanou Y, Astruc D (1996) Chem Commun 17:2047
37. Jacob S, Majoros I, Kennedy JP (1996) Macromolecules 29:8631
38. Shohi H, Sawamoto M, Higashimura T (1991) Polym Bull 25:529
39. Kaszas G, Puskas JE, Kennedy JP, Hager WG (1991) J Polym Sci: Part A: Polym Chem 29:427
40. Storey RF, Chisholm BJ, Lee Y (1993) Polymer 34:4330
41. Jacob S, Majoros I, Kennedy JP (1997) ACS PMSE Prepr 77:185
42. Fodor Zs, Faust R (1995) J Macromol Sci -Pure Appl Chem A32:575
43. Gadkari A, Kennedy JP (1989) J Appl Polym Sci 44:19
44. Feldhusen J, Ivan B, Müller AHE (1998) Macromolecules 31:578
45. Sawamoto M, Enoki T, Higashimura T (1987) Macromolecules 20:1
46. Fukui H, Sawamoto M, Higashimura T (1993) J Polym Sci: Part A: Polym Chem 31:1531
47. Fukui H, Sawamoto M, Higashimura T (1993) Macromolecules 26:7315
48. Fukui H, Sawamoto M, Higashimura T (1994) Macromolecules 27:1297
49. Fukui H, Sawamoto M, Higashimura T (1994) J Polym Sci: Part A: Polym Chem 32:2699
50. Omura N, Lubnin AV, Kennedy JP (1997) ACS Symp Ser 665:178
51. Majoros I, Marsalko TM, Kennedy JP (1997) Polym Bull 38:15
52. Omura N, Kennedy JP (1997) Macromolecules 30:3204
53. Fukui H, Sawamoto M, Higashimura T (1995) Macromolecules 28:3756
54. Fukui H, Yoshihashi S, Sawamoto M, Higashimura T (1996) Macromolecules 29:1862
55. Bae YC, Fodor Zs, Faust R (1997) Macromolecules 30:198
56. Bae YC, Faust R (1998) Macromolecules (in press)
57. Lemaire C, Tessier M, Maréchal E (1997) Macromol Symp 122:371
58. Sinai-Zingde G, Verma A, Liu Q, Brink A, Bronk J, Allison D, Goforth A, Patel N, Marand H, McGrath JE, Riffle JS (1990) ACS Polym Prepr 31(1):63
59. Merrill NA, Powers KW, Wang HC (1992) ACS Polym Prepr 33(1):962
60. Nuyken O, Sanchez JR, Voit B (1997) Macromol Rapid Commun 18:125
61. Matyjaszewski K, Hrkach JS (1995) J Inorg Organomet Polym 5(2):183
62. Sogah DY, Webster OW (1986) Macromolecules 19:1775
63. Sogah DY, Webster OW (1987) In: Fontanille M, Guyot A (eds) Recent advances in mechanistic and synthetic aspects of polymerization, p 61
64. Jiang Y, Fréchet JMJ (1989) ACS Polym Prepr 30(1):127
65. (a) Puskas JE (1997) International Symposium on Ionic Polymerization, Paris (to be published in Macromol Symp); (b) Puskas JE, Wilds CJ (1998) J Polym Sci: Part A: Polym Chem 36:85
66. Gadkari A, Farona MF (1987) Polym Bull 17:299
67. Schappacher M, Deffieux A (1997) International Symposium on Ionic Polymerization, Paris (to be published in Macromol Symp)
68. Labeau MP, Cramail H, Deffieux A (1996) Polym Int 41:453
69. Gyor M, Kitayama T, Fujimoto N, Nishiura T, Hatada K (1994) Polym Bull 32:155
70. Peng YX, Dai HS (1997) J Macromol Sci Pure Appl Chem A34:1285
71. Aoshima S, Ebara K, Higashimura T (1985) Polym Bull 14:425

72. Higashimura T, Aoshima S, Sawamoto M (1986) Makromol Chem Macromol Symp 3:99
73. Sawamoto M, Aoshima S, Higashimura T (1988) Makromol Chem Macromol Symp 13/14:513
74. Higashimura T, Aoshima S, Sawamoto M (1988) ACS Polym Prepr 29(2):1
75. Higashimura T, Ebara K, Aoshima S (1989) J Polym Sci:Part A: Polym Chem 27:2937
76. Kawakamie Y, Aoki T, Yamashita Y (1987) Polym Bulletin 18:473
77. Charleux B, Pichot C (1993) Polymer 34:195
78. Charleux B, Pichot C, Llauro MF (1993) Polymer 34:4352
79. Miyashita K, Kamigaito M, Sawamoto M, Higashimura T (1994) Macromolecules 27:1093
80. Sawamoto M, Hasebe T, Kamigaito M, Higashimura T (1994) J Macromol Sci Pure Appl Chem A31:937
81. Fukui H, Deguchi T, Sawamoto M, Higashimura T (1996) Macromolecules 29:1131
82. Lu J, Kamigaito M, Sawamoto M, Higashimura T (1997) Macromolecules 30:22
83. Lu J, Kamigaito M, Sawamoto M, Higashimura T, Deng YX (1997) J Polym Sci: Part A: Polym Chem 35:1423
84. Kennedy JP, Lo CY (1982) ACS Polym Preprint 23:99
85. Kennedy JP, Lo CY (1982) Polym Bull 8:63
86. Kennedy JP, Lo CY (1985) Polym Bull 13:441
87. Balogh L, Takacs A, Faust R (1992) ACS Polym Prepr 33(1):958
88. Takacs A, Faust R (1996) J Macromol Sci Pure Appl Chem A33:117
89. Sawamoto M, Enoki T, Higashimura T (1986) Polym Bull 16:117
90. Kanaoka S, Sueoka M, Sawamoto M, Higashimura T (1993) J Polym Sci: Part A: Polym Chem 31:2513
91. Goethals EJ, Haucourt N, Verheyen AM, Habimana J (1990) Makromol Chem, Rapid Commun 11:623
92. Percec V, Lee M, Tomazos D (1992) Polym Bull 28:9
93. Lievens SS, Goethals EJ (1996) Polym Int 41:277
94. Goethals EJ, Roose P, Reyntjens W, Lievens S (1997) International Symposium on Ionic Polymerization, Paris (to be published in Macromol Symp) 64
95. Minoda M, Sawamoto M, Higashimira T (1990) Polym Bull 23:133
96. Minoda M, Sawamoto M, Higashimura T (1990) Macromolecules 23:4890
97. Minoda M, Sawamoto M, Higashimura T (1993) J Polym Sci: Part A: Polym Chem 31:2789
98. Shohi H, Sawamoto M, Higashimura T (1992) Macromolecules 25:53
99. Shohi H, Sawamoto M, Higashimura T (1992) Makromol Chem 193:1783
100. Miyashita K, Kamigaito M, Sawamoto M (1994) J Polym Sci: Part A: Polym Chem 32:2531
101. Wilczek L, Kennedy JP (1987) J Polym Sci: Polym Chem Ed 25:3255
102. Kennedy JP, Hiza M (1983) J Polym Sci: Polym Chem Ed 21:1033
103. Chen D, Kennedy JP, Allen AJ (1988) J Macromol Sci Chem A25:389
104. Puskas JE, Kaszas G, Chen CC, Kennedy JP (1988) Polym Bulletin 20:253
105. Maenz K, Stadermann D (1996) Angew Makromol Chem 242:183
106. Kennedy JP, Midha S, Gadkari A (1990) ACS Polym Prepr 31(2):655
107. Kennedy JP, Midha S, Gadkari A (1991) J Macromol Sci Chem A28:209
108. Nemes S, Pernecker T, Kennedy JP (1991) Polym Bull 25:633
109. Fréchet JMJ, Henmi M, Gitsov I, Aoshima S, Leduc MR, Grubbs RB (1995) Science 269:1080
110. Grubbs RB, Liu MJ, Fréchet JMJ (1997) ACS PMSE Preprint 77:197

Received: May 1998

Asymmetric Star Polymers: Synthesis and Properties

Nikos Hadjichristidis, Stergios Pispas, Marinos Pitsikalis, Hermis Iatrou and Costas Vlahos

Department of Chemistry, University of Athens, Panepistimiopolis, Zografou, 15771 Athens, Greece E-mail: *nhadjich@atlas.uoa.gr*

The synthesis and the properties, both in bulk and in solution, of asymmetric star polymers are reviewed. Asymmetry is introduced when arms of different molecular weight, chemical nature or topology are incorporated into the same molecule. The phase separation, aggregation phenomena, dilute solution properties etc. are examined from a theoretical and experimental point of view. Recent applications of these materials show their importance in modern technologies.

Keywords. Asymmetry, Miktoarm stars, Synthesis, Morphology, Aggregation, Chain conformation

List of Symbols and Abbrevations		72
1	Introduction	74
2	Synthesis	75
2.1	Stars with Molecular Weight Asymmetry	75
2.2	Stars with Chemical Asymmetry	78
2.2.1	Miktoarm Stars	78
2.2.1.1	General Strategies and Methods	78
2.2.1.2	Synthesis of A_2B Miktoarm Star Copolymers	82
2.2.1.3	Synthesis of A_3B Miktoarm Star Copolymers	85
2.2.1.4	Synthesis of A_nB (n>5) Miktoarm Star Copolymers	85
2.2.1.5	Synthesis of A_2B_2 Miktoarm Star Copolymers	88
2.2.1.6	Synthesis of A_nB_n (n>2) Miktoarm Star Copolymers	89
2.2.1.7	Synthesis of ABC Miktoarm Star Terpolymers	93
2.2.1.8	Synthesis of ABCD Miktoarm Star Quaterpolymers	96
2.2.2	Asymmetric ω–Functionalized Polymers	97
2.3	Stars with Topological Asymmetry	98
3	Properties	100
3.1	Solution Properties	100
3.1.1	Theory	100
3.1.2	Experimental Results	104
3.2	Bulk Properties	110

3.2.1 Theory 110
3.2.2 Experimental Results 115

4   **Applications** 123

5   **Concluding Remarks** 124

6   **References and Notes** 125

List of Symbols and abbreviations

A	adsorbed amount
AIBN	N,N'-azobisisobutyronitrile
Bd	butadiene
c_{gel}	gelation concentration
d	critical dimensionality
D	diffusion coefficient
D_{inter}	interchain distance
DMAPLi	3-dimethylaminopropyllithium
DOP	diisooctylphthalate
DSC	differential scanning calorimetry
DVB	divinylbenzene
f	volume fraction
f_i	number of precursors of the i-th kind in a miktoarm star
g	the ratio $R_{g,star}/R_{g,linear}$
g'	the ratio $[\eta]_{star}/[\eta]_{linear}$
$\vec{G}_{A_n}, \vec{G}_{B_m}$	orientation vectors
$<G^2_{A_nB_m}>$	mean square distance between the centers of mass of the two homopolymer parts in a miktoarm star
$<G^2_k>$	mean square distance between the center of mass of a homopolymer part and the star common origin
I	molecular weight distribution index, M_w/M_n
IMDS	intermaterial dividing surface
K	constant in the Mark-Houwink-Sakurada equation
k	Boltzmann constant
k_H	Huggins constant
l	chain packing parameter
LALLS	low angle laser light scattering
LAS	asymmetric three-arm homostar with two identical arms and a third arm with double the molecular weight of the others
M	molecular weight
M_e	molecular weight between entanglements
MMA	methyl methacrylate
M_n	number average molecular weight
m_{ni}	number average molecular weight of the precursors

MO	membrane osmometry
MW	molecular weight
M_w	weight average molecular weight
m_{wi}	weight average molecular weight of the precursors
N	total number of unit in a miktoarm star
N_w, N_n	weight and number average aggregation number
N_A	Avogadro number
N_A, N_B	number of unit A or B in a miktoarm star
n_A, n_B	number of branches of A and B kind in a miktoarm star
N_e	entanglement length: number of segments between two entanglements
NMR	nuclear magnetic resonance
OBDD	ordered bicontinuous double diamond structure
ODT	order-disorder transition
$P(z_o, z)$	probability distribution function
P2VP	poly(2-vinyl pyridine)
P2VPK	poly(2-vinyl pyridinyl) potassium
P4MeS	poly(4-methyl styrene)
PB	polybutadiene
PBuMA	poly(*n*-butyl methacrylate)
PCL	poly(ε-caprolactone)
PDDPE	1,4-bis(1-phenylethenyl)benzene
PDMS	poly(dimethyl siloxane)
PE	polyethylene
PEO	poly(ethylene oxide)
PEP	poly(ethylene-*co*-propylene)
PI	polyisoprene
PILi	polyisoprenyllithium
PMMA	poly(methyl methacrylate)
POX	polyoxazoline
PPO	poly(2,6-dimethyl phenylene oxide)
PS	polystyrene
PtBuA	poly(*tert*-butyl acrylate)
PtBuMA	poly(*tert*-butyl methacrylate)
PtBuS	poly(*tert*-butyl styrene)
PVN	poly(2-vinyl naphthalene)
RG	renormalization group theory
R_g	radius of gyration
R_h	hydrodynamic radius
R_i	end to end distance
S	entropy
SANS	small angle neutron scattering
SAS	asymmetric three-arm homostar with two identical arms and a third arm with half the molecular weight of the others
SAXS	small angle X-ray scattering

SEC	size exclusion chromatography
sec-BuLi	secondary-butyllithium
SLS	static light scattering
$<S^2_{A_nB_m}>$	radius of gyration of a miktoarm star copolymer
$<S^2_k>$	radius of gyration of homopolymer
T	absolute temperature
tBuA	tert-butyl acrylate
TEM	transmission electron microscopy
T_g	glass transition temperature
THF	tetrahydrofuran
u	interaction parameter between segments
UV	ultraviolet
VN	2-vinyl naphthalene
VS	(4-vinylphenyl)dimethyl vinyl silane
$w(z_o)$	statistical weight of a macrostate
x_A, x_B	fractions of components A and B
z	distance perpendicular to the interface
α	exponent in the Mark-Houwink-Sakurada equation
Γ	decay rate of a correlation function
γ_{Gk}	the ratio $<G^2_k>/<G^2_{k,star}>$
γ_{sk}	the ratio $<S^2_k>/<S^2_{k,star}>$
ϵ	asymmetry parameter
ϵ-CL	ϵ-caprolactam
$[\eta]$	intrinsic viscosity
θ	theta conditions
μ_2	second cumulant
ν	critical exponent
σ	grafting density
σ_G	the ratio $<G^2_{A_nB_m}>/<G^2_{A_n,star}>+<G^2_{B_m,star}>$
τ	fractional position of the branch along the backbone
φ	volume concentration
$\Phi_{A,B}$	Flory parameter for components A and B
Φ_e	volume concentration for entanglements to occur
Φ	length fraction
χ	Flory-Huggins interaction parameter
$\Omega(z_o)$	density profile of chain end

1
Introduction

Asymmetric star polymers are megamolecules [1] emanating from a central core. In contrast to the symmetric stars very little was known, until recently, about the properties of the asymmetric stars. This was due to the difficulties associated with the synthesis of well-defined architectures of this class of polymeric materials. The synthesis, solution and bulk properties, experimental and theoretical, of the following categories of asymmetric stars will be considered in this review:

(a) Stars with molecular weight asymmetry
 The arms are chemically identical but differ in molecular weight.
(b) Stars with chemical asymmetry
 The arms differ in chemical nature. The term miktoarm stars (coming from the Greek word μικτός meaning mixed) is used for these polymers. The term heteroarm star polymers (hetero from the Greek word έτερος meaning other), used by others for this class of polymers, is not appropriate since it does not convey the concept of a group of dissimilar objects. Stars having similar chemical nature but different end-groups also belong to this category.
(c) Stars with topological asymmetry
 The arms are block copolymers which may or not have the same composition and molecular weight but differ with respect to the polymeric block which is attached to the central point.

Schematically the above structures are depicted in Fig. 1.

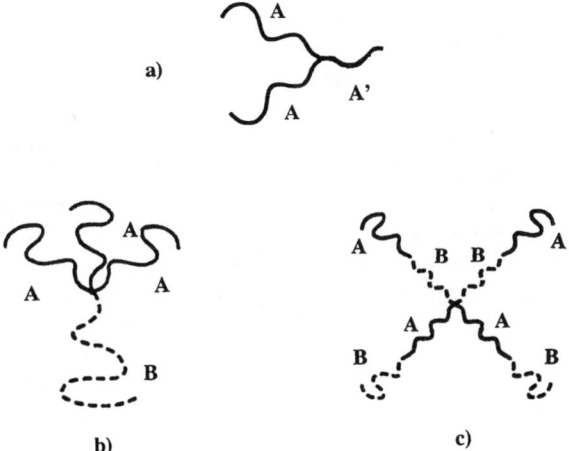

Fig. 1a–c. Asymmetric stars with: **a** molecular weight asymmetry; **b** chemical asymmetry; **c** topological asymmetry

2
Synthesis

2.1
Stars with Molecular Weight Asymmetry

Three-arm polystyrene (PS) stars having two arms of equal molecular weight, and a third one with molecular weight either half or twice that of the identical arms, were prepared by Pennisi and Fetters [2]. Their approach involves the reaction of living PS chains with a ten-fold excess of methyltrichlorosilane for the preparation of the methyldichlorosilane end-capped PS. The addition of the linking agent's solution to the dilute living polymer solution in benzene, under vigorous stirring proved to be efficient for the preparation of the desired product. No coupled byproduct, i.e., the two-arm "star" with a remaining Si-Cl bond was detected following this procedure.

The excess silane was removed after freeze-drying the end-capped PS under dynamic high vacuum and then heating the resulting porous material at 50 °C for at least 72 h. Purified benzene was re-introduced into the reaction vessel to dissolve the polymer. The methyldichlorosilane end-capped PS acted as a macromolecular coupling agent when it was added to a solution containing a small excess of living PS chains having molecular weight half or twice that of the end-capped PS. It is well known [3-5] that PSLi cannot undergo complete reaction with methyltrichlorosilane due to the steric hindrance of the polystyryllithium anions. Therefore the living chains were end-capped with a small amount of butadiene. The reduced steric hindrance of the butadienyllithium chain ends facilitates the completion of the reaction with the methyldichlorosilane-capped PS.

The complete linking reaction of low molecular weight PS ($<4\times10^4$) to the macromolecular coupling agent was achieved using small quantities (2–10 vol.%) of triethylamine instead of the butadiene capping method. Triethylamine is known to disrupt the association of the polystyryllithium in hydrocarbon media [6, 7], thus facilitating the linking reaction. The synthesis of the asymmetric PS stars is outlined in Scheme 1.

The key step of the synthetic procedure is the preparation of the methyldichlorosilane end-capped PS. This is achieved by choosing the suitable reaction conditions, i.e., excess of Si-Cl bonds over living polymer chains, use of dilute polymer solutions, vigorous stirring during the addition of the excess linking agent to the polymer solution. Size exclusion chromatography (SEC) was used to monitor the reaction sequence. After elimination of the excess of the second PS

$$PS_A^- Li^+ + excess\,(CH_3)SiCl_3 \longrightarrow PS_A Si(CH_3)Cl_2 + LiCl + (CH_3)SiCl_3 \uparrow$$

$$PS_A Si(CH_3)Cl_2 + 2\,PS_B^- Li^+ \longrightarrow PS_A Si(CH_3)(PS_B)_2 + 2\,LiCl$$

Scheme 1

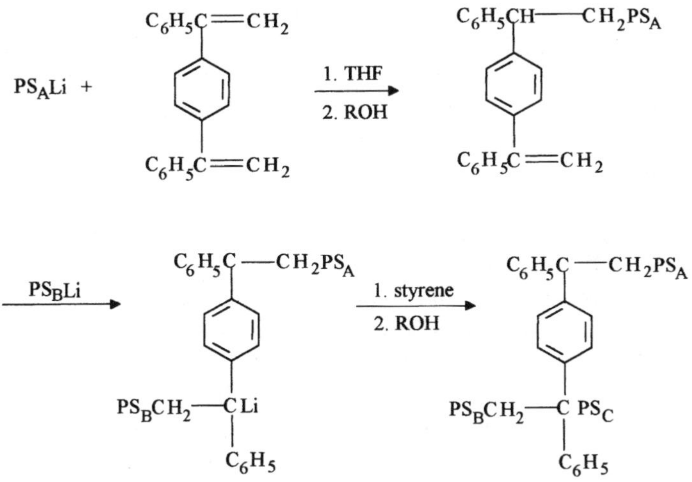

Scheme 2

arms by fractionation, the final products and the different kind of arms, which were isolated before the coupling reaction, were characterized by membrane osmometry (MO) and static light scattering (SLS), revealing that well defined star polymers were prepared.

Using the same procedure Pennisi and Fetters prepared a series of asymmetric polybutadiene (PB) stars in which the third arm was of variable molecular weight [2]. It was found more efficient to add the living PB solution to the methyltrichlorosilane linking agent in order to reduce the formation of the coupled byproduct. Similar characterization techniques were also employed in this case.

Asymmetric polyisoprene (PI) three-arm stars with variable length of the third arm were synthesized using the same method [8]. The reaction of the living PI chains with excess methyltrichlorosilane was performed at 5 °C. This low temperature was selected in an effort to minimize the coupled byproduct. Nevertheless the reduced steric hindrance of the PILi chain end in association with the low molecular weight of the polydienes used (M_n=5500 and 1100) led to the formation of an appreciable amount of the coupled byproduct, which was later separated by fractionation, with the excess of the last coupled arm, using a solvent-precipitant system. Pure products were finally isolated as evidenced by the molecular characterization techniques used (SEC, MO, SLS).

Asymmetric PS stars of the type $(PS_A)_n(PS_B)_n$ were also prepared by the divinylbenzene (DVB) method [9]. Living PS chains, prepared by sec-BuLi initiation, were reacted with a small amount of DVB producing star homopolymers. The DVB core of the stars contains active anions which, if no accidental deactivation occurs, are equal to the number of the arms that have been linked to this core. These active sites are available for the polymerization of an additional quantity of monomer. Consequently further addition of styrene produced asymmetric star polymers having n branches with molecular weight A and n branches with molecular weight B. A small quantity of THF was used to accelerate the second polymerization step.

The last method for the synthesis of asymmetric stars suffers from the disadvantages that characterizes the DVB method: the broad molecular weight distributions, compared to stars prepared by chlorosilane chemistry, molecular heterogeneity, since n is an average value, absence of complete control over the final product etc. More details will be given in Sect. 2.2.1.1. SEC analysis revealed the existence of high molecular weight species. This was attributed to the formation of linked stars. These structures can be produced when active anionic living arms react with other DVB-linked cores. It is evident from the above that the products are not as pure as those produced by suitable chlorosilane chemistry.

Asymmetric three-arm PS stars, possessing chains of different molecular weights were also prepared by Quirk and Yoo [10] using 1,4-bis(1-phenylethenyl)benzene (PDDPE) as the linking agent. It was observed that the addition reaction of polystyryllithium with PDDPE in THF leads primarily to the formation of the monoadduct product, due to the ability of the negative charge to be delocalized into the phenyl rings and the remaining vinyl group. The formation of this product was then followed by the addition of the second polystyryllithium chain in order to obtain the coupled product. The efficiency of the coupling reaction depends on the control of the stoichiometry between the reactants. Finally the addition of styrene in the presence of THF to promote the crossover reaction leads to the formation of the asymmetric PS stars, as shown in Scheme 2. Unreacted monoadduct product and PS_B homopolymer (the second arm) were also observed in the SEC trace of the final product due to incomplete linking reactions.

2.2
Stars with Chemical Symmetry

2.2.1
Miktoarm Stars

Star polymers of chemically different arms are usually called miktoarm stars. Although there are several individual methods for the synthesis of miktoarm stars four general methodologies have been developed. Three of them are based on anionic polymerization and the fourth on cationic polymerization. In all of them the use of appropriate linking agents is necessary.

2.2.1.1
General Strategies and Methods

2.2.1.1.1
Anionic Polymerization Method with Divinylbenzene (DVB)

The synthesis of miktoarm stars by the DVB method is a three step procedure. The first step involves the preparation of the living arm by anionic polymeriza-

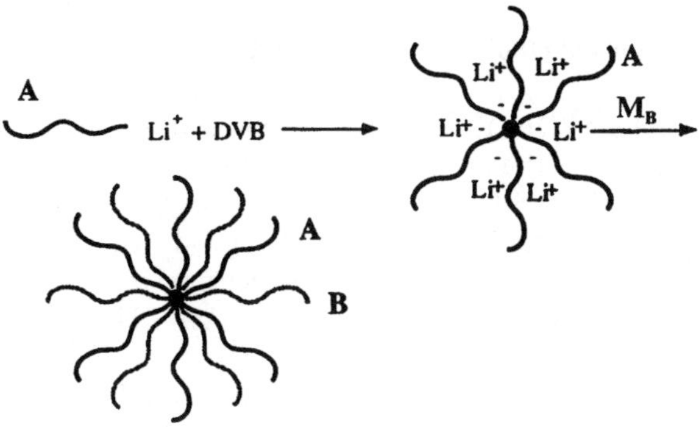

Scheme 3

tion using a suitable initiator. The living precursor then reacts in the second step with a small amount of DVB, leading to the formation of a star molecule bearing within its core a number of active sites, which is theoretically equal to the number of the A arms of the star polymer. Subsequent addition of another monomer, in the third step, results in the growth of B arms of the miktoarm stars, since the active star, prepared at the second step, acts as a multifunctional initiator for the polymerization of the second monomer.

The growing B arms have anionic sites at their outer ends thus providing the possibility of reacting with electrophilic compounds or other monomers towards the preparation of end-functionalized stars or star-block copolymers. This method can be carried out in inert atmosphere, avoiding the use of the highly demanding and time consuming vacuum technique. It was first reported by Okay and Funke [11] and by Eschwey and Burchard [12] and developed by Rempp and collaborators [13–16]. Scheme 3 illustrates the DVB method.

Despite the advantages mentioned above the DVB method is characterized by several disadvantages, the foremost being the architectural limitations. Only stars of the type A_nB_n can be prepared and even in this case there is no absolute control of the number of arms, n. In fact, n is an average value and is influenced by several parameters. Specifically, n is increased by decreasing molecular weight of the precursor A and by increasing the molar ratio of the DVB to living chains. Another major problem is that a fraction of the living chains A are not incorporated into the star structure due to accidental deactivation, the high molecular weight of the chains (steric reasons) or the low molar ratio of DVB to living chains. The unreacted living arms A can act as initiators after the addition of the second monomer. Another disadvantage is that the B arms cannot be isolated and characterized independently. Finally, reaction of the living ends with the remaining double bonds of the DVB nodule can lead to the formation of loops (intramolecular reaction) or networks (intermolecular reaction). From the above, it is clear that the miktoarm stars prepared by this method are characterized by rather poor molecular and compositional homogeneity.

2.2.1.1.2
Anionic Polymerization with Diphenylethylenes (DPE)

1,1-Diphenyl ethylene (DPE) derivatives were used for the synthesis of miktoarm stars according to the method developed by Quirk [17, 18]. Two moles of living polymer A react with one mole of 1,3-bis(1-phenylethenyl) benzene, DDPE, leading to the formation of the coupled product having two active sites. These active sites can act as initiators for the polymerization of another monomer, thus producing miktoarm stars of the type A_2B_2. The reaction sequence is given in Scheme 4.

It is a three step procedure, using a divinyl compound in a similar manner as the DVB method. Stars of predetermined architectures can be prepared by this method but only polymers of the type A_2B_2 and ABC have been produced so far. More complicated structures such as AB_3, AB_5, A_nB_n (with n>2)or ABCD have not appeared in the literature.

The crucial point of the procedure is the control of the stoichiometry of the reaction between the living A chains and the DPE derivative, otherwise a mixture of stars is produced. A major problem is the fact that the rate constants for the reaction of the first and second polymeric chain with the DPE derivative are different. This results in bimodal distributions because of the formation of both the monoanion and dianion. In order to overcome this problem polar compounds have to be added, but it is well known that they affect dramatically the microstructure of the polydienes that are formed in the last step. However the addition of lithium sec-butoxide to the living coupled DPE derivative, prior to the addition of the diene monomer, was found to produce monomodal well defined stars with high 1,4 content. Finally another weak point of the method is that, as in the case of the DVB route, the B arms cannot be isolated from the reaction mixture and characterized separately. It is therefore difficult to obtain unambiguous information about the formation of the desired products.

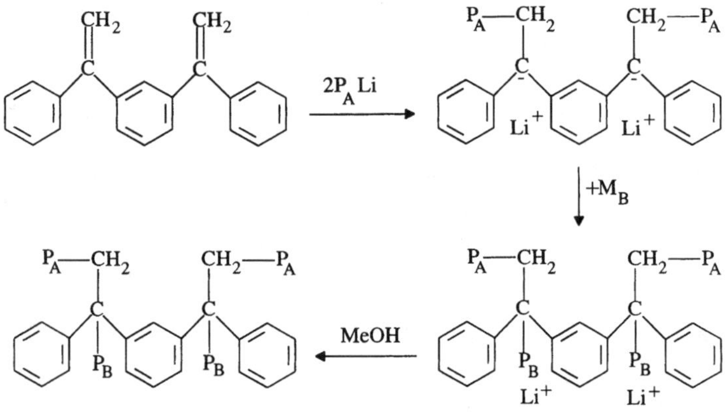

Scheme 4

2.2.1.1.3
Anionic Polymerization Followed by Chlorosilane Coupling

The method is based on consecutive reactions of living, anionically prepared polymer chains with multifunctional chlorosilane compounds, which act as linking agents. Miktoarm star copolymers, terpolymers and quaterpolymers of the type A_2B, A_3B, A_5B, A_8B_8, $(AB)_2B$, $(AB)_3B$, A_2B_2, ABC and ABCD have been prepared by this method. Using the suitable chlorosilane and the appropriate reaction sequence it is rather easy to predetermine the structure of the final product. The synthesis sequence can be monitored by SEC and all the arms and the intermediate and final products can be characterized, thus providing unambiguous proof for the formation of the desired products. The disadvantage of this method is that it is time consuming compared with the other ones. Nevertheless this limitation is a small price to pay given the potential of this method for generating true model compounds of a wide variety of macromolecular architectures.

2.2.1.1.4
Living Cationic Polymerization Method

The recent development of living cationic polymerization systems has opened the way to the preparation of rather well defined star homopolymers and miktoarm star polymers [19 and see the chapter in this volume]. Divinyl ether compounds were used as linking agents in a manner similar to the DVB method for anionic polymerization. Typically the method involves the reaction of living polymer chains with a small amount of the divinyl compound. A star polymer is formed carrying at the core active sites capable of initiating the polymerization of a new monomer. Consequently a miktoarm star copolymer of the type A_nB_n is produced.

Several experimental parameters influence the value of n making it difficult to have precise control over the structure [20]. The value of n increases by decreasing the length of A arms, by increasing the feed ratio, i.e., the molar ratio of the divinyl ether to the living ends and by increasing the concentration of the living ends. The structure of the monomers and the linking agent plays an important role in determining the quality of the produced stars. Monomers having bulky groups lead to the formation of a large amount of low molecular weight polymers at the beginning of the reaction sequence, thus leading to mixed products, which are difficult to separate. Divinyl ethers having long and rigid spacers between the two vinyl groups proved to be more efficient coupling agents. Due to the analogy with the DVB method this approach is characterized by similar disadvantages. Nevertheless the possibility of using monomers that cannot be polymerized anionically makes the method attractive and susceptible to several applications.

2.2.1.1.5
Other Methods

Individual methods have also been devised for the preparation of miktoarm stars. One of these approaches involves the preparation of macromonomers possessing either central or end vinyl groups which can be used to produce miktoarm stars either by copolymerization of the double bonds or by reacting the double bonds with living polymer chains, thus creating active centers able to initiate the polymerization of another monomer. All these methods are limited to specific synthetic problems and cannot be used for the preparation of a wide range of different structures.

2.2.1.2
Synthesis of A_2B Miktoarm Star Copolymers

One miktoarm star copolymer of the A_2B type was prepared by Mays using a method similar to the one adopted by Pennisi and Fetters for the synthesis of asymmetric stars [21]. According to this method living PS chains were reacted with excess CH_3SiCl_3 to produce the monosubstituted linking agent, followed by the removal of the excess silane by the addition of a slight excess of living PI chains. Fractionation was performed to remove the excess PI after the linking

$$PS^-Li^+ + \text{excess } (CH_3)SiCl_3 \longrightarrow PS\text{-}Si(CH_3)Cl_2 + LiCl + (CH_3)SiCl_3 \uparrow$$

$$PS\text{-}Si(CH_3)Cl_2 + \text{excess } PI^-Li^+ \longrightarrow PS\text{-}Si(CH_3)(PI)_2$$

Scheme 5

$$2\,PS^-Li^+ + (CH_3)SiCl_2H \longrightarrow (PS)\underset{H}{\overset{CH_3}{\underset{|}{\overset{|}{Si}}}}(PS) \quad (I)$$

$$P2VP^-Li^+ + CH_2{=}CH{-}CH_2Br \longrightarrow P2VPCH_2CH{=}CH_2 \quad (II)$$

$$(I) + (II) \xrightarrow{Pt} (PS)_2(P2VP)$$

Scheme 6

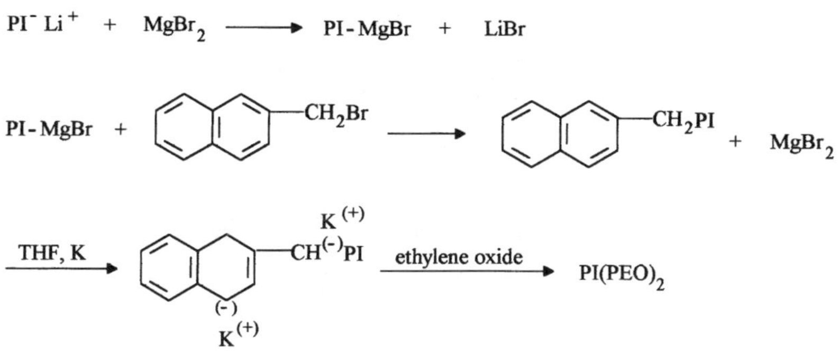

$$PI^-Li^+ + MgBr_2 \longrightarrow PI\text{-}MgBr + LiBr$$

Scheme 7

reaction was completed. The reaction sequence used for the synthesis of the $(PI)_2PS$ miktoarm star is given in Scheme 5.

The method takes advantage of the steric hindrance of the polystyryllithium living end, which in combination with the excess silane used for the linking reaction, reduces the possibility for the formation of the coupled byproduct. The reaction sequence was monitored by SEC and the reaction products were characterized by MO, differential laser refractometry and LALLS, revealing that well defined polymers were prepared.

This method was further developed by Iatrou et al. [22]. All possible combinations of A_2B polymers with A and B being PS, PI or PB were prepared. A more sophisticated and complicated high vacuum technique was used to ensure the formation of well defined products. High degrees of molecular, structural and compositional homogeneity were achieved by this technique, as was evidenced by the combination of all the molecular and spectroscopic characterization data. In a more recent study stars having deuterated PS arms, $(PI)_2(d\text{-}PS)$ were also prepared [23].

An A_2B star having two PS arms and one poly(2-vinyl pyridine) (P2VP) arm, $(PS)_2(P2VP)$ was prepared by Eisenberg et al. using a different approach [24]. Living PS chains were linked to dichloromethylsilane, CH_3SiCl_2H to produce the two arms of the star. In another reactor living P2VP was reacted with allyl bromide. A hydrosilylation addition of the Si-H group of the two-arm star to the vinyl group of the end-functionalized P2VP was performed to produce the final miktoarm star. The reaction sequence is outlined in Scheme 6.

The miktoarm stars were characterized by medium polydispersities $(M_w/M_n=I=1.33-1.50)$ probably due to incomplete hydrosilylation. It is characteristic that only small molecular weight P2VP arms were used to facilitate the linking reaction. This is evidence of the limitations of the hydrosilylation reaction for the preparation of miktoarm stars.

Anionic polymerization techniques and naphthalene chemistry were used by Teyssié et al. to prepare A_2B miktoarm stars, where A is poly(ethylene oxide) (PEO) and B is PS, PI, poly(α-methyl styrene) or poly(*tert*-butyl styrene) [25]. The reaction sequence is shown in Scheme 7.

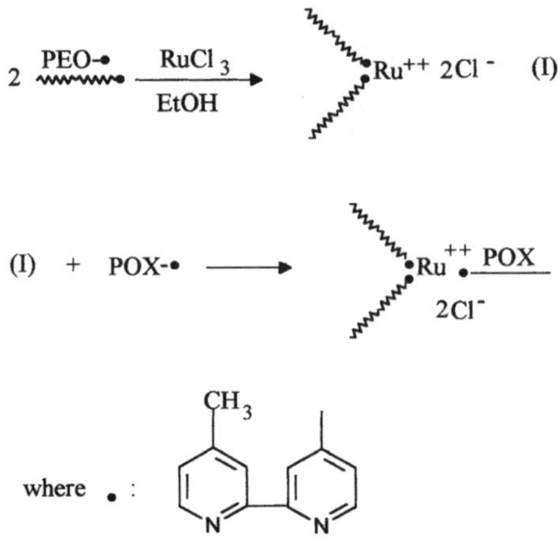

$$2 \text{ PEO-}\bullet \xrightarrow[\text{EtOH}]{\text{RuCl}_3} \text{>Ru}^{++} \text{ 2Cl}^- \quad \text{(I)}$$

$$\text{(I)} + \text{POX-}\bullet \longrightarrow \text{>Ru}^{++} \cdot \underline{\text{POX}} \quad 2\text{Cl}^-$$

where •: [4,4'-dimethyl-2,2'-bipyridine structure]

Scheme 8

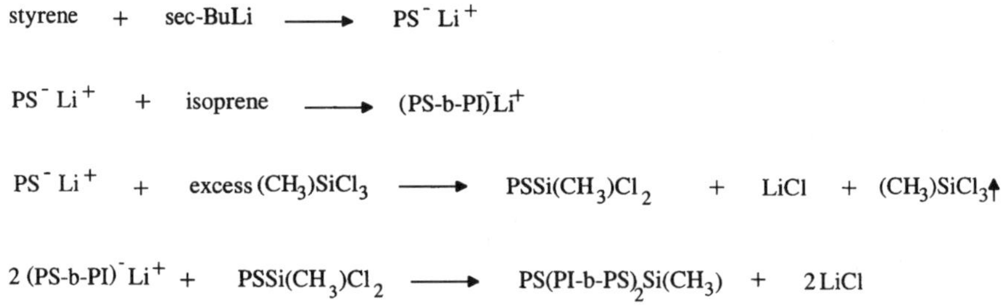

$$\text{styrene} + \text{sec-BuLi} \longrightarrow \text{PS}^- \text{Li}^+$$

$$\text{PS}^- \text{Li}^+ + \text{isoprene} \longrightarrow \text{(PS-b-PI)}^- \text{Li}^+$$

$$\text{PS}^- \text{Li}^+ + \text{excess (CH}_3\text{)SiCl}_3 \longrightarrow \text{PSSi(CH}_3\text{)Cl}_2 + \text{LiCl} + \text{(CH}_3\text{)SiCl}_3\uparrow$$

$$2 \text{ (PS-b-PI)}^- \text{Li}^+ + \text{PSSi(CH}_3\text{)Cl}_2 \longrightarrow \text{PS(PI-b-PS)}_2\text{Si(CH}_3\text{)} + 2 \text{ LiCl}$$

Scheme 9

Living polymer chains were reacted with bromomethyl naphthalene but a fairly large amount of the coupled byproduct was formed. The byproduct was minimized to 5–10% by using the Grignard reagent, leading to the formation of A_2B stars. The final polymer was contaminated by the starting homopolymers and traces of PEO homopolymer. Polydispersity indices as high as 1.2–1.3 were obtained with this procedure.

A special technique was employed by Naka et al. for the preparation of A_2B stars, A being PEO and B polyoxazoline (POX) [26] according to Scheme 8. Ru(III) complexes with bipyridyl terminated polymers were utilized in this method. Characterization data were not provided for these miktoarm star copolymers.

The chlorosilane method was adopted for the synthesis of miktoarm stars of the type $B(A-b-B)_2$ where A is PI and B is PS [27]. The synthetic procedure was similar to that used for the preparation of A_2B stars except that two of the arms are diblock copolymers, as shown in Scheme 9. Extensive characterization data were given to confirm the synthesis of well defined copolymers.

$$\text{PS}^- \text{Li}^+ + \text{excess SiCl}_4 \longrightarrow \text{PS-SiCl}_3 + \text{LiCl} + \text{SiCl}_4\uparrow$$

$$\text{PS-SiCl}_3 + \text{excess PI}^- \text{Li}^+ \longrightarrow \text{PS-SiPI}_3 + 3\,\text{LiCl}$$

Scheme 10

2.2.1.3
Synthesis of A_3B Miktoarm Star Copolymers

The work on the A_2B stars was expanded to the synthesis of miktoarm stars of the type A_3B, where A is PI and B is PS [28]. $SiCl_4$ was used as the linking agent. Living PS chains were reacted with an excess of the linking agent for the preparation of the trichlorosilane end-capped PS. After the evaporation of excess $SiCl_4$ the macromolecular linking agent was added to a slight excess of the living PI chains to obtain the A_3B star. The reaction sequence, given in Scheme 10, was monitored by SEC. The products after extensive characterization were found to have high degree of molecular and compositional homogeneity.

A similar procedure was followed for the synthesis of miktoarm stars of the type $B(A-b-B)_3$, A being PI and B PS [27].

A different approach but still in the frame of the chlorosilane method was adopted by Tsiang for the synthesis of $(A-b-B)B_3$ miktoarm star copolymers, where A is PS and B is PB [29]. Living PB chains were reacted with $SiCl_4$ in a molar ratio 3:1, followed by the addition of the living diblock PS-b-PBLi. The key step of the method is the succesfull synthesis of the $(PB)_3SiCl$ intermediate product. The reduced steric hindrance of the PBLi chain end poses questions about the purity of this polymer, since several byproducts, such as $(PB)_2SiCl_2$, $(PB)_4Si$, $PBSiCl_3$ can be formed in the first step of the synthesis. SEC analysis was performed to monitor the reaction sequence.

It is obvious that the method developed by Tsiang is very demanding with regard to the stoichiometry of the reagents. The byproducts are almost impossible to separate.

2.2.1.4
Synthesis of A_nB ($n \geq 5$) Miktoarm Star Copolymers

Miktoarm star copolymers of the type A_5B were prepared [30] in a similar manner to the A_2B and A_3B type stars. The reaction sequence is outlined in Scheme 11.

Living PS chains were reacted with 1,2-bis(trichlorosilyl)ethane in a ratio Li:Cl=1:6. Dropwise addition of the living polymer solution into the vigorously stirred solution of the linking agent was performed to minimize the coupling product. Under these conditions 15% of the coupled product was formed. The

styrene + sec-BuLi ⟶ PS⁻Li⁺

PS⁻Li⁺ + Cl₃SiCH₂CH₂SiCl₃ ⟶ Cl₃SiCH₂CH₂SiCl₂(PS) + LiCl

isoprene + sec-BuLi ⟶ PI⁻Li⁺

Cl₃SiCH₂CH₂SiCl₂(PS) + 5 PI⁻Li⁺ ⟶ (PI)₃SiCH₂CH₂Si(PI)₂(PS) + 5 LiCl

Scheme 11

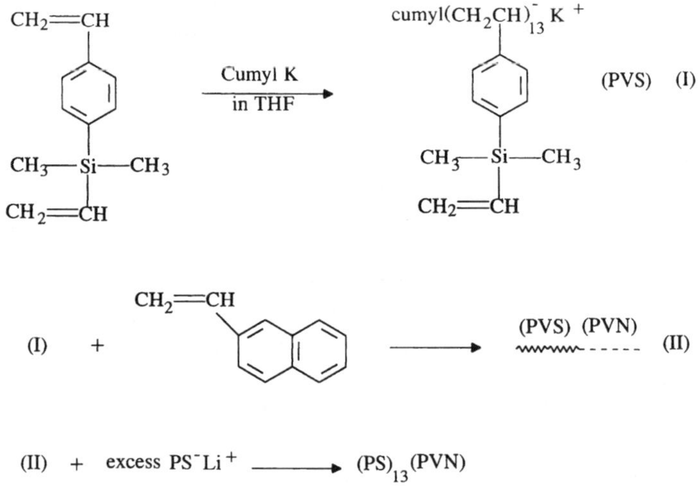

Scheme 12

pentachlorosilane-capped PS was then reacted with excess PILi for the preparation of the A_5B copolymer. Fractionation techniques were employed to isolate the desired polymer. The stoichiometric addition of the living PS to the silane was chosen in this case instead of using excess silane, since this hexafunctional silane is solid and consequently its excess cannot easily be removed. The progress of the reaction was monitored by SEC. The combined characterization results proved the narrow distribution in molecular weight and composition of the final products.

Star polymers having several PS branches and only one poly(2-vinyl naphthalene), PVN branch were prepared by Takano et al. using anionic polymerization techniques [31]. Sequential anionic block copolymerization of (4-vinyl-phenyl) dimethylvinylsilane (VS) and VN was employed. The double bonds attached to silicon have to remain unaffected during the polymerization of VS. This was ac-

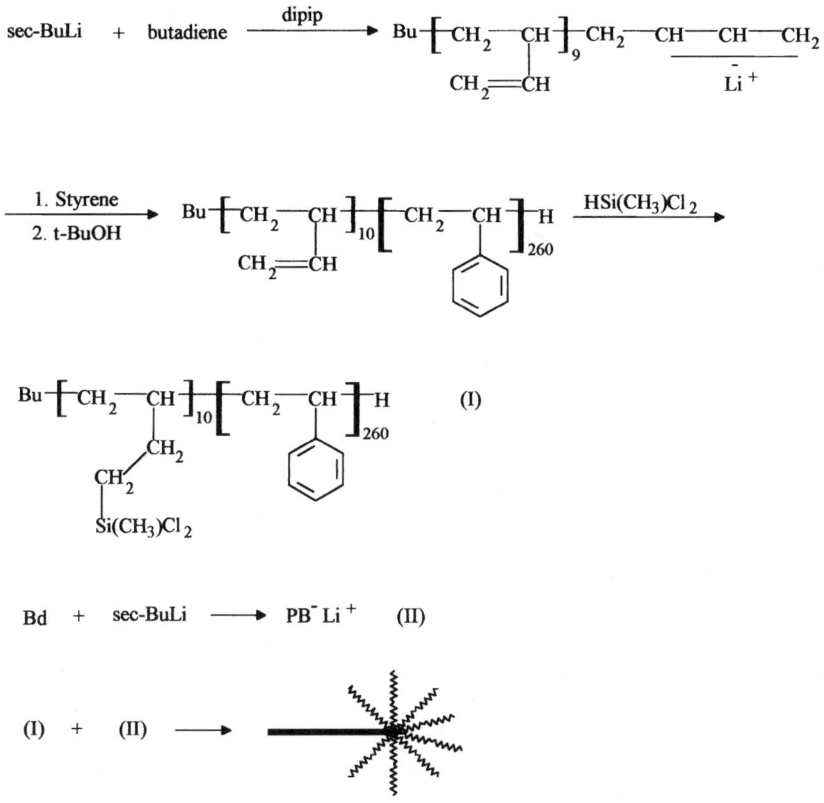

Scheme 13

complished using THF as a solvent and short polymerization times. The PVS block with the unreacted double bonds was used as a multifunctional linking agent. Subsequent addition of living PS chains produced miktoarm stars of the type $(PS)_n PVN$, as shown in Scheme 12. Characterization studies revealed that n=13.

The method adopted for the synthesis of these stars can be considered as a macromonomer method, since end-reactive vinyl groups were used for the linking of the PS arms. There is a possibility that the silyl vinyl anion formed after the addition of the living PS chains reacts with silyl vinyl groups; this effect was minimized using short VS blocks and a large excess of PS anions.

Similar structures of the AB_n type were prepared by Wang et al., A being PS and B PB or P2VP [32, 33]. Due to the much higher molecular weight of the PS arm the polymers were called umbrella copolymers. The reaction sequence for the preparation of the $PS(PB)_n$ copolymers is given in Scheme 13. Butadiene was polymerized anionically in the presence of dipiperidinoethane (dipip) followed by the addition of styrene. A diblock copolymer having a PS chain and a short

1,2 PB block was thus prepared. Hydrosilylation chemistry was employed for the incorporation of the -Si(CH$_3$)Cl$_2$ or -Si(CH$_3$)Cl groups to the 1,2-PB double bonds. Subsequent addition of 1,4 PBLi or P2VPK leads to the formation of the umbrella copolymers. The limited control exercized over the hydrosilylation reaction means that the number of the arms cannot be accurately predicted. Roovers and collaborators succeeded to prepare umbrella-star copolymers [(PS-u-P2VP)$_n$]$_m$. The synthesis is based on the reaction of (PB-1,2)-b-PSLi with chlorosilane having 32 peripheral Si-Cl bonds followed by hydrosilylation of the PB and reaction with P2VPLi.

2.2.1.5
Synthesis of A$_2$B$_2$ Miktoarm Star Copolymers

The synthesis of miktoarm star copolymers of the type A$_2$B$_2$ was first reported by Xie and Xia [34]. A chlorosilane method was employed to prepare PS$_2$PEO$_2$ stars according to Scheme 14. Living PS chains were reacted with SiCl$_4$ in a molar ratio 2:1 leading to the formation of the two-arm product. The remaining Si-Cl bonds can be used for the linking reaction of living PEO chains. The process is facilitated by the increased steric hindrance of the living PS chain ends. It is, indeed, very difficult to prepare the three- or four-arm PS stars. From this point of view the control of the stoichiometry is not very important in this specific case. Using CH$_3$SiCl$_3$ instead of SiCl$_4$ A$_2$B miktoarm stars were also prepared.

A different approach was followed by Iatrou and Hadjichristidis for the synthesis of PS$_2$PB$_2$ miktoarm stars [35] according to Scheme 15. The first step involved the reaction of living PS chains with excess SiCl$_4$, followed by the evaporation of the excess silane in a similar manner as was described in the case of the A$_3$B stars. The second PS arm was incorporated by slow stoichiometric addition (titration) of one living PS chain to each PSSiCl$_3$. This procedure was monitored by SEC taking samples from the reactor during the titration. The last step involved the addition of a small excess of living PB chains for the preparation of the (PS)$_2$(PB)$_2$ miktoarm star.

$$2\,PS^-Li^+ + SiCl_4 \xrightarrow{C_6H_6/THF} (PS)_2SiCl_2 \xrightarrow{excess\,PEO^-K^+} (PS)_2Si(PEO)_2$$

Scheme 14

$$PS^-Li^+ + excess\,SiCl_4 \longrightarrow PSSiCl_3 + LiCl + SiCl_4\uparrow$$

$$PSSiCl_3 + PS^-Li^+ \xrightarrow{titration} (PS)_2SiCl_2 + LiCl$$

$$(PS)_2SiCl_2 + excess\,PB^-Li^+ \longrightarrow (PS)_2(PB)_2 + 2\,LiCl$$

Scheme 15

Using a method similar to that of Xie and Xia, Young et al. prepared $(PS)_2(PI)_2$ miktoarm stars [36]. Living PS chains were reacted with $SiCl_4$ in a molar ratio 2:1 for the formation of the two-arm product. Subsequent addition of living PI chains led to the formation of the miktoarm star.

A_2B_2 stars, A being PI and B PB were also prepared by two different methods [37]. The first method involved the end-capping reaction of living PI chains with 2–3 units of styrene in order to increase the steric hindrance of the active chain end, followed by titration with $SiCl_4$ and finally reaction with an excess of PBLi. According to the second method living PI chains were reacted with $SiCl_4$ in a molar ratio 2:1 at –40 °C. This low temperature route was performed in order to reduce the reactivity of the living chain end, thus avoiding the formation of macromolecular linking agents with different functionalities. Subsequent addition of excess PILi led to the formation of the A_2B_2 miktoarm stars.

Stars of the type A_2B_2 were also prepared by the method developed by Quirk et al. In this case A was PS and B PI or PB [38, 39]. Stars of the type $A_2(B-b-A)_2$, where A was PS and B PB were also synthesized by this method. The disadvantages of the method have already been mentioned. In order to overcome these problems the reaction of the living PS chains with the divinyl compound were monitored by SEC and UV spectroscopy, by observing the increase in absorbance of the diphenyl alkyllithium species at 438 nm. It is obvious that the method is very demanding experimentally and a lot of effort has to be exercised for the preparation of well defined products.

2.2.1.6
Synthesis of A_nB_n (n>2) Miktoarm Star Copolymers

Multiarm miktoarm stars have been prepared by a variety of methods. Model miktoarm stars, called Vergina star copolymers, bearing 8 PS and 8 PI branches, PS_8PI_8 were synthesized using chlorosilane chemistry [40]. A silane with 16 Si-Cl bonds $Si[CH_2CH_2Si(CH_3)(CH_2CH_2Si(CH_3)Cl_2)_2]_4$ was used as linking agent. Living PS chains were reacted with the linking agent in a molar ratio 8:1 for the preparation of the eight-arm star. Even a slight excess of PSLi (~5%) can be used without the incorporation of more than eight arms due to the steric hindrance of the already attached chain per Si atom. A small excess of PILi was finally added to prepare the desired product.

The most widely used method for the preparation of miktoarm stars of the type A_nB_n is the DVB method, which has already been mentioned. The polymers prepared by this method have PS as A arms and PtBuMA, PtBuA, PBuMA, PEO or P2VP as B arms [41–43]. SEC was used to monitor the reaction steps and the molecular characterization data showed that the products were not of the same degree of homogeneity as those prepared by the chlorosilane method, due to the disadvantages inherent of the method.

Structures of the same type have also been prepared by cationic polymerization techniques, as can be seen in Scheme 16. Vinyl ethers having isobutyl-, acetoxy ethyl-, and malonate ethyl- pendant groups have been used. Hydrolysis of

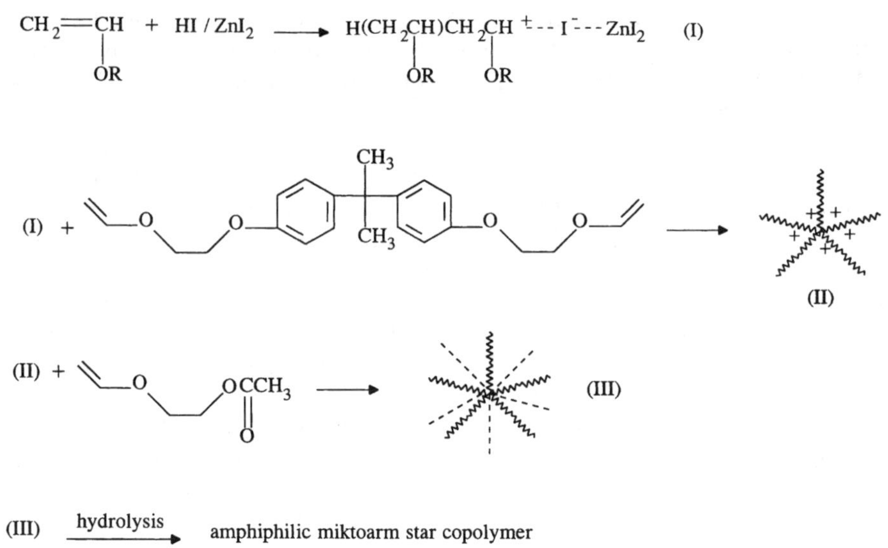

CH$_2$=CH + HI / ZnI$_2$ $\longrightarrow$ H(CH$_2$CH)CH$_2$CH $^\pm$--- I$^-$---ZnI$_2$ (I)
| | |
OR   OR   OR

(I) + [diagram: CH$_2$=CH–O–CH$_2$CH$_2$–O–C$_6$H$_4$–C(CH$_3$)$_2$–C$_6$H$_4$–O–CH$_2$CH$_2$–O–CH=CH$_2$] $\longrightarrow$ (II)

(II) + [diagram: CH$_2$=CH–O–CH$_2$CH$_2$–OC(=O)CH$_3$] $\longrightarrow$ (III)

(III) $\xrightarrow{\text{hydrolysis}}$ amphiphilic miktoarm star copolymer

Scheme 16

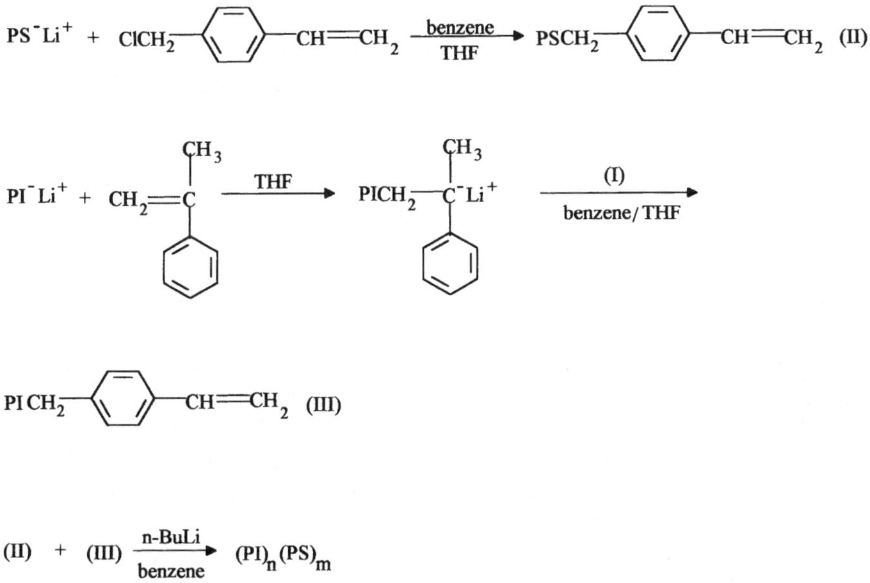

PS$^-$Li$^+$ + ClCH$_2$–C$_6$H$_4$–CH=CH$_2$ $\xrightarrow[\text{THF}]{\text{benzene}}$ PSCH$_2$–C$_6$H$_4$–CH=CH$_2$ (II)

PI$^-$Li$^+$ + CH$_2$=C(CH$_3$)(C$_6$H$_5$) $\xrightarrow{\text{THF}}$ PICH$_2$–C$^-$Li$^+$(CH$_3$)(C$_6$H$_5$) $\xrightarrow[\text{benzene/THF}]{\text{(I)}}$

PICH$_2$–C$_6$H$_4$–CH=CH$_2$ (III)

(II) + (III) $\xrightarrow[\text{benzene}]{\text{n-BuLi}}$ (PI)$_n$(PS)$_m$

Scheme 17

the esters led to the preparation of amphiphilic miktoarm star copolymers [44–46]. The products were mainly characterized by SEC and NMR spectroscopy, whereas only limited molecular characterization data were given in these studies.

Ishizu and Kuwahara have developed a macromonomer technique for the synthesis of miktoarm stars of the type A_nB_n. PS and PI macromonomers having end vinyl groups were prepared by the coupling reaction of the corresponding living anions with p-chloromethylstyrene. Anionic copolymerization of the PS and PI macromonomers was performed in benzene solutions using n-BuLi as initiator. The products can be considered miktoarm stars of the type A_nB_m, as evidenced by their solution and solid state properties [47]. The reaction series is given in Scheme 17.

It was found that the reactivity ratios of the copolymerization system greatly influences the number of the arms of the star polymer.

Diblock macromonomers having central vinyl groups were used for the synthesis of $(PS)_n(PtBuMA)_n$ miktoarm stars [48, 49]. The macromonomers were prepared by sequential anionic polymerization of styrene, 1,4-divinyl benzene (DVB) and tert-butyl methacrylate. The DVB monomer was left to react with the living PS chains for short times (~5 min) so that a few DVB units can be incorporated at the end of the PS chains and the formation of PS stars can be avoided. Free radical polymerization in solution and in bulk using AIBN as initiator, tetramethylthiuram as a photosensitizer and ethylene glycol dimethacrylate as a cross linking agent was carried out for the synthesis of the miktoarm stars. A similar experiment was performed using PS-b-P2VP diblocks having central isoprene units [49].

A cyclophosphazene derivative was used as a linking agent to produce miktoarm stars consisting of PS and Nylon-6 branches [50], according to Scheme 18.

The linking agent was prepared by reacting the hexachlorocyclotriphosphazene with 4-hydroxy benzoic acid ethyl ester and subsequent hydrolysis with NaOH. The acid groups thus prepared were transformed into acid chloride by treatment with $SOCl_2$. PS chains were linked to this linking agent by two methods. The first method involved the addition of anionically living PS chains to the linking agent. For the second method radical polymerization of styrene in the presence of 2-aminoethanethiol was performed producing amine terminated PS. These end-capped polymers were reacted with the linking agent. In both cases the coupling was not complete. Hydrolysis of the remaining acid chloride groups and titration of the resulting acid groups showed that less than 2 groups remained unreacted. These acid groups were used for the ring opening polymerization of ε-caprolactam (ε-CL) giving rise to the formation of miktoarm stars. It is obvious that there is poor control over the coupling reaction of the PS chains to the linking agent. The molecular weight of the PS arm affects the degree of displacement of the acid chloride groups. The higher the molecular weight of the PS the lower the number of arms incorporated at the star's center. These results in combination with the use of ring opening and radical polymerization in one of the possible routes leads to products with broad molecular weight distributions and poor control over the final structure.

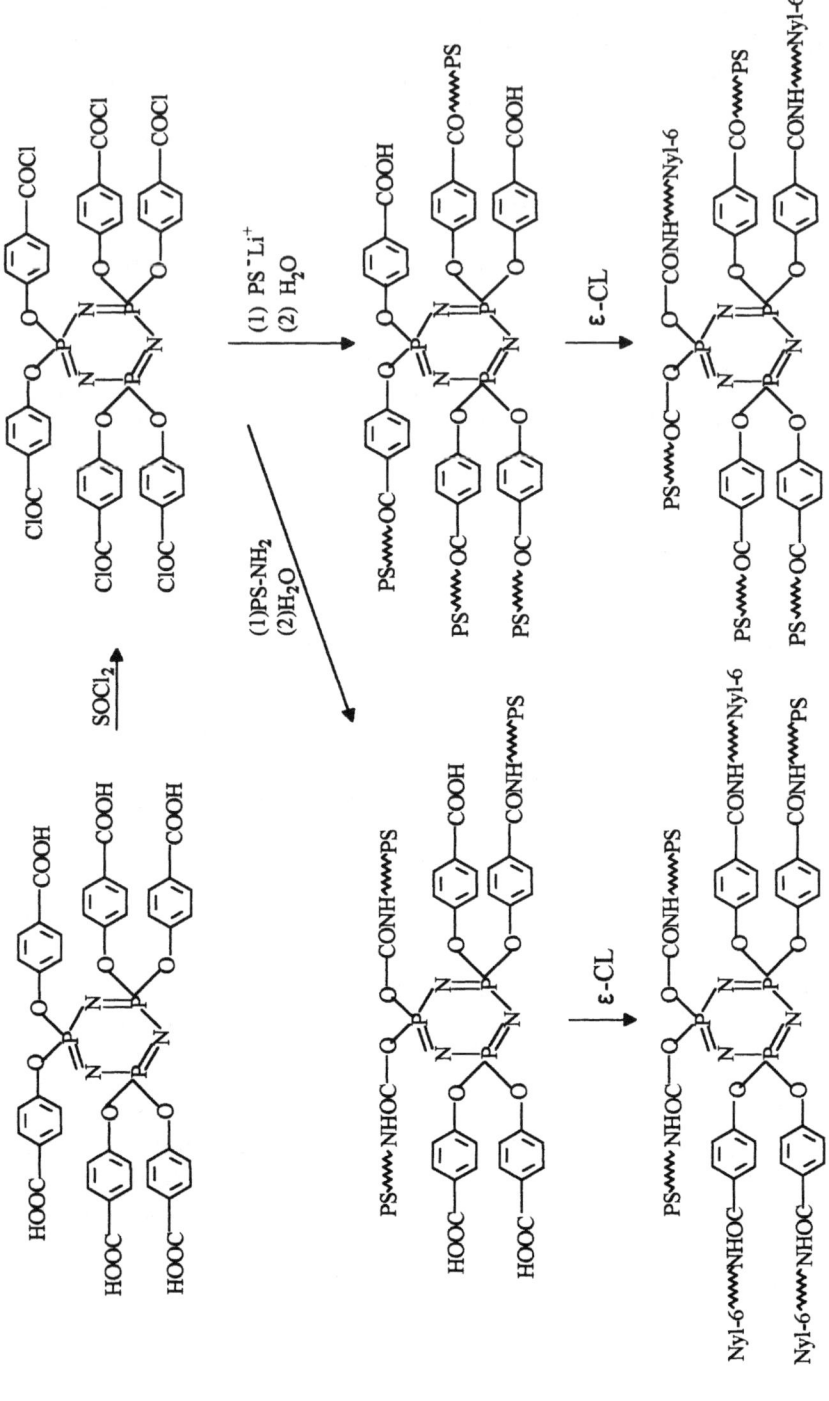

Scheme 18

2.2.1.7
Synthesis of ABC Miktoarm Star Terpolymers

Miktoarm star terpolymers of the type ABC have been prepared by several methods. Iatrou and Hadjichristidis reported the synthesis of a miktoarm star consisting of PS, PI and PB branches radiating from the star center [51]. This was achieved using the chlorosilane method and the step by step linking of the different branches to the trichloromethylsilane, which was the linking agent, as shown in Scheme 19.

A solution of living PI chains was added to a large excess of the silane, followed, after the evaporation of the excess trichloromethylsilane, by the slow stoichiometric addition (titration) of the living PS chains, exactly as was described in the case of the A_2B_2 star copolymers. The formation of the desired product, (PS)(PI)(CH$_3$)SiCl was monitored by SEC taking samples from the reactor during the titration process. The ABC star was finally prepared by the addition of a slight excess of PBLi.

The order of linking of the different branches to the silane plays an essential role, since the less sterically hindered chain end, namely PBLi, has to be added at the end of the procedure and the most sterically hindered chain end, namely the PSLi has to be added at the titration step in order to prevent the complete reaction with the macromolecular difunctional linking agent. The final products were characterized by low molecular weight distributions and high structural, compositional and molecular homogeneity indicating that this step by step addition of the different branches provides excellent control during the synthetic procedure.

Using the same route asymmetric AA'B miktoarm stars were also prepared [52]. These are stars having two chemically identical A arms but of different molecular weights. In other words the B chain is not grafted at the middle of the A chain as in the case of the symmetric A_2B stars. A was deuterated PS and B PI in that case.

The macromonomer method was used by Fujimoto et al. for the preparation of (PS)(PDMS)(PtBuMA) stars [53], as described in Scheme 20. The lithium salt of the p-(dimethylhydroxy)silyl-α-phenyl styrene was synthesized and used as initiator for the polymerization of hexamethylcyclotrisiloxane (D$_3$). Living PS chains were reacted with the end double bond of the macromonomer, followed by the anionic polymerization of the t-BuMA.

$$PI^-Li^+ + \text{excess } (CH_3)SiCl_3 \longrightarrow PI(CH_3)SiCl_2 + LiCl + (CH_3)SiCl_3 \uparrow$$

$$PI(CH_3)SiCl_2 + PS^-Li^+ \xrightarrow{\text{titration}} (PI)(PS)(CH_3)SiCl + LiCl$$

$$(PI)(PS)(CH_3)SiCl + \text{excess } PB^-Li^+ \longrightarrow (PI)(PS)(PB) + LiCl$$

Scheme 19

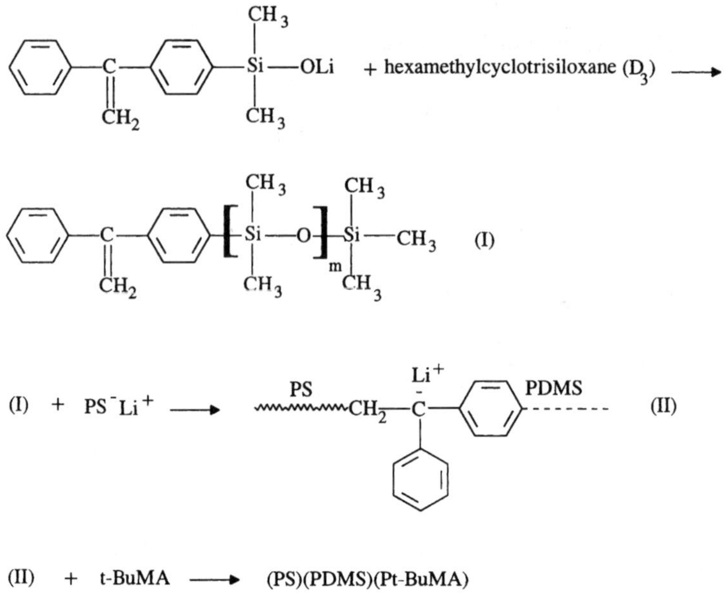

Scheme 20

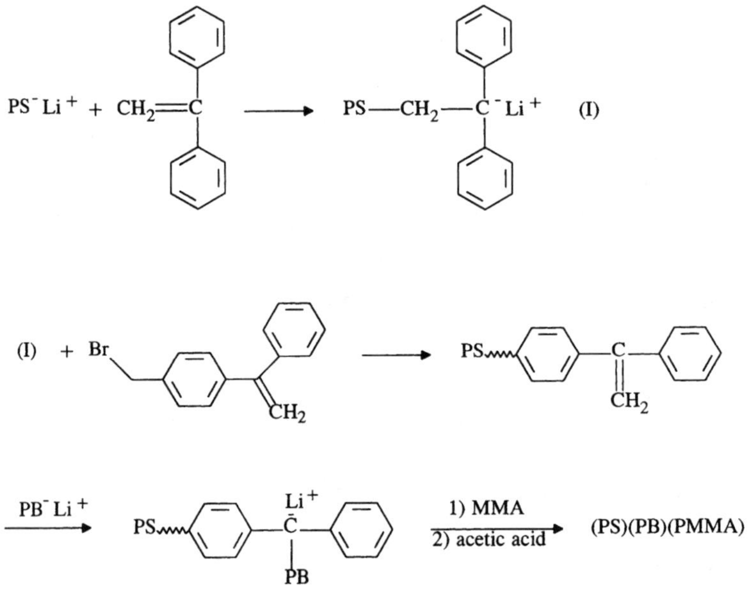

Scheme 21

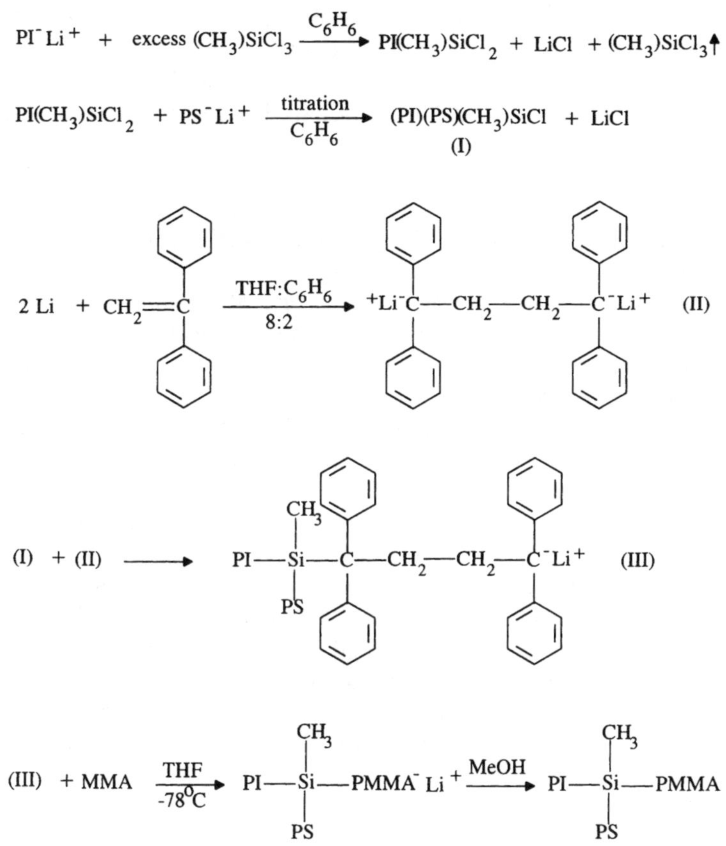

$$PI^-Li^+ + \text{excess } (CH_3)SiCl_3 \xrightarrow{C_6H_6} PI(CH_3)SiCl_2 + LiCl + (CH_3)SiCl_3\uparrow$$

$$PI(CH_3)SiCl_2 + PS^-Li^+ \xrightarrow[C_6H_6]{\text{titration}} \underset{(I)}{(PI)(PS)(CH_3)SiCl} + LiCl$$

$$2\,Li + CH_2=C(C_6H_5)_2 \xrightarrow[8:2]{THF:C_6H_6} {}^+Li^-C(Ph)_2-CH_2-CH_2-C(Ph)_2{}^-Li^+ \quad (II)$$

$$(I) + (II) \longrightarrow PI-\underset{PS}{\overset{CH_3}{Si}}-C(Ph)_2-CH_2-CH_2-C(Ph)_2{}^-Li^+ \quad (III)$$

$$(III) + MMA \xrightarrow[-78^\circ C]{THF} PI-\underset{PS}{\overset{CH_3}{Si}}-PMMA^-Li^+ \xrightarrow{MeOH} PI-\underset{PS}{\overset{CH_3}{Si}}-PMMA$$

Scheme 22

The PDMS was characterized by rather broad molecular weight distributions (I~1.4) and so fractionation was performed, before continuing to the following synthetic steps, in order to reduce the polydispersity of the final product. As in the previous two cases the polymethacrylate branch cannot be isolated and checked independently.

A similar synthetic route was adopted by Stadler et al. for the synthesis of (PS)(PB)(PMMA) stars [54] as shown in Scheme 21. Living PS chains were end-capped with 1-(4-bromomethylphenyl)-1-phenyl ethylene to produce the macromonomer. The capping reaction with DPE was employed in order to reduce the reactivity of the PSLi chain ends thus avoiding several side reactions (transmetallation, addition to the double bond of the DPE derivative). The next step involved the linking of living PB chains, prepared in THF at −10 °C to the end double bond of the macromonomer. This produces a new active center which was used to initiate the polymerization of MMA leading to the formation of the desired product.

The chlorosilane method was also used by Hadjichristidis et al. for the synthesis of miktoarm stars having PS, PI and PMMA arms [55, 56]. The reaction sequence is presented in Scheme 22. The monofunctional macromolecular linking agent (PS)(PI)(CH$_3$)SiCl was prepared using procedures similar to those described above, followed by reaction with a dilute solution of a dilithium initiator, formed by the reaction between 1,1-diphenylethylene (DPE) and Li. This route was carried out to ensure that only one of the initiator's active centers reacts with the linking agent. The remaining active center was used to polymerize MMA in THF at −78 °C yielding the desired miktoarm star. According to this procedure the PMMA branch cannot be isolated and cannot be characterized independently. It was observed that during the synthesis of the difunctional initiator a rather large amount (as high as ~30%) of monofunctional species was also formed. These active species do not affect the synthesis of the ABC star but they give PS-b-PI diblocks by reacting with the (PS)(PI)(CH$_3$)SiCl linking agent, thus reducing the yield of the desired polymer and making its isolation from the crude product more difficult. Nevertheless this was the first attempt to combine the chlorosilane method with the polymerization of methacrylates.

A similar technique was employed for the synthesis of miktoarm stars having PS, PEO, poly(ε-caprolactone) (PCL) or PMMA branches [57]. A PS-b-PMMA diblock copolymer possessing a central DPE derivative, bearing a protected hydroxyl function was prepared. After deprotection and transformation of the hydroxyl group to an alkoxide the anionic ring opening polymerization of the third monomer (EO or ε-CL) was initiated. Only limited characterization data were given in this communication.

2.2.1.8
Synthesis of ABCD Miktoarm Star Quaterpolymers

Only one case concerning the synthesis of a miktoarm star quaterpolymer has appeared in the literature. It consists of four different branches, namely PS, poly(4-methyl styrene) (P4MeS), PI and PB [35]. The reaction sequence for the preparation of this miktoarm star is presented in Scheme 23. The procedure was similar to the one adopted for the synthesis of the ABC-type terpolymers by the chlorosilane method. The characteristic of this method is that two of the arms were incorporated to the linking agent by titration. Consequently the order of addition plays an important role for the preparation of well defined products. PS was chosen to react first with an excess of SiCl$_4$, followed after the evaporation of the excess silane, by the titration with the more sterically hindered P4MeS so that only one arm can be incorporated in the star. The moderately hindered PILi anion was then added by titration, followed by the addition of the fourth arm, which is the least sterically hindered PBLi anion so that complete linking can be achieved. The reaction sequence was monitored by SEC and these results in combination with the molecular and spectroscopic characterization data showed that well defined quaterpolymers were prepared.

Scheme 23

PS⁻Li⁺ + excess SiCl₄ ⟶ (PS)SiCl₃ + LiCl + SiCl₄↑

(PS)SiCl₃ + P4MeS⁻Li⁺ —titration→ (PS)(P4MeS)SiCl₂ + LiCl

PI⁻Li⁺ + (PS)(P4MeS)SiCl₂ —titration→ (PS)(P4MeS)(PI)SiCl + LiCl

(PS)(P4MeS)(PI)SiCl + excess PB⁻Li⁺ ⟶ (PS)(P4MeS)(PI)(PB) + LiCl

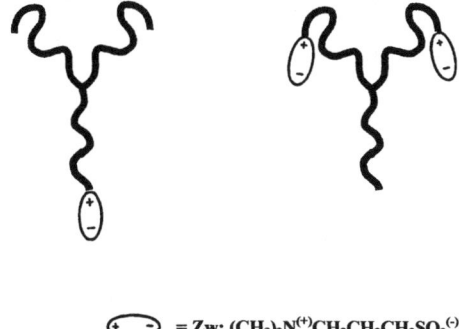

⊖⊕ = Zw: $(CH_3)_2N^{(+)}CH_2CH_2CH_2SO_3^{(-)}$

Fig. 2. Asymmetric PB stars with functional end-groups

2.2.2
Asymmetric ω-Functionalized Polymers

Three-arm PB stars having arms of equal molecular weight but with different functional end-groups were prepared [58]. One or two functional dimethylamine groups were introduced using the functional initiator 3-dimethylaminopropyllithium (DMAPLi) [59–61]. Post polymerization reactions were carried out to transform the dimethylamine groups to zwitterions of the sulfobetaine type (Fig. 2).

The method used for the synthesis was similar to the one developed by Pennisi and Fetters for the synthesis of asymmetric stars, having different molecular weight arms. The synthesis of the three-arm stars with one of them end-functionalized with dimethylamine end-group is outlined in Scheme 24. A solution of living amine-functionalized PB, prepared using DMAPLi as initiator was added to a large excess of methyltrichlorosilane (Si-Cl/C-Li~100/1) in order to synthesize the methyldichlorosilane end-capped N-functionalized PB. The excess silane was removed under reduced pressure. The polymer was repeatedly redissolved and dried. Purified benzene was distilled into the reactor to redissolve the silane-capped arm. Finally a slight excess of the unfunctionalized arms, prepared using sec-BuLi as initiator, was reacted with the macromolecular linking

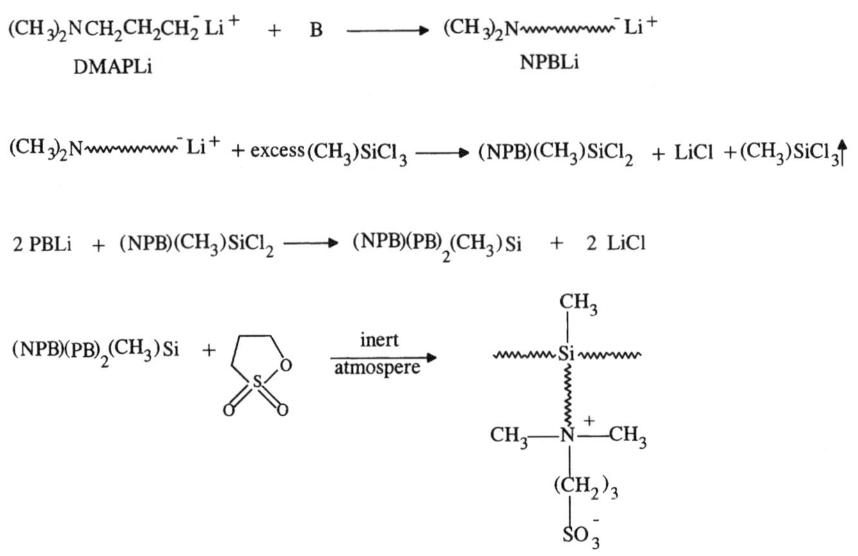

$(CH_3)_2NCH_2CH_2CH_2^- Li^+$ + B $\longrightarrow$ $(CH_3)_2N\text{wwwww}^- Li^+$

DMAPLi NPBLi

$(CH_3)_2N\text{wwwww}^- Li^+$ + excess $(CH_3)SiCl_3$ $\longrightarrow$ $(NPB)(CH_3)SiCl_2$ + LiCl + $(CH_3)SiCl_3\uparrow$

2 PBLi + $(NPB)(CH_3)SiCl_2$ $\longrightarrow$ $(NPB)(PB)_2(CH_3)Si$ + 2 LiCl

$(NPB)(PB)_2(CH_3)Si$ + [cyclopropane sultone] $\xrightarrow{\text{inert atmospere}}$ \text{wwww}Si\text{wwww} with CH$_3$ substituent and $CH_3-N^+-CH_3$, $(CH_2)_3$, SO_3^- group

Scheme 24

agent leading to the formation of three-arm stars with one dimethylamine end-group. Subsequent reaction with cyclopropane sultone transformed the dimethylamine groups into zwitterions.

The synthesis of stars with two functional end-amine groups, 2N-3-PB were prepared in a similar manner. The living unfunctionalized arm was reacted with an excess of methyltrichlorosilane followed after the removal of the excess silane by addition of a small excess of the functionalized living arms to the unfunctionalized chlorosilane-capped arm. When the arm molecular weight was lower than ca. 10^4 the amount of coupled byproduct of the reaction with the excess silane was very large (>10%). This result, together with the fact that all the arms have the same molecular weight, makes the separation of the byproduct from the desired star impossible. In order to minimize the coupling reaction the living polymer was end-capped with 1,1-diphenylethylene (DPE). This capping reaction, accelerated by the addition of a small quantity of THF, reduces the amount of the coupled product to acceptable levels (<3.5%).

The reaction sequence was monitored by SEC. The molecular characterization was carried out by MO and SLS for both the intermediate and final products, revealing that well defined stars were prepared.

2.3
Stars with Topological Asymmetry

A new class of asymmetric stars, the so-called inverse star block copolymers, were recently reported by Tselikas et al. [62]. These polymers are four-arm stars,

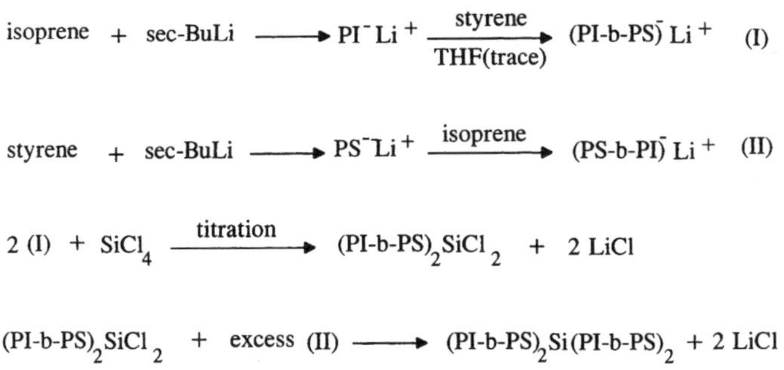

$$\text{isoprene} + \text{sec-BuLi} \longrightarrow \text{PI}^-\text{Li}^+ \xrightarrow[\text{THF(trace)}]{\text{styrene}} (\text{PI-b-PS})^-\text{Li}^+ \quad (I)$$

$$\text{styrene} + \text{sec-BuLi} \longrightarrow \text{PS}^-\text{Li}^+ \xrightarrow{\text{isoprene}} (\text{PS-b-PI})^-\text{Li}^+ \quad (II)$$

$$2\,(I) + \text{SiCl}_4 \xrightarrow{\text{titration}} (\text{PI-b-PS})_2\text{SiCl}_2 + 2\,\text{LiCl}$$

$$(\text{PI-b-PS})_2\text{SiCl}_2 + \text{excess (II)} \longrightarrow (\text{PI-b-PS})_2\text{Si}(\text{PI-b-PS})_2 + 2\,\text{LiCl}$$

Scheme 25

with each of these arms being a diblock copolymer of isoprene and styrene. Two of these blocks are connected to the star's center with their PS end, while the other two blocks with their PI end. Consequently the asymmetry is due to the different topology of the arms.

The synthetic procedure for the synthesis of the inverse starblock copolymers is given in Scheme 25. Diblock arms (I) having the living end at the PS chain end were prepared by anionic polymerization with sequential addition of monomers. In order to accelerate the crossover reaction from the PILi to the PSLi chain end a small quantity of THF was added prior the addition of styrene. The living diblock (I) solution was added dropwise to a stoichiometric amount of SiCl$_4$ until two arms are linked to the silane. This step was monitored by SEC and is similar to a titration process. The end point of the titration was determined by the appearance of a small quantity (~1%) of trimer in the SEC trace. The diblock (I) was selected over the diblock (II) due to the increased steric hindrance of the styryl anion over the isoprenyl anion, which makes easier the control of the incorporation of only two arms into the silane.

The difunctional macromolecular linking agent was then reacted with a small excess of the living diblock (II) for the preparation of the inverse star block copolymers. In order to facilitate the reaction with the macromolecular linking agent the living diblock was end-capped with 3–4 units of butadiene.

Extensive molecular characterization studies of the arms and the final products by MO and SLS and compositional characterization studies by ^{1}H NMR, UV-SEC and differential refractometry lead to the conclusion that the copolymers are characterized by a high degree of molecular and compositional homogeneity.

3
Properties

3.1
Solution Properties

3.1.1
Theory

Relatively few theoretical studies have been devoted to the conformational characteristics of asymmetric star polymers in solution. Vlahos et al. [63] studied the conformational properties of A_nB_m miktoarm copolymers in different solvents. Analytical expressions of various conformational averages were obtained from renormalization group calculations at the critical dimensionality d=4 up to the first order of the interaction parameters u_A, u_B, and u_{AB} between segments of the same or different kind, among them the radii of gyration of the two homopolymer parts $<S_k^2>$ (k=A_n or B_m) and the whole miktoarm chain $<S_{A_nB_m}^2>$, the mean square distance between the centers of mass of the two homopolymer parts A and B $<G_{A_nB_m}^2>$ and the mean square distance between the center of mass of a homopolymer part and the star common origin $<G_k^2>$ (k=A_n or B_m). The critical exponents, ν, of these conformational averages ($<A>\sim M^{2\nu}$) were found to be unaffected by the cross interactions between dissimilar units and are determined only by the prevailing solvent conditions within the homopolymer arms. Thus in common theta solvents, ν is equal to 1/2, while for common good and selective solvents, $\nu=1/2+\varepsilon/16$, where $\varepsilon=4-d$. From the analytical expressions and those corresponding to the homopolymer precursors (i.e., a homopolymer star chain with the same number of branches and units as the k homopolymer part in the miktoarm star copolymer) they arrived at the following ratios

$$\sigma_G = \left\langle G_{A_nB_m}^2 \right\rangle / \left[\left\langle G_{A_n\text{star}}^2 \right\rangle + \left\langle G_{B_n\text{star}}^2 \right\rangle \right] \tag{1}$$

$$\gamma_{S_k} = \left\langle S_k^2 \right\rangle / \left\langle S_{k,\text{star}}^2 \right\rangle \quad (k = A_n \text{ or } B_m) \tag{2}$$

$$\gamma_{G_k} = \left\langle G_k^2 \right\rangle / \left\langle G_{k,\text{star}}^2 \right\rangle \quad (k = A_n \text{ or } B_m) \tag{3}$$

These ratios are the most important means of quantitatively characterizing the effects of heterointeractions in copolymers, since their analytical formulas depend only on the chain composition and the cross excluded volume parameter $u_{AB}*$. For a particular miktoarm chain the ratio σ_G increases from a common good to selective and then to a common theta solvent since the intensity of heterointeractions increases. In Fig. 3 is illustrated the dependence of the chain topology as function of the length fraction of the B branch $\Phi_B=N_B/(N_A+N_B)$. Similarly, the ratios γ have higher values in a common theta than in a common good

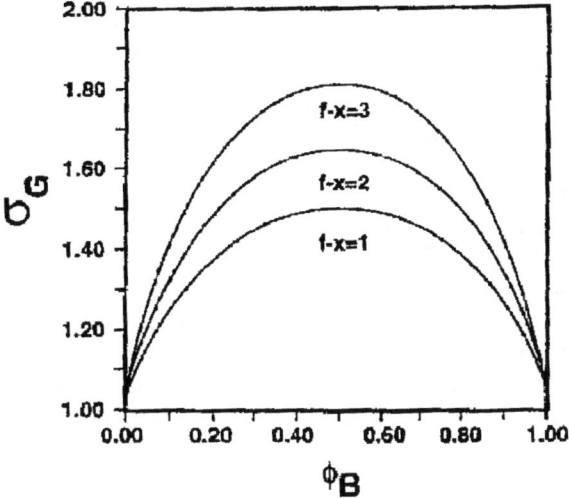

Fig. 3. Expansion factor σ_G of the mean square distance between the two centers of mass as a function of Φ_B for miktoarm copolymers A_xB_{f-x} with x=2 and various values of f-x. $u_A^* = u_B^* = 0, u_{AB}^* = \varepsilon/8$ (reproduced with permission from [63])

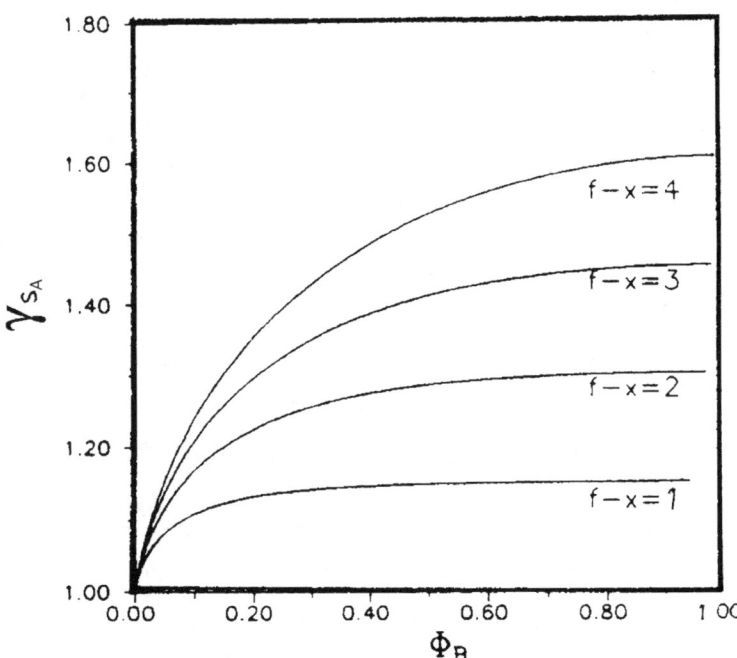

Fig. 4. Ratio γ_{SA} as a function of Φ_B for miktoarm copolymers A_xB_{f-x} with x=3 and various values of f-x. $u_A^* = u_B^* = 0, u_{AB}^* = \varepsilon/8$ (reproduced with permission from [63])

solvent. In a selective solvent however the cross interactions are as effective as in the macroscopic state of common theta or a common good solvent depending on the solvent conditions within the homopolymer part. The dependence of length fraction of B arms on the ratios γ is shown in Fig. 4. Miktoarm stars A_nB_m, where A or B are deuterated are necessary in order to evaluate these theoretical predictions.

The same group also calculated the dimensionless ratios σ_G and γ for miktoarm star copolymers by means of off-lattice Monte Carlo simulations [64]. The model considers Gaussian units with Lennard-Jones intramolecular interactions whose parameters have been set to mimic the various macroscopic states. Polymers of the A_2B, A_2B_2, A_2B_4, A_3B_3 and A_6B_6 type with equal length branches and total number of units N=24, 48, 84, 108, 144 and 204 have been generated. The Pivot algorithm was used for the generation of the Monte Carlo sampling. The values of the ratios extrapolated to the limit of large molecular weights are presented in Table 1.

Table 1. The extrapolated values of the dimensionless ratios σ_G, γ_{S_A} and γ_{G_A} by Monte Carlo calculations. The values in brackets are data by Renormalization Group analysis

	A_2B	A_2B_2	A_2B_4	A_3B_3	A_6B_6
		common theta			
σ_G	1.518±0.013 (1.498)	1.833±0.024 (1.648)	2.207±0.016 (1.997)	2.454±0.022 (1.971)	5.021±0.050 (2.943)
γ_{S_A}	1.074±0.15 (1.127)	1.103±0.013 (1.254)	1.140±0.020 (1.508)	1.120±0.009 (1.413)	1.119±0.013 (1.872)
		selective solvent (theta for A)			
σ_G	1.419±0.005 (1.374)	1.605±0.015 (1.486)	1.920±0.021 (1.748)	1.980±0.018 (1.729)	3.054±0.038 (2.457)
γ_{S_A}	1.061±0.009	1.107±0.018	1.173±0.014	1.150±0.009	1.207±0.010
γ_{G_A}	1.258±0.020	1.371±0.014	1.584±0.018	1.590±0.027	2.093±0.035
		common good			
σ_G	1.294±0.005 (1.249)	1.424±0.009 (1.324)	1.562±0.015 (1.498)	1.616±0.017 (1.486)	2.102±0.037 (1.971)
γ_{S_A}	1.037±0.004	1.068±0.004	1.129±0.008	1.102±0.004	1.151±0.003
γ_{G_A}	1.086±0.007	1.171±0.009	1.321±0.015	1.285±0.011	1.529±0.007
		artificially segregated			
σ_G	3.497±0.004			6.053±0.003	9.311±0.002
γ_{G_A}	2.868±0.004			3.484±0.003	5.141±0.002

According to the simulation results the mean conformation achieved by the miktoarm stars in solution is characterized by a significant deformation along the arms, mainly near the star center where the possibility of interactions is high and a change in the relative orientation of the vectors $\vec{G}_{A_n}$ and $\vec{G}_{B_m}$ (determined from their angle) extends the center of mass separation in order to diminish the repulsions. The results show that both mechanisms have a similar influence on the increase of the $<G^2_{A_nB_m}>$. The values of the ratio σ however corresponding to an artificial segregated model (also considered in this work) are very large compared to the value of a single chain in various solvents. It is concluded that in dilute solutions the miktoarm chain does not segregate intramolecularly. As in the previous case experimental results are not available.

Vlahos et al. [65] have determined the dimensionless ratio σ_G by intrinsic viscosity analysis for the A_2B and A_3B miktoarm stars in various solvent conditions. Considering the ratios γ_{S_A} and γ_{S_B} to be close to 1 for the particular miktoarms studied in this work they arrived at the following equations:

$$[\eta]^{2/3}_{A_2B} = x_A^{5/3}(1+x_B\sigma_G)[\eta]^{2/3}_{A_2}(\frac{\Phi_{A_2B}}{\Phi_{A_2}})^{2/3} + x_B^{5/3}(1+x_A 2\sigma_G)[\eta]^{2/3}_B(\frac{\Phi_{A_2B}}{\Phi_B})^{2/3} \quad (4)$$

$$[\eta]^{2/3}_{A_3B} = x_A^{5/3}(1+\frac{2}{7}x_B\sigma_G)[\eta]^{2/3}_{A_3}(\frac{\Phi_{A_3B}}{\Phi_{A_3}})^{2/3} + x_B^{5/3}(1+x_A 2\sigma_G)[\eta]^{2/3}_B(\frac{\Phi_{A_3B}}{\Phi_B})^{2/3} \quad (5)$$

Monte Carlo calculations based on the lower bound method were applied to estimate the Flory parameter Φ for asymmetric stars, whereas, for the symmetric stars, experimental results on three- and four-arm homopolymer stars were used. The obtained values of σ_G for different macroscopic states together with those obtained from RG and MC calculations are presented in Table 2. p is the ratio of the lengths of the B and A arms ($p=N_B/N_A$).

Table 2. The ratio σ_G of miktoarm stars [22, 65]

Common good	MC	RG	Exp.
A_2B (p=1)	1.294±0.005	1.249	1.236
A_2B (p=1.7)	1.289±0.006	1.238	1.241
A_3B (p=2)	1.342±0.010	1.316	1.402
A_2B (p=3.9)	1.241±0.017	1.262	1.402
Selective (θ for A)			
A_2B (p=1)	1.419±0.005	1.374	1.331

Finally Kosmas and Hadjichristidis [66] by using a single molecule argument predicted the analytical relations

$$M_n = \sum_i f_i m_{ni}, \quad M_w = M_n + \frac{\sum_i f_i m_{ni}(m_{wi} - m_{ni})}{M_n} \tag{6}$$

which links the average molecular weights of the homopolymer precursors and those of the final miktoarm macromolecule; here, f_i is the number of the precursors of the i-th kind (i=1, 2 ,...n), m_{ni}, m_{wi}, their number and weight average molecular weight and M_w, M_n the molecular weights corresponding to the miktoarm chain. Expressing m_{ni}, m_{wi} and M_w, M_n in terms of the Laplace tranform of the distributions of molecular weights of the precursors $P_i(m)$ and the whole molecule $P_c(M)$ they found that both equations at Eq. 6 are valid for any $P_i(m)$. Experimental values of M_w based on LALLS and a combination of MO and SEC are found to be close to those calculated from the theory.

3.1.2
Experimental Results

Star polymers, with a structure where chains of different molecular weight and/or chemical nature radiate from a common junction point, are expected to have different solution behavior compared to regular or symmetric star polymers. By forcing chemically different arms to be joined, expansion of the molecular dimensions is predicted (see Sect. 3.1.1) as a consequence of the increase in the number of heterocontacts in good and theta solvents or new kinds of structures can be formed in selective solvents.

The dilute solution properties of asymmetric three-arm polystyrene stars were studied by Fetters and coworkers [67] under different solvent conditions. These materials comprised two series of samples, the first having a short third arm with half the molecular weight of the two identical arms (SAS) and the second a long arm having twice the molecular weight of the other two arms (LAS). In toluene, a good solvent for PS, the two types of asymmetric stars exhibited identical values of g', defined as

$$g' = [\eta]_{star}/[\eta]_{linear} \tag{7}$$

where $[\eta]_{linear}$ is the intrinsic viscosity of the linear polymer with the same molecular weight. Values of g' are a little higher than values of g' for a regular three-arm star. In the theta solvent cyclohexane at 35 °C the value of g' for SAS is lower than that of LAS which was equal to g' value of regular three-arm stars. It has to be mentioned that in both solvents the value of α in the Mark-Houwink-Sakurada equation, $[\eta] = KM^\alpha$, was higher for the SAS than LAS. The latter has a similar α value as the symmetric stars.

The dilute solution characteristics of asymmetric stars, with equal number of short and long arms, prepared by the DVB method, were studied by Mays et al.

[9]. They used size exclusion chromatography with a multiangle laser light scattering detector and viscometry in an attempt to characterize the structure of the material produced by this two step method. Depending on the flow rate, different forms of chromatograms were obtained. At very low flow rates, i.e., of 0.1 ml/min a true size separation could be achieved in contrast to the normal flow rates, where the separation mechanism could be influenced by diffusion and/or entanglement effects. Plots of the weight average molecular weight, M_w, or radius of gyration, R_g, against elution volume for the asymmetric stars had a "U" shape, in contrast to the expected linear decrease of M_w or R_g with increasing elution volume. It was concluded that the material was a complex mixture of asymmetric and linked stars. In the case where the asymmetric star peak was resolved the branching factors g and g', ($g=R_{g,\,star}/R_{g,\,linear}$) were obtained. Their values were found to be lower than the ones obtained for regular star precursors, indicating a more compact structure for the asymmetric molecules.

The hydrodynamic properties of miktoarm stars of the types A_2B and A_2B_2 where A, B=PS, PI, PBd were studied by Hadjichristidis and coworkers [22] in solvents good for both kinds of arms or theta for one of them and good for the other. Using the techniques of dilute solution viscometry and dynamic light scattering they tried to probe the conformation of these molecules in solution and extract some information on the effect of heterocontacts on the overall size of the molecule. The experimentally determined [η] and R_h values for the copolymers were compared with the theoretical values. The latter were calculated from data on linear homopolymers in the same solvents, corrected for the star structure. The final theoretical value was an average value weighted according to composition of those calculated for the homostars.

Experimental [η] and R_h values were found to be consistently larger than the theoretical ones, irrespective of the solvent quality, the differences being larger for the A_2B_2 case. It was suggested that a small expansion of the miktoarm stars occurs either in a good or in a selectively θ solvent, due to the increased repulsive interactions between A and B chains when they are tethered to the same point.

As an extension of this work viscometry was used to study the conformation of A_3B stars (A=PI and B=PS) in the common good solvent toluene [65]. In this case the experimental data, together with the ones obtained from the previous work, were used to extract the dimensionless ratio σ_G. This ratio expresses quantitatively the effects of heterointeractions between unlike segments on the conformational properties of the copolymers. The σ_G values for the A_3B case were higher than for the A_2B case; it seems that this is due to the increased segment density of A units in the vicinity of B units for A_3B. The experimental values compared rather well with the ones obtained from renormalization group theory and Monte Carlo calculations taking into account the uncertainty in the asymmetry correction coefficient used in the calculations (see Table 2).

Tsitsilianis et al. [14] also reached the conclusion, from size exclusion chromatography measurements, that A_nB_n type copolymers, where A is PS and B is poly(*tert*-butyl acrylate), are more expanded than the corresponding homostars, due to the increased density of two different kinds of segments which leads to an increased number of heterocontacts. They noticed that the expansion was greater as the molecular weight of the arms increased.

Teyssié and coworkers [68, 69] have studied the surface, interfacial and emulsifying properties of A_2B stars, where B is a polydiene, PS or poly(*tert*-butyl styrene), (PtBuS) and A poly(ethylene oxide). The miktoarm stars were shown to be better emulsifying agents for water-organic solvents mixtures than linear block copolymers. Saturation of the interface was reached more quickly with the miktoarm polymers. The differences were ascribed to the miktoarm architecture which produces higher segment density near the interface of the two components. These results are in agreement with the experiments conducted by Xie and Xia on PS_2PEO_2 stars [34].

Higashimura et al. [70] tried to elucidate the interactions between amphiphilic miktoarm star molecules produced by cationic polymerization and small molecules using NMR techniques. In their comparison between two different architectures no distinct differences were observed for star-blocks and miktoarm stars, both species being sufficiently capable of accommodating hydrophilic molecules within their hydrodynamic volume.

Anastasiadis and coworkers [71] investigated the effect of macromolecular architecture on the dynamic behavior of three-arm miktoarm stars of the types A_2B and AA'B where A is PI and B is PS. AA'B stands for asymmetric miktoarm stars where the two arms of identical chemical constitution have different molecular weights. From a different point of view these two series of samples can be considered as symmetric simple grafts (A_2B) and asymmetric simple grafts (AA'B) where in the second case the B branch is situated at a position $\tau=0.25$ on the backbone, where τ is the ratio of the distance of the branching point from one backbone end to the length of the backbone. They used the pulsed-field-gradient NMR technique to probe the dependence of the self-diffusion coefficient, D, on the concentration in semidilute and concentrated solutions in the common good solvent toluene. For both types of samples the diffusion coefficient for the polyisoprene rich grafts were consistently lower than those for the copolymers with high styrene contents when compared at the same concentration due to the different entanglement characteristics of the two components. In a $DN^{0.59}N_e^{0.41}$ vs ϕ/ϕ_e representation, where N is the number of segments, N_e the entanglement length, ϕ the volume fraction of copolymer and ϕ_e the volume fraction for entanglements to occur, the data for the asymmetric I_2S stars were slightly higher than those for the symmetric I_2S samples. Further analysis of the data showed that the D dependence on c in both I_2S copolymers and P(S-*b*-I) diblock copolymers could be collapsed onto a master curve when plotted as $D(gN)^{0.59}N_e^{0.41}$ vs ϕ/ϕ_e, a procedure that takes into account the difference in the entanglement characteristics of the parent homopolymers and the different radii of gyration of stars vs linear polymers. The exponential slowing down of the diffusivities expected for star molecules was not observed due to the moderately high molecular weights and concentrations studied.

Tsitsilianis and coworkers [41] presented results on the micellization behavior of anionically synthesized amphiphilic miktoarm star copolymers with PS and PEO branches in THF. Although THF is not very selective for PS increased values of apparent weight average molecular weights were obtained. The micelles formed in THF had molecular weights two orders of magnitude larger than the M_w of the unimer. The authors observed an increase in the depolarization ratio by increasing polymer concentration indicative of the formation of large multimolecular micelles with PEO cores.

Micelle formation was also observed, by the same authors [72], in the case of miktoarm stars with equal number of polystyrene and poly(*tert*-butyl acrylate) branches in methanol a good solvent for PtBuA and a precipitant for PS. For a sample of the type $PS_{15}PtBuA_{15}$ of molecular weight $M_w=218,000$ and 70 wt% tBuA, the association number was estimated to be 27 by light scattering. It was assumed that the multimolecular micelles were spherical in shape with PtBuA chains in the corona and PS in the core. The DVB cores of the unimers were assumed to be located at the core-corona interface forming a third domain with high segment density.

In the same study viscometric experiments in acetone indicated the existence of a conformational transition around 35 °C. In that temperature region a sudden decrease in the intrinsic viscosity and a maximum for the Huggins constant, k_H, were observed. Although static light scattering envelopes were curved downwards at low angles, indicating the existence of a small number of high molecular weight associated species, extrapolation from higher angles gave molecular weights very close to the unimer molecular weight. The authors concluded that these phenomena could only be interpreted by a transition from a segregated to a nonsegregated conformation.

The micellar structures formed in solutions of well defined $(PE)_2(PEP)_2$, $PE(PEP)_2$ and $(PE)_2PEP$ type polyethylene (PE)-poly(ethylene-*alt*-propylene) (PEP) miktoarm stars in *n*-decane were investigated by Richter et al. [73] by small angle neutron scattering using the contrast variation technique. The solvent is selective for the PEP branches. The formation of multimolecular structures was also influenced by the ability of PE to crystallize. The PE chains were 75% deuterated to enhance scattering contrast and their MW was kept constant and the MW of PEP was varied in order to study the influence of composition and architecture on the micellar properties. The miktoarm stars formed multimolecular micelles with flat dense, disk-like, crystalline PE cores and soft coronas of PEP "hairs" sticking out on both sides of the core. Using model fitting procedures structural parameters like core and corona thickness were estimated.

It was confirmed that the molecular architecture does not modify the global characteristics (shape) of the micelles but only its geometrical parameters. The average thickness of the core was nearly constant and independent of the MW of the PEP brush for the PE_2PEP_2 case. For $PE(PEP)_2$ and $(PE)_2PEP$ the thickness decreased with increasing MW of PEP. It was found that for all PEP molecular weights the chains are more extended if the PE chains are connected to two PEP arms instead of one. The difference was more pronounced when the PEP molecular weights were higher. On the other hand the thicknesses of the cores were higher for the system $(PE)_2PEP$. In comparing the systems $(PE)_2(PEP)_2$ and

PE(PEP)$_2$ it was suggested that the average extension of the brush arms was higher for the latter architecture except for the samples with medium molecular weight (M$_{PEP}$~9000). At the same time the thickness of the core and the blob size were lower for the PE(PEP)$_2$ samples. From measurements at different temperatures a critical micelle temperature of about 60 °C was found. At room temperature the equilibrium was found to be shifted completely in favor of the micelles. The authors also developed a thermodynamic model to describe the system at low temperature, based on the free energy of a simple micelle. Although this model predicts scaling relations dependent on the star architecture and was found to agree well with previous data on PE-*b*-PEP diblocks it could not describe the core thickness variation correctly for the miktoarm systems.

The solution properties of umbrella star poly(styrene-*u*-butadiene) copolymers, with PS cores in solvents with decreasing solvating power for polystyrene, were studied by viscometry and light scattering [33]. Intrinsic viscosities decreased in the order toluene (common good solvent)>cyclohexane (θ for PS at 35 °C)>methylcyclohexane (θ solvent for PS at 70 °C). Decreasing the temperature from the θ point produced only a slight contraction of the size of the molecule most likely due to the contraction of the PS core. Intrinsic viscosities of a sample in *n*-decane are about two times lower than in the θ solvents and the umbrella star was found, by light scattering, to be molecularly dissolved in the same concentration range, forming monomolecular micelles with collapsed PS star cores and marginally solvated PB coronas.

The association behavior of three-arm polybutadiene stars with one or two dimethylamino or sulfozwitterionic groups (see Scheme 24 for structures) was studied by Hadjichristidis and coworkers [58, 74] in dilute solutions and compared with the behavior of monofunctional linear and regular three-arm star polybutadienes with all ends functionalized. The amine-capped samples showed no evidence of association in cyclohexane, a good nonpolar solvent for polybutadiene. Strong association was observed for the zwitterionic materials. The weight average aggregation number, N_w, determined by low angle laser light scattering and the number average aggregation number, N_n, by membrane osmometry, were found to decrease by increasing the number of functional groups in the star polymers, at the same total molecular weight, whereas N_w and N_n were decreasing function of molecular weight for the same number of end groups. The difunctional and trifunctional stars formed gels even at low concentrations. Taking into account the low aggregation numbers for these samples at concentrations lower than c_{gel} the authors postulated that in very dilute solutions intramolecular association dominates. The association numbers obtained for the monofunctional stars were lower than those determined for the linear monofunctional species due to the steric hindrance of the unfunctionalized arms.

Studies on the hydrodynamic properties of the same samples by dynamic light scattering and viscometry confirmed the conclusions of the static methods. Decreased diffusion coefficients and k_D values, where k_D is the coefficient of the concentration dependence of D ($k_D=2A_2M-k_f v$), were observed for the zwitterionic polymers compared to the amine precursors. Increased values of the polydispersity parameter μ_2/Γ^2 in dynamic light scattering experiments indicated that the aggregates were polydisperse supporting the high values of polydispersity calculated from osmometry and static light scattering. The strongly negative k_D values for the trifunctional stars indicated the existence of strong hydrodynamic interactions. These samples showed also low aggregation numbers and only slight increase in the hydrodynamic radius, R_h, compared to their precursors. This behavior is the result of intramolecular association at low concentrations. For the difunctional stars the phenomena were less pronounced, although intramolecular association is also possible in this case.

Assuming that the aggregates of the monofunctional stars can be described as regular stars (more precisely as umbrella star polymers) and the precursors as the arms the authors calculated aggregation numbers from the ratio of $(R_h)_{Zw}/(R_h)_N$, where $(R_h)_{Zw}$ is the hydrodynamic radius of the zwitterion-capped polymers in cyclohexane and $(R_h)_N$ is the same radius for the amine precursor in the same solvent, using the literature data for stars of different functionalities. The results showed that the aggregates behave hydrodynamically as star polymers with functionality equal to $2N_w$, where N_w is the weight average aggregation number determined by static light scattering. They concluded that the two unfunctionalized arms anchored at the periphery of the aggregates are responsible for the overall size of the associates.

Low intrinsic viscosities and high k_H (Huggins constant) values were determined by dilute solution viscosity measurements, indicative of large hydrodynamic interactions, supporting the conclusions drawn from light scattering that intramolecular association does occur at low concentration. Comparative examination of R_v and R_h values showed that $R_v<R_h$ for the zwitterionic polymers, which was attributed to dissociation of the aggregates, to some extent, under the applied shear rate in the capillary. Since these forces are generally not very strong it was concluded that the critical shear rate should be small. Compared to linear ω-functionalized polymers the lower stability of the aggregates formed by the end functionalized stars was attributed to the steric repulsion of the unfunctionalized arms.

In another publication [75] the adsorption behavior of the same materials from dilute solutions on silicon wafers was studied by ellipsometry. A mixture of cyclohexane and toluene (50:50 by volume) was used in order to provide enough refractive index contrast for the measurements and also to inhibit association of the end groups, which would influence the adsorption process.

The adsorbed amount, A, was found to increase with decreasing molecular weight of the stars and increasing the number of the zwitterionic groups per molecule. The grafting density, σ, defined as $\sigma=AN_A/M_w$, where N_A is the Avogadro number, presented a stronger dependence on the molecular weight than on the functionality of the stars. The σ values of two samples of comparable molecular weights but having two and three zwitterionic groups were very close, indicating that, despite the fact that the adsorption energy is high, the entropic loss involved in the attachment of the third arm when two arms have already been adsorbed may be very high and does not occur. This conclusion was also supported by the similar D_{inter} values, where $D_{inter}=1/\sigma^{1/2}$ is the interchain distance, obtained for stars with two and three polar groups. The adsorbed stars were found to be less stretched than the linear chains. The adsorption energy was calculated to be (9±1) kT per group from the linear polymer data.

The adsorption kinetics from time resolved ellipsometry measurements showed two regimes: (a) a diffusion controlled process at the initial stages and (b) at longer times, an exponential behavior where the arriving chain must penetrate the barrier formed by the already adsorbed chains. The experimental data indicated that the stars penetrate this barrier faster than the linear chains.

3.2
Bulk Properties

3.2.1
Theory

Few theoretical studies deal with the behavior of miktoarm star copolymers in the solid state. Issues like the phase diagram and the order-disorder transitions however have been studied in considerable detail.

Olvera de la Cruz and Sanchez [76] were first to report theoretical calculations concerning the phase stability of graft and miktoarm A_nB_n star copolymers with equal numbers of A and B branches. The static structure factor $S(q)$ was calculated for the disordered phase (melt) by expanding the free energy, in terms of the Fourier transform of the order parameter. They applied path integral methods which are equivalent to the random phase approximation method used by Leibler. For the copolymers considered $S(q)$ had the functional form $S(q)^{-1}= (Q(q)/N)-2\chi$ where N is the total number of units of the copolymer chain, χ the Flory interaction parameter and Q a function that depends specifically on the copolymer type. $S(q)$ has a maximum at q^* which is determined by the equation $\partial Q/\partial Q=0$.

For the graft copolymers they found that $q^*(graft) \geq q^*(diblock)$ and there is no symmetry around the composition f=1/2 due to the inherent asymmetry in a graft copolymer. The spinodal curves where $S(q^*)$ diverges at the spinodal temperature or at the characteristic value of $(\chi N)_s$ are plotted in Figs. 5 and 6 as functions of the composition. They predicted that a simple graft has no critical point for any f. Irrespective of the position of the branch point, the minimum value of the spinodal appears at volume fraction f=0.5. For the special case of the symmetric graft (an A_2B miktoarm star) $(\chi N)_s=13.5$. This means that it is more difficult for the chain to phase separate in this kind of architecture than it is for linear diblock copolymers. Miktoarm copolymers of the A_nB_n type are predicted to have critical point at $(\chi N_0)_c=10.5$ (the same value as for the diblocks when f= 0.5) which implies that the microphase separation of nearly symmetrical miktoarms depends solely on the molecular weight of the span diblock copolymer N_0 and not on that of the whole star chain.

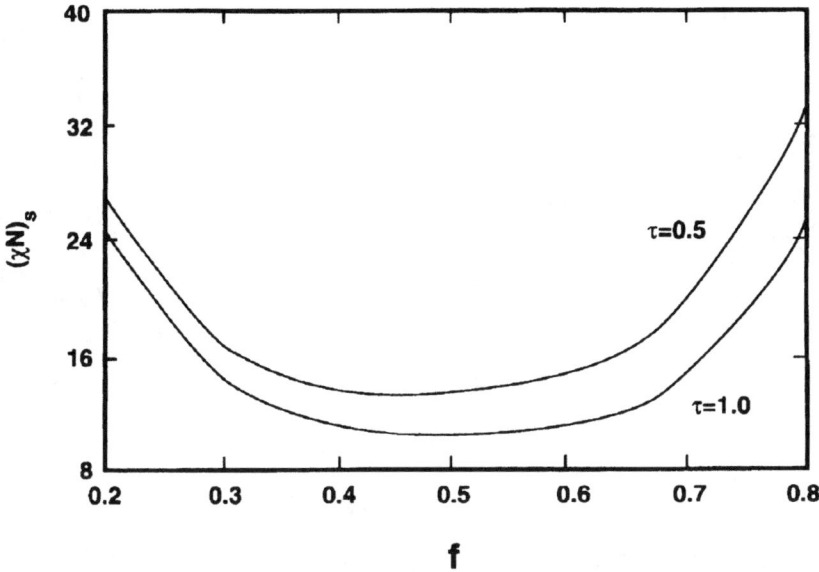

Fig. 5. Comparison of the variation of $(\chi N)_s$ with composition for graft and diblock copolymers. The number monomers (N) in each copolymer is the same. τ is the fractional position of the branch. (reproduced with permission from [76])

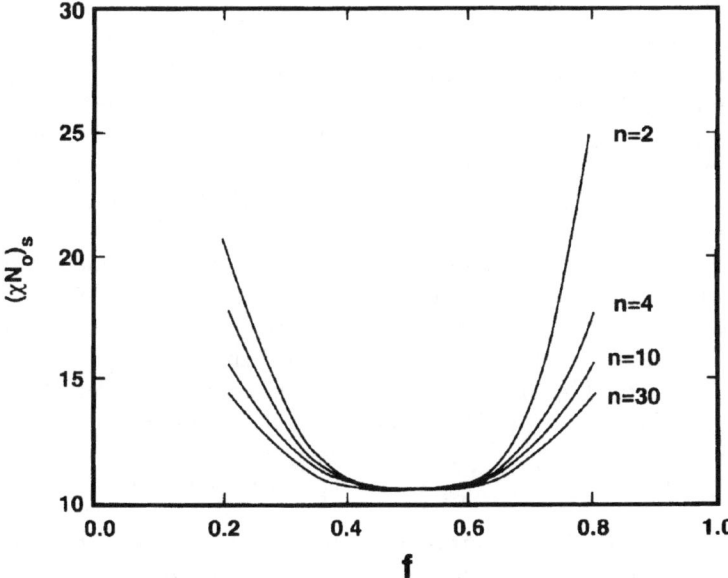

Fig. 6. Variation of $(\chi N_0)_s$ with composition and arm number for $A_n B_n$ star copolymers. N_0 is the number of monomers in the diblock (A_1-b-B_1). (reproduced with permission from [76])

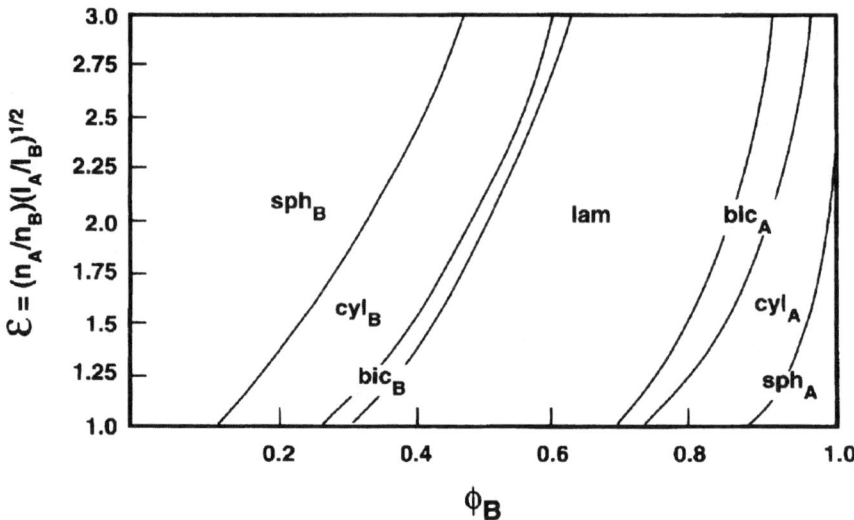

Fig. 7. Phase diagram in the strong- segregation limit for starblock copolymers with n_A A arms and n_B B arms as a function of the volume fraction of the B monomer (reproduced with permission from [77])

Milner [77] studied the phase diagram of A_nB_m miktoarm copolymers in the strong segregation limit in which the interface between A and B chains is sharp. The morphology of microphases and their length scale are determined by a competition of the interfacial tension and the increase in the stretching free energy as the copolymer arms stretch away from the interface. Calculations are made in terms of the cross sectional area relative to the outer surface from different "wedges" corresponding to spherical, cylindrical, and OBDD phases. The surface tension contribution is the area of the dividing surface times the surface tension γ, while the stretching energy was calculated by considering the miktoarm stars as brushes. The copolymer chains are added one by one and the work to add each one is summed. The total free energy of each structure (the sum of the stretching and interfacial energy) is then minimized. The phase diagram is determined by the cross lines of the free energy of different structures and is presented in Fig. 7 as a function of the volume fraction of the B units Φ_B and the parameter $\varepsilon = \dfrac{n_A (l_A)^{1/2}}{n_B (l_B)^{1/2}}$ which expresses the polymer asymmetry. n_A, n_B are the numbers of A or B arms and l_A, l_B are material parameters expressing the elasticity of the A and B homopolymers. The consideration of miktoarm stars as brushes (branches without common origin) with A and B arms crossing the interface will make no sense for small number of arms but in the other case,

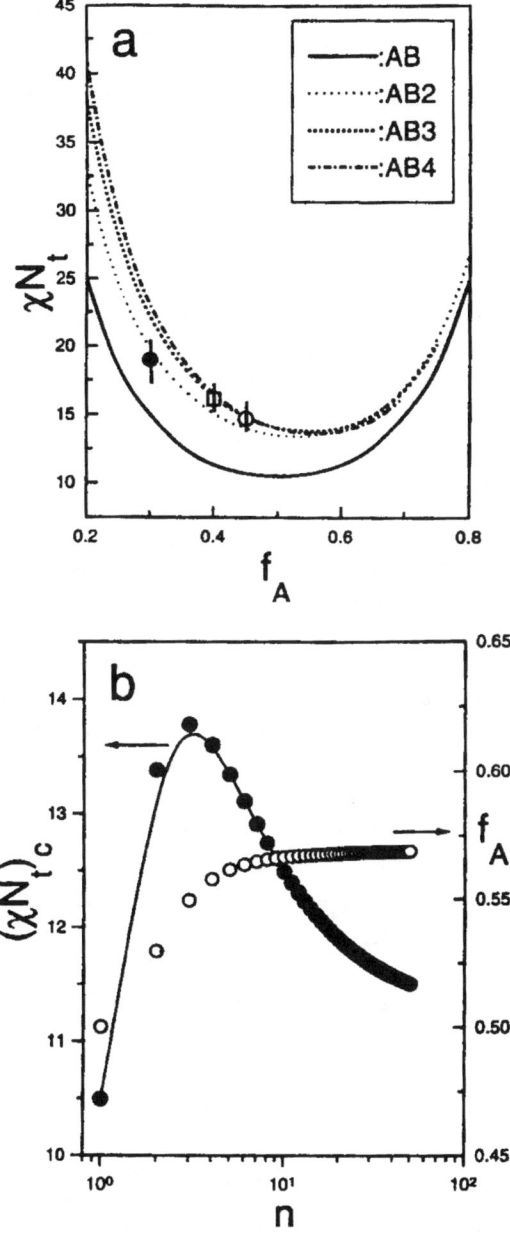

Fig. 8. a Comparison of the spinodal curves $\chi N_t(f_A)$ for diblock and different AB_n miktoarm copolymers with n=2, 3, and 4. The experimental results from an AB_2 (●) and the two AB_3 (□, ○) miktoarm stars are shown together with the investigated temperature range (*vertical lines*). **b** Critical values of χN_t plotted as a function of the number of arms of the B blocks. The dependence of the optimal composition corresponding to the minima of the spinodal curves are also shown. The line is only a guide to the eye (reproduced with permission from [78])

because of the presence of the core, the arms will be stretched more than the present theory predicts.

Floudas et al. have developed [78] a mean field theory to treat the phase stability criteria for the most general case of miktoarm star copolymers, namely, A_nB_m, with asymmetric number of branches m≠n. The case of AB_n was treated in detail. The static structure factor in the disordered phase was calculated and the resulting spinodal curves for the series of AB_n miktoarms are plotted in Fig. 8a. They found that these copolymers are more difficult to phase separate than linear diblocks since the critical values of χN_t ($N_t=N_A+nN_B$) are higher than the diblock case at any composition. The spinodal curves are asymmetric due to the inherent asymmetry of miktoarm stars. The critical values corresponding to the minima of the spinodal and the respective compositions are plotted in Fig. 8b for miktoarms with different n. For n up to 3 there is a considerable increase in the value of $(\chi N_t)_c$ which indicates an increasing compatibility between A and B blocks. However for n>3 there is a reversal in $(\chi N_t)_c$ which, for n>10, differs only slightly from that of the A-*b*-B diblock copolymers. The maximum value of $(\chi N_t)_c$ at n=3 results from a delicate balance between the stretching free energies of the A and B blocks while for higher number of arms the free energy of the system is mainly determined by the B blocks forming the star. These theoretical predictions have been tested experimentally by SAXS measurments.

Kosmas [79] studied the behavior of miktoarm stars at the interface of two different phases using a simple mesoscopic model that incorporates the thermodynamic parameters into the size of the Kuhn length. The latter depends on the distance z perpendicular to the interface by the simple discontinuous function $l(z)=l_+$ for $z≥0$, $l(z)=l_-$ for $z≤0$. From the discontinuity of the Kuhn length only the P_Z component of the probability distribution function $P(z_0,z)=P_xP_yP_z$ of finding the end of an homopolymer chain at z distance from the origin (z_0) is affected. The weight of a macrostate $W(z_0)$ of finding the one end at z_0 and the other anywhere then is obtained from the entropy $S=-kT\Sigma P(z_0,z)\ln P(z_0, z)$ which connects $P(z_0,z)$ with $W(z_0)$, $S=kT\ln W(z_0)$. The density profiles of the chain end as a function of the distance z_0 from the interface in the limit of large available volume is given then by $\Omega(z_0) = W(z_0) \Big/ \int_{-L}^{L} dz_0 W(z_0)$. In the case of miktoarm stars consisting of i arms (i=1, 2, ..., f) the weight of macrostate W_{mikto} is the product of the weights of the macrostates of the f different arms $W_{mikto}= \prod_i Wi$. The density profile of the star common junctions is given by the relation

$$\Omega_{mikto} = \frac{2\rho}{1+\prod_i r_i} \prod_i \left[\sqrt{r_i}\right]^{erfc(\sqrt{3z_0}/R_i(z_0)\sqrt{2})}$$ where $R_i(z_0)=R_{i+}$ for $z≥0$ and $R_i(z_0)=R_{i+}$

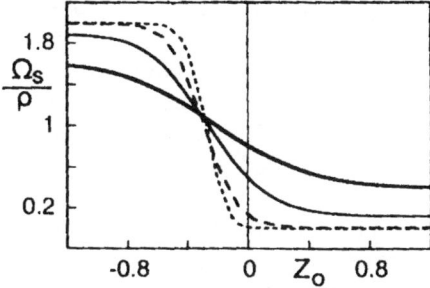

Fig. 9. Density profiles of the cores of stars of the same total molecular weight but different number of equal branches as a function of the reduced distance Z_0, linear chain f=2 (*heavy*), f=4 (*normal*), f=8 (*dashed*), f=15 (*dotted*). The ratio r=2, defined in text (Sect. 3.2.1) represents the ratio of the sizes of a branch at the left and right phase (reproduced with permission from [79])

for z≤0 are the sizes of the i-th branch of the miktoarm at the two phases and r_i their ratios. The concentration of miktoarm chains is ρ (number of chains per volume n/V). The density profile of a homopolymer star which is the simplest case of miktoarm star is illustrated in Fig. 9 as a function of the reduced distance

$$Z_0 = \sqrt{3z_0} / \sqrt{f} R(z_0)\sqrt{2}.$$

3.2.2
Experimental Results

Only a few studies have been devoted to the bulk properties of asymmetric homopolymer stars. The main issue under investigation was, up to now, the self-diffusion and viscoelastic behavior of three-arm stars where the molecular weight of the third arm was varied in order to observe the transition in the diffusion from linear polymer to a polymer with a star architecture.

In the earlier study Fetters et al. [80] utilized a series of deuterated polybutadienes ranging from a linear 2A-mer to a three-arm with a centrally located third arm which had a molecular weight A' varying from 920 to 4.56×10^3 g/mol. A linear hydrogenous 1,4 PBd of Mw=2×10^5 g/mol was used as the matrix. Diffusion coefficients were measured using the thin film IR microdensitometry technique. A monotonic decrease of the diffusion coefficient was observed as the centrally placed side-arm mol. weight increased. The decrease was sharp for MW(A')<M_e, the critical entanglement molecular weight. For MW(A')>$2M_e$ the behavior could be qualitatively described by a simple model assuming independence of the arm retraction probability for each of the arms.

In a recent investigation along the same lines, Graessley and coworkers [8] used a series of asymmetric three-arm poly(ethylene-*alt*-propylene) stars, obtained by hydrogenation or deuteration of the corresponding asymmetric poly(isoprene) stars, in order to study the crossover from linear to branched polymer relaxation dynamics in highly entangled melts. The molecular weight of the two equal arms, constituting the linear backbone was ~38 times larger than M_e, the critical molecular weight for entanglements, where the M_w of the third arm was varied between 0.5 M_e and 18 M_e. Dynamic moduli were measured in shear over a wide range of frequencies and temperatures. Time-temperature superposition was observed for all samples. However, in the branched samples a significantly larger modulus shift was required than in the linear ones. The shift factor for the time scale depended more strongly on temperature for the asymmetric stars. The viscosity, η_o, increased by nearly a factor of three going from the linear backbone sample to the sample with the shortest arm. The change in recoverable compliance across the series was much smaller than the change in viscosity. The diffusion coefficient, determined by forward recoil spectroscopy, decreases even more rapidly. For a sample with only 2.2 entanglements on the third arm D was reduced by a factor of 100. The transition from linear to branched dynamics could be seen more clearly in the product $\eta_o D$. It was concluded that even a single mid-backbone arm with two to three entanglements is quite sufficient to generate a branched polymer relaxation behavior.

Much experimental work has appeared in the literature concerning the microphase separation of miktoarm star polymers. The issue of interest is the influence of the branched architectures on the microdomain morphology and on the static and dynamic characteristics of the order-disorder transition, the ultimate goal being the understanding of the structure-properties relation for these complex materials in order to design polymers for special applications.

In the first seminal morphological study of A_2B miktoarm stars [81] where A and B are PS or PI, by Hadjichristidis and collaborators, an SI_2 sample with 37 vol.% PS was found, by TEM, to be microphase separated as PS cylinders in PI matrix in contrast to a lamellae structure expected for linear diblock copolymer of the same composition. Later, more thorough studies [23, 82, 83] with a larger number of samples and covering a wider range of compositions showed that differences exist in the phase diagram for miktoarm star copolymers in comparison to the linear diblocks. The boundaries for the microstructures usually encountered in diblocks are shifted to higher compositions of the single arm. At constant composition the miktoarm star copolymer has a morphology with greater curvature than the diblock. This provides a means of relieving overcrowding of the A_2 component and overstretching of the B component of the miktoarm star. Analogous shifts were also observed in the case of A_3B (A=PI and B=PS) miktoarm stars [83]. These findings are in qualitative agreement with the theoretical predictions of Milner, which was developed after the first experimental findings [81]. However some discrepancies in terms of the exact location of the morphology boundaries do exist between theoretical prediction and experimental results. An I_2S sample with 53 vol.% PS formed a tricontinuous cubic structure where a lamellae structure was predicted by theory (Fig. 10). Another I_2S sample with high PS content, (81% by vol.), showed microphase separation without long range order. This randomly oriented wormlike morphology of

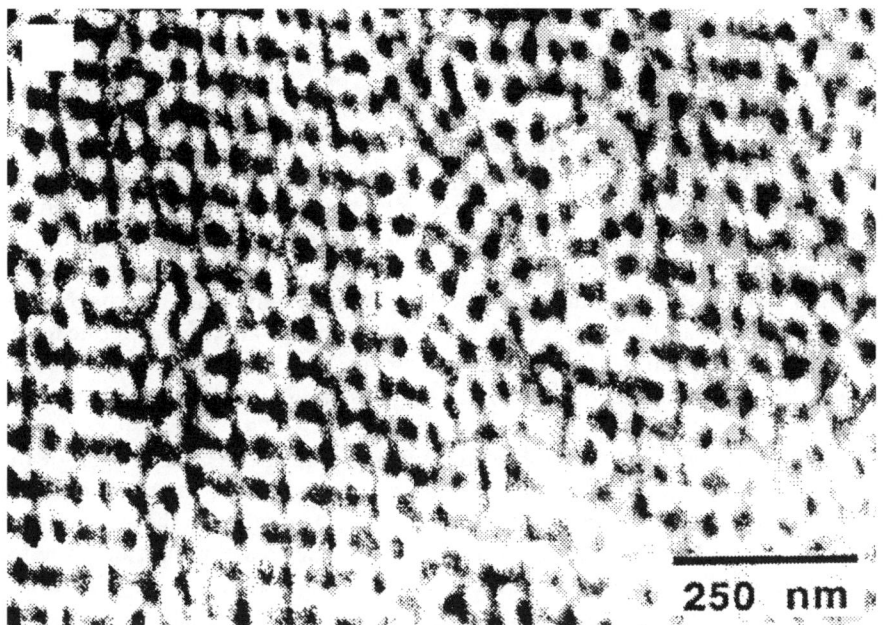

Fig. 10. Bright field TEM micrograph of I_2S miktoarm with 53 vol.% PS (reproduced with permission from [83])

polyisoprene "cylindrical" micelles in a matrix of polystyrene, was shown to be an equilibrium one by selective solvent casting and annealing experiments. On careful annealing the authors were able to observe the transition from a kinetically trapped non-equilibrium folded layer morphology into the equilibrium randomly oriented wormlike phase through an intermediate "fold-lace" morphology. Furthermore the experimentally determined phase diagram for the I_3S stars showed absence of the spherical morphology in the high PS volume fraction (92%) region. The micrographs of an I_3S sample with 39 vol.% PS had the structure of PS cylinders instead of spheres, a sample with 86% PS, PI cylinders instead of lamellae and a sample having 92 vol.% PS consists of disordered PI cylinders in a PS matrix. It is obvious that as the number of arms in A_nB type samples increases differences between the present theory and experiment become more apparent. The theory may be brought "in line" with the experimental findings if the theoretical boundary curves (see Fig. 7) could be "bend back" towards the low volume fraction side of the phase diagram. Presently the theory omits the multiple domain effects, i.e., effects of crowding of chains emanating from different but neighboring interfaces [83].

The microdomain morphology of asymmetric miktoarm stars of the type AA'B with A, A'=PI of different molecular weight and B deuterated PS, covering a wide range of compositions, was investigated by Gido and coworkers [52] by means of TEM and SANS. The morphologies of the asymmetric miktoarm stars are identical to those of diblock copolymers with the same composition. However, the lateral crowding and chain stretching is partially alleviated in the asymmetric miktoarms on the PI side of the interface, by changes in the lattice dimensions, relative to the symmetric ones.

The morphology of I_8S_8 (Vergina) miktoarm stars was studied by Gido et al. [84] by TEM, SAXS and SANS. The three samples studied had compositions in the range 37–47% PS and all formed well ordered lamellae as expected for diblock polymers with the same composition. From calculations of the area per junction point it was concluded that in the case of Vergina stars the spacing between molecules at the interface is larger than in the I_2S and I_3S cases due to the increased functionality of these star molecules. This may also result in more spreading of at least some of the arm trajectories parallel to the interface, relative to linear diblocks.

Thomas and coworkers reported the morphological characterization of the type $(PS-b-PI)_n PS$ miktoarm stars where n=2, 3 [85]. Due to their composition (volume fraction of PS between 0.51 and 0.56) the neat samples showed lamellar microdomains like their corresponding diblocks. The area per junction point for the complex architectures was found to be very close to that obtained for simple diblock and triblock copolymers, leading to the conclusion that in any case the relative amounts of looped and bridged conformations of the PI blocks remains the same on changing the architecture. Blending with PS or PI homopolymers of low MW induced a transition to hexagonally packed cylinders without the observation of an intermediate cubic morphology.

Teyssié and coworkers [86] studied the effect of macromolecular architecture on the lamellar structure of the poly(ethylene oxide) crystallizable arms in (poly *tert*-butyl styrene)(poly(ethylene oxide))$_2$ [PtBuS(PEO)$_2$] miktoarm stars by using SAXS and differential scanning calorimetry (DSC). The results were compared with the ones obtained on poly(tBuS-*b*-EO) materials. At the same total molecular weight and composition the melting temperature, the degree of crystallinity and the number of folds of PEO chains were found to be lower for the branched samples.

The morphological investigation of a three miktoarm star terpolymer was first reported by Hadjichristidis et al. [81]. A SIB miktoarm star with MWs of the arms PS: 20.7 K/PI: 15.6 K/PBd: 12.2 K and a volume fraction of PS equal to 0.4 formed PS cylinders in a polydiene matrix. Due to the low molecular weights the polydiene arms formed one phase. The cylinders are in contrast to the lamellae expected for diblock copolymers but the behavior is analogous to the behavior of I_2S and S_2I miktoarm stars of the same composition.

Hashimoto et al. [87] recently reported on the microdomain morphology of three component miktoarm stars, consisting of polystyrene, poly(dimethylsiloxane) and poly(*tert*-butyl methacrylate) with nearly the same fraction of each component. Three distinct microdomains were observed by TEM, a result that was confirmed by DSC experiments which showed three distinct T_gs. The two-dimentional images of the unstained sections showed triangular patterns exhib-

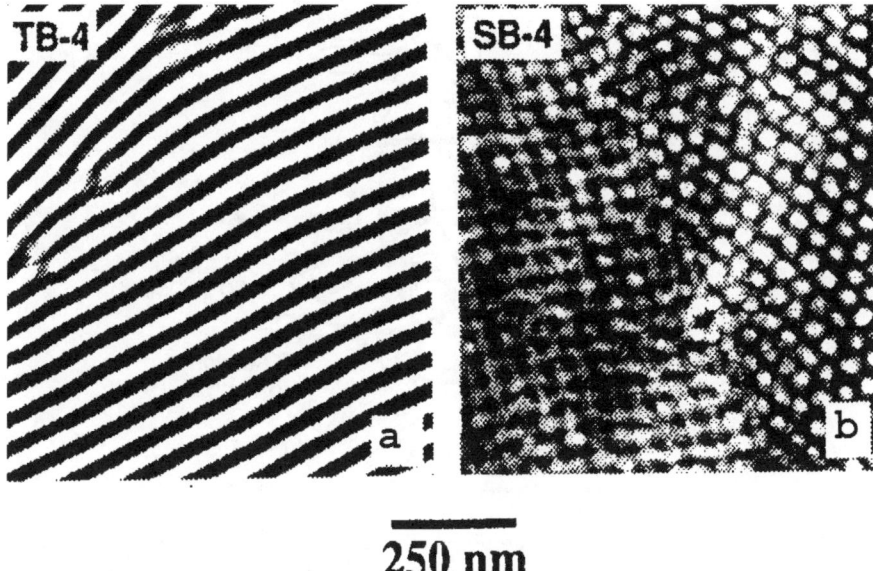

Fig. 11a,b. Bright field TEM micrographs of: **a** a linear tetrablock copolymer; **b** a topologically asymmetric miktoarm starblock copolymer of the same composition and molecular weight (see text) (reproduced with permission from [62])

iting a three-fold symmetry. The authors suggested that the PDMS phase was continuous in three dimensions. The other two phases were not distinguishable. Because of the equal fractions of the three components the authors hypothesized that the other two microphases should also be continuous in three dimensions. All the examined simple models failed to describe quantitatively the observed morphology because of the molecular complexity. It has to be mentioned that tilting experiments gave images of other kinds of symmetry, which could not be identified in terms of simple geometrical arrangements. The crystallinity of the PDMS phase was found by DSC to be lower in the miktoarm case than in the PS-PDMS case, due to the increased steric hindrance close to the central branch point.

Thomas and coworkers [62] studied the structures formed by strongly segregated compositionally symmetric four-arm inverse star block copolymers and their linear counterparts. These macromolecules having the general formula $(PS_M\text{-}b\text{-}PI_{aM})_n\text{-}(PI_M\text{-}b\text{-}PS_{aM})_n$ where $M\sim20,000$, $n=1, 2$ and $a=1, 2, 4$ (a is the ratio of the outer block molecular weight to that of the inner block, see Fig. 1c) can be considered as topologically asymmetric star copolymers. Transmission electron microscopy (TEM) and small angle X-ray scattering (SAXS) were used to characterize the microdomain morphology of these materials. Samples with $n=1, 2$ $a=1, 2$ formed highly ordered lamellae structures. For $a=4$ (sample SB4) a transition from a lamellar structure for $n=1$ to a triconnected cubic structure ($n=2$) was observed (Fig. 11). Comparison of TEM data with simulated images

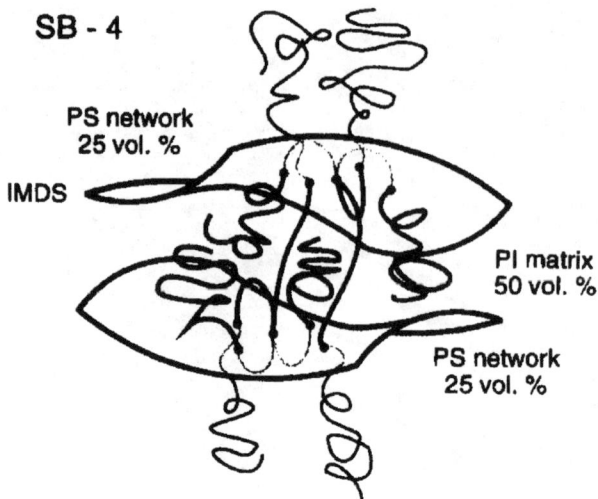

Fig. 12. Schematic representation of proposed arrangement of A/B junctions on a triply periodic IMDS of constant mean curvature for SB-4 miktoarm copolymer of Fig. 11 (reproduced with permission from [62])

of three dimensional models of level surfaces confirmed that the cubic continuous structure is best described as an ordered bicontinuous double diamond (OBDD) structure. The area per junction point was found to increase with molecular weight in the star architecture compared to the linear tetrablocks. From d spacing measurements it was concluded that as "a" increases the ability of the shorter interior blocks to bridge the domains decreases and more loop conformations are formed. These loop conformations force the A/B junctions further apart on average and the long exterior blocks can assume less stretched conformations (Fig. 12). For the SB-4 sample the short interior blocks would be too overcrowded as loops on flat intermaterial dividing surface (IMDS). In order to avoid overcrowding the IMDS becomes curved, allowing the peripheral junctions to spread further apart than on a flat interface and the microdomain structure becomes triply periodic.

Floudas and coworkers [88] investigated the static and kinetic aspects of the order-disorder transition in SI_2 and SIB miktoarm stars using SAXS and rheology. At temperatures above the order-disorder transition (ODT) the mean field theory describes the experimental results quite well. Near the ODT, SAXS profiles gave evidence for the existence of fluctuations. Both samples separated into cylindrical microdomains below the ODT. The ODT was determined on shear oriented samples and found, by SAXS, to be 379 K in both cases. This was confirmed by rheology. The discontinuities in SAXS peak intensity and in the storage modulus near the ODT were more pronounced for the miktoarm stars than for the diblocks. The χN values, where χ is the interaction parameter and N the overall degree of polymerization, of sample SI_2 near the ODT were higher than in the case of diblocks of the same composition and almost independent of temperature. The broad distribution of relaxation times (longer than the relaxation times of individual chains) observed by rheology was attributed to a broad distribution of grain sizes. At very long times the rheological behavior of the samples was proposed to be governed by the relaxation of grains which rearrange with rates related to their sizes. Quenching experiments were used to elucidate the mechanism governing the ordering process. The rheological measurements after each quench from the melt indicated that ordering proceeds via heterogeneous nucleation. Quenching to final temperatures near the ODT made the kinetics very slow with times which depended strongly on the final quench temperature. Deeper quenches produced only moderate changes in the kinetics. Compared with diblocks the kinetic times were slower for miktoarm stars in deeper quenches. For shallow quenching there were no measurable differences. The width of the kinetically accessible metastable region was also larger for the star copolymers. External perturbations, i.e., continuous oscillating shear, produced no changes in the ordering process.

Dielectric spectroscopy was also used by the same group in order to study the local and global dynamics of the PI arm of the same miktoarm star samples [89]. Measurements were confined to the ordered state, where the dynamics of the PI chain tethered on PS cylinders were observed in different environments since in the SIB case the faster moving PB chains are tethered in the same point as the PI arm. The distribution of segmental relaxation times were broader for SI_2 than SIB. The effect was less pronounced at higher temperatures. The PI normal mode time was found to be slower in SIB, when compared to SI_2 although both arms had the same molecular weight. Additionally, the normal mode relaxation time distributions of the PI chains tethered to PS cylinders in the miktoarm samples were narrower than in P(S-b-I) systems of lamellar structure.

The characteristics of the order-disorder transition in two SI_3 miktoarm stars were also investigated by Floudas et al. using SAXS and rheology [78]. Due to the high molecular weights diisooctylphthalate (DOP) was used as the nonpreferential solvent to bring the transition into the experimentally accessible temperature window. All SAXS parameters were discontinuous at the ODT revealing a first-order transition. The interaction parameter exhibited a weak temperature dependence consistent with fluctuation effects near the transition. The experimental results agreed qualitatively with the theoretical spinodal curves $\chi N_{total} = F(f_A)$ for AB_n miktoarm copolymers. For shallow quenches rheological measurements indicated a nucleation and growth ordering mechanism with an Avrami exponent of about three, reminiscent of undiluted block copolymers.

Floudas and coworkers [90] employed dielectric spectroscopy to probe the interfacial width in lamellae forming non-linear block copolymers of the type (B-b-A)$_3$B and (B-b-A)$_3$B(A-b-B)$_3$, where A is PI and B is PS. Their experiments were conducted at temperatures below the ODT and below the glass transition of the PS "hard phase". In this temperature region the global chain dynamics of PI bridges were used to provide an estimate of the "dynamic" interface between the incompatible blocks. A broad relaxation process in the vicinity of the normal mode relaxation time of PI could be identified. It was shown to arise from the mobile junction points at the interface. By fitting the experimental data with Havriliak-Negami functions the parameters characterizing the process were determined and the width of the interphase was estimated. The experimentally determined value of 5–6 nm for the interfacial width compared well with the theoretically calculated one (3.6 nm), based on a theory developed for linear block copolymers.

The dependence of the experimentally observed ODT on the ordering temperature (final quenching temperature) was shown recently by rheology measurements on SI_3 and SI low molecular weight samples [91]. This dependence resembles the dependence of the melting temperature in semicrystalline homopolymers on the crystallization temperature. The existence of an equilibrium transition temperature ($T_{ODT}°$) was demonstrated and a procedure, analogous to the known Hofmann-Weeks plot for crystalline polymers, was proposed as a means to calculate $T_{ODT}°$. The phenomenon was attributed to the fact that grains when produced at different temperatures have different sizes.

The rheological behavior of S_2I_2 miktoarm star copolymers was studied by McLeish and coworkers [92]. An independence of the rheological characteristics on temperature, for the temperature range between 100 and 150 °C was observed for a sample having 20 wt% PS. In this system stress relaxation may occur via molecular diffusion along the interface in two ways. First the entangled polyisoprene arms can retract as in the case of the star polymers, with the junction points remaining at the interface. Second, the PS arms can collapse and then the whole molecule can reptate in the PI phase. From the available results the authors could not distinguish between the two mechanisms.

Tsitsilianis [93] used DSC to study the phase behavior of (polybutylmethacrylate)$_n$(polystyrene)$_n$ and (polybutylacrylate)$_n$(polystyrene)$_n$ miktoarm stars with varying n. The T_g of the PS phase was found to be much lower than in the homopolymer and the width of the transition was increased. In some cases an intermediate T_g was observed due to the extended interphase region in these materials, as a result of mixing of the different arms in the vicinity of the star cores.

The compatibilizing ability of miktoarm stars comprised of polystyrene and Nylon-6 arms emanating from a phosphazene core were investigated in detail by Miyata et al. [50]. For binary blends of poly(2,6-dimethylphenylene oxide) (PPO) and a miktoarm star with about four PS arms ($M_{n,arm}$~62,000) and two poly(ε-caprolactam) arms ($M_{n,arm}$~26,000), having more than 10 wt% miktoarm star, the PPO T_g values were apparently shifted to lower temperatures and the T_g decreased with increasing the amount of miktoarm star polymer. For all binary blends a single T_g was observed indicating miscibility of the two polymers (PS is miscible with PPO over a considerable range of molecular weights in all proportions). The T_g values were lower for the cases of high PS/Nylon-6 ratios in the miktoarms. For a ternary blend of PPO/Nylon-6/miktoarm with 45/45/10 composition respectively only one T_g was observed corresponding to the

PPO/PS phase. Scanning electron microscopy experiments showed that the different domains were smaller in the blends containing miktoarm material. The size of the domains decreased and their uniformity increased with increasing miktoarm content in the blends. Miktoarm stars with short PS arms were less effective in compatibilizing PPO and Nylon-6. Ternary blends were found to have higher modulus, strength and elongation than binary PPO/miktoarm blends. They also showed yield behavior. These improvements were attributed to the increased adhesion at the PPO/Nylon-6 interface due to the presence of polystyrene-containing copolymer.

4
Applications

Although the synthesis of asymmetric star polymers was rather recently achieved, several efforts have been made to use these polymers for specific applications. An increased number of patents have appeared in recent years. Sometimes ill-defined and not well characterized polymers were used in these studies.

Conventional star block copolymers bearing PS-b-PB or PS-b-PI arms have been employed in adhesive compositions. These copolymers have the disadvantage that the polydiene fraction degrades in processing and over time. The unsaturation is also susceptible to possible grafting and crosslinking reactions which lead to increased molecular weights, thus making the polymer ineffective in applications such as pressure sensitive adhesives and removable tapes. These applications require an adhesive that possesses a fourfold balance of tack, cohesive strength and the proper balance between adhesion and resistance to low stress peel. Asymmetric stars have been useful in the formulation of adhesive compositions for pressure sensitive adhesives, removable tape applications etc., showing improved properties compared to conventional star block copolymers.

Miktoarm stars of the type (PS-b-PI)$_n$(PI)$_m$ prepared using DVB as linking agent were employed in adhesive compositions, showing a good balance of tack, peel strength and hold power near room temperature due to their reduced cohesive strength and viscosity without any reduction in the service temperature [94].

Stars of the type [(PS-co-PI)-b-PI]$_n$(PI-b-PS)$_m$ were also used in adhesives showing high resistance to low-stress peel while maintaining moderate adhesion [95]. This was achieved by controlling the T_g of the end-block to relatively low levels.

Tapes made from adhesives containing asymmetric star block copolymers of the type (PS-b-PI)$_n$ does not represent an asymmetric starblock copolymer [96]. By using asymmetric rather than the conventional symmetric star block copolymers the preparation of tapes with moderate adhesion and good resistance to low stress peel which are easily removed from the substrate without leaving adhesive residue was achieved.

High adhesive performance at low molecular weight and viscosity was achieved using [PS-b-P(ethylene/butylene)]$_2$PI$_2$ and [PS-b-P(ethylene/butylene)]$_2$PI$_4$ miktoarm stars [97]. This behavior was probably obtained due to the superior phase separation between the PS end blocks and P(ethylene/butylene) rubber blocks compared to PS and PI blocks.

(PS-b-PB)$_n$PB$_m$ miktoarm stars having n+m=4 proved to be able to produce good adhesives after crosslinking by radiation [98]. Improved high temperature resistance was obtained without chemical crosslinking agents, such as dimethacrylates or acrylates that have been proven to be dangerous for the human health and the environment.

Asymmetric stars have also been added to a variety of oils including crude oils, lubricating and fuel oils to produce oil compositions, generally having improved viscosity index characteristics. The newer engines place increased demands on the engine lubricants. Asymmetric stars are effective viscosity index improvers.

Miktoarm stars of the type (PI)$_n$(PtBuMA)$_n$ provided lower viscosity at low temperatures and better engine oil pumpability compared with common dispersant viscosity index improvers [99]. Asymmetric stars possessing PI arms of different molecular weights have been shown to produce oil films with increased thickness [100]. This is a result of the increased elasticity of the polymer due to the combined longer and shorter arms. Stars having the higher ratio of molecular weights between long and short arms (~2:1 to 3:1) and >12% long arms had lower viscosity values, thus providing improved fuel efficiency. Improved low temperature pumpability was also achieved using hydrogenated (PS-b-PI)$_n$PI$_m$ stars [101]. The thickening efficiency in this case varies mainly with the MW of the hydrogenated PI block of the copolymer arm and secondly with the PS molecular weight.

Asymmetric stars have also been used or have the potential to be used in many other applications as compatibilizers [50], impact modifiers and in sealant [100] and molding compositions etc.

5
Concluding Remarks

Living polymerization methods have been proven very efficient in preparing asymmetric star polymers. These materials include stars having arms of different molecular weight, chemical nature or topology. Most of these products are well defined and molecularly and compositionally homogeneous, thus providing the ability to correlate the molecular structure with the properties. Several theories exist that try to predict the properties of the asymmetric stars. However in most cases there are no experimental results to evaluate these theories. Partly deuterated miktoarm stars and new architectures have to be designed for this purpose. The growing interest of the industry for products involved in special applications will expand the use of asymmetric stars as adhesives, melt viscosity index modifiers, compatibilizers, impact modifiers etc.

6
References and Notes

1. Macromolecules (from the Greek word μακρύς meaning long) is not applicable for non-linear polymers. Megamolecules (from the Greek word μέγας meaning big) better represents the non-linear polymeric species
2. Pennisi RW, Fetters LJ (1988) Macromolecules 21:1094
3. Morton M, Helminiak TE, Dadkary SD, Bueche F (1962) J Polym Sci 57:471
4. Roovers JEL, Bywater S (1972) Macromolecules 5:385
5. Tung L, Gatzke A (1983) J Polym Sci Polym Phys Ed 21:1839
6. Morton M, Fetters LJ (1964) J Polym Sci, Part A 2:3311
7. Worsfold DJ, Bywater S (1972) Macromolecules 5:393
8. Gell CB, Graessley WW, Efstratiadis V, Pitsikalis M, Hadjichristidis N (1997) J Polym Sci :Part B: Polym Phys Ed 35:1943
9. Frater DJ, Mays JW, Jackson C (1997) J Polym Sci Polym Phys Ed. 35:141
10. Quirk RP, Yoo T (1993) Polym Bull 31:29
11. Okay O, Funke W (1990) Macromolecules 23:2623
12. Eschwey H, Burchard W (1975) Polymer 16:180
13. Tsitsilianis C, Chaumont P, Rempp P (1990) Makromol Chem 191:2319
14. Tsitsilianis C, Graff S, Rempp P (1990) Eur Polym J 27:243
15. Tsitsilianis C, Lutz P, Graff S, Lamps J-P, Rempp P (1991) Macromolecules 24:5897
16. Rein D, Rempp P, Lutz PJ (1993) Makromol Chem Macromol Symp 67:237
17. Yamagishi A, Szwarc M, Tung L, Lo GY-S (1978) Macromolecules 11:607
18. Quirk RP, Hoover FI (1987) In: Hogen-Esch TE, Smid J (eds) Recent advances in anionic polymerization. Elsevier, New York, p 393
19. Kawakami Y (1994) Prog Polym Sci 19:203
20. Kanaoka S, Sawamoto M, Higashimura T (1991) Macromolecules 24:2309
21. Mays JW (1990) Polym Bull 23:247
22. Iatrou H, Siakali-Kioulafa E, Hadjichristidis N, Roovers J, Mays JW (1995) J Polym Sci Polym Phys Ed 33:1925
23. Pochan DJ, Gido SP, Pispas S, Mays JW, Ryan AJ, Fairclough JPA, Hamley IW, Terrill NJ (1996) Macromolecules 29:5091
24. Khan IM, Gao Z, Khougaz K, Eisenberg A (1992) Macromolecules 25:3002
25. Ba-Gia H, Jerome T, Teyssié P (1980) J Polym Sci Polym Chem Ed 18:3483
26. Naka A, Sada K, Chujo Y, Saegusa T (1991) Polym Prepr Jp 40(2):E1161
27. Avgeropoulos A, Hadjichristidis N (1997) J Polym Sci Polym Chem Ed 35:813
28. Tselikas Y, Hadjichristidis N, Iatrou H, Liang KS, Lohse DJ (1996) J Chem Phys 105:2456
29. Tsiang RCC (1994) Macromolecules 27:4399
30. Velis G, Hadjichristidis N (1997) 4th Conference of the Greek Polymer Society, Patras, Greece; Macromolecules, submitted.
31. Takano A, Okada M, Nose T, Fujimoto T (1992) Macromolecules 25:3596
32. Wang F, Roovers J, Toporowski PM (1995) Macromol Symp 95:205
33. Wang F, Roovers J, Toporowski PM (1995) Macromolecular Reports A32(Suppls 5/6):951
34. Xie H, Xia J (1987) Makromol Chem 188:2543
35. Iatrou H, Hadjichristidis N (1993) Macromolecules 26:2479
36. Wright SJ, Young RN, Croucher TG (1994) Polym International 33:123
37. Algaier J, Young RN, Efstratiadis V, Hadjichristidis N (1996) Macromolecules 29:1794
38. Quirk RP, Lee B, Schock LE (1992) Makromol Chem Macromol Symp 53:201
39. Quirk RP, Yoo T, Lee B (1994) JM-Pure Appl Chem A31:911
40. Avgeropoulos A, Poulos Y, Hadjichristidis N, Roovers J (1996) Macromolecules 29:6076
41. Tsitsilianis C, Papanagopoulos D, Lutz P (1995) Polymer 36:3745

42. Tsitsilianis C, Boulgaris D (1995) Macromol Reports A32 (Suppls 5/6):569
43. Tsitsilianis C, Voulgaris D (1997) Macromol Chem Phys 198:997
44. Kanaoka S, Sawamoto M, Higashimura T (1991) Macromolecules 24:5741
45. Kanaoka S, Omura T, Sawamoto M, Higashimura T (1992) Macromolecules 25:6407
46. Kanaoka S, Sawamoto M, Higashimura T (1993) Macromolecules 26:254
47. Ishizu K, Kuwahara K (1994) Polymer 35:4907
48. Ishizu K, Yikimasa S, Saito R (1991) Polym Commun 32:386
49. Ishizu K, Yikimasa S, Saito R (1992) Polymer 33:1982
50. Miyata K, Watanabe Y, Itaya T, Tanigaki T, Inoue K (1996) Intern Symp Ionic Polym Paris, p 161
51. Iatrou H, Hadjichristidis N (1992) Macromolecules 25:4649
52. Lee C, Gido SP, Pitsikalis M, Mays JW, Beck Tan N, Trevino SF, Hadjichristidis N (1997) Macromolecules 30:3738
53. Fujimoto T, Zhang H, Kazama T, Isono Y, Hasegawa H, Hashimoto T (1992) Polymer 33:2208
54. Huckstadt H, Abetz V, Stadler R (1996) Macromol Rapid Commun 17:599
55. Sioula S, Tselikas Y, Hadjichristidis N (1997) Macromol Symp 117:167
56. Sioula S, Tselikas Y, Hadjichristidis N (1997) Macromolecules 30:1518
57. Lambert O, Reutenauer S, Hurtrez G, Riess G, Dumas P (1997) Intern Symp Ionic Polym Paris, p 161
58. Pitsikalis M, Hadjichristidis N (1995) Macromolecules 28:3904
59. Eisenbach CD, Schnecko H, Kern W (1975) Eur Polym J 11:699
60. Stewart MJ, Shepherd N, Service D (1990) Br Polym J 22:319
61. Pispas S, Pitsikalis M, Hadjichristidis N, Dardani P, Morandi F (1995) Polymer 36:3005
62. Tselikas Y, Hadjichristidis N, Lescanec RL, Honeker CC, Wohlgemuth M, Thomas EL (1996) Macromolecules 29:3390
63. Vlahos CH, Horta A, Freire J (1992) Macromolecules 25:5974
64. Vlahos CH, Horta A, Hadjichristidis N, Freire J (1995) Macromolecules 28:1500
65. Vlahos CH, Tselikas Y, Hadjichristidis N, Roovers J, Rey A, Freire J (1996) Macromolecules 29:5599
66. Kosmas MK, Hadjichristidis N (1994) Macromolecules 27:5216
67. Khasat N, Pennisi RW, Hadjichristidis N, Fetters LJ (1988) Macromolecules 21:1100
68. Gia H-B, Jerome R, Teyssié Ph (1981) J Appl Polym Sci 26:343
69. Gia H-B, Jerome R, Teyssié Ph (1980) J Polym Sci Part B:Polym Phys 18:2391
70. Kanaoka S, Sawamoto M, Higashimura T (1992) Macromolecules 25:6414
71. Anastasiadis SH, Chrissopoulou K, Fytas G, Fleischer G, Pispas S, Pitsikalis M, Mays JW, Hadjichristidis N (1997) Macromolecules 30:2445
72. Tsitsilianis C, Kouli O (1995) Makromol Rapid Commun 16:591
73. Ramzi A, Prager M, Richter D, Efstratiadis V, Hadjichristidis N, Young RN, Allgaier JB (1997) Macromolecules 30:7171
74. Pitsikalis M, Hadjichristidis N, Mays JW (1996) Macromolecules 29:179
75. Siqueira DF, Pitsikalis M, Hadjichristidis N, Stamm M (1996) Langmuir 12:1631
76. Olvera de la Cruz M, Sanchez IC (1986) Macromolecules 19:2501
77. Milner ST (1994) Macromolecules 27:2333
78. Floudas G, Hadjichristidis N, Tselikas Y, Erukhimovich I (1997) Macromolecules 30:3090
79. Kosmas MK (1996) J Chem Phys 104:405
80. Jordan EA, Donald AM, Fetters LJ, Klein J (1989) Polym Prep (Am Chem Soc Polym Chem Div) 30(1):63
81. Hadjichristidis N, Iatrou H, Behal SK, Chludzinski JJ, Disko MM, Garner RT, Liang KS, Lohse DJ, Milner ST (1993) Macromolecules 26:5812
82. Pochan DJ, Gido SP, Pispas S, Mays JW (1996) Macromolecules 29:5099
83. Tselikas Y, Iatrou H, Hadjichristidis N, Liang KS, Mohanty K, Lohse DJ (1996) J Chem Phys 105:2456

84. Beyer FL, Gido SP, Poulos Y, Avgeropoulos A,. Hadjichristidis N (1997) Macromolecules 30:2373
85. Avgeropoulos A, Dair BJ, Thomas EL, Hadjichristidis N (1998) Macromolecules (submitted)
86. Gervais M, Gallot B, Jerome R, Teyssié Ph (1986) Makromol Chem 187:2685
87. Okamoto S, Hasegawa H, Hashimoto T, Fujimoto T, Zhang H, Kazama T, Takano A, Isono Y (1997) Polymer 38:5275
88. Floudas G, Hadjichristidis N, Iatrou H, Pakula T, Fischer EW (1994) Macromolecules 27:7735
89. Floudas G, Hadjichristidis N, Iatrou H, Pakula T (1996) Macromolecule 29:3139
90. Floudas G, Alig I, Avgeropoulos A, Hadjichristidis N (1998) J Non Crystalline Solids 235–237:485
91. Floudas G, Pakula T, Velis G, Sioula S, Hadjichristidis N (1998) J Chem Phys 108(15)
92. Johnson J, Young RN, Wright ST, McLeish T (1994) Polym Prep (Am Chem Soc Polym Chem Div) 35(1):600
93. Tsitsilianis C (1993) Macromolecules 26:2977
94. StClair DJ (1983) US Pat 4,391,949 A
95. Ma J (1994) Eur Pat 632,073 A
96. Nestegard MK, Ma J (1994) US Pat 5,296,547 A
97. Himes GR, Spence BA, Hoxmeier RJ, Chin SS (1995) US Pat 5,393,841 A
98. Debier ERS (1993) WO 9,324,547 A
99. Sutherland RJ, Dubois DA (1994) Eur Pat 603,955 A
100. Rhodes RB, Bean AR (1994) US Pat 5,302,667 A
101. Rhodes RB, Bean AR (1989) Eur Pat 298,578 A

Received: April 1998

Poly(macromonomers): Homo- and Copolymerization

Koichi Ito[1], Seigou Kawaguchi

Department of Materials Science, Toyohashi University of Technology, Tempaku-cho, Toyohashi 441-8580, Japan [1]E-mail: itoh@tutms.tut.ac.jp

Syntheses and characterization of branched polymers prepared by homo- and copolymerization of macromonomers are reviewed. A number of macromonomers have so far been available as potential building blocks to design a variety of well-defined, branched homo- and copolymers including comb, star, brush, and graft types. Recent progress in macromonomer syntheses, macromonomers' homo- and copolymerization, characterization of the branched polymers obtained, as well as application to design of polymeric microspheres are described. Macromonomers and their homo- and copolymerization appear to provide continuing interest in designing and characterizing a variety of branched polymers and in their unique applications.

Keywords. Poly(macromonomers), Graft copolymers, Comb, Star, Brush, Polymeric microspheres

List of Symbols and Abbreviations		130
1	Introduction	133
2	Survey of Macromonomer Techniques	134
3	Syntheses of Macromonomers	136
3.1	Polyolefins	136
3.2	Polystyrenes	137
3.3	Polyacrylates	139
3.4	Poly(ethylene oxide)	139
3.5	Some Other New Macromonomers	141
4	Homopolymerization and Copolymerization of Macromonomers	141
4.1	Homopolymerization	142
4.2	Copolymerization	145
5	Characterization of Star and Comb Polymers	148
5.1	Characterization and Solution Properties of Poly(macromonomers)	149

| 5.2 | Bulk Properties | 154 |
| 5.3 | Some Properties of Graft Copolymers | 156 |

6	**Design of Polymeric Microspheres Using Macromonomers**	157
6.1	Dispersion Polymerization	157
6.2	Mechanistic Model of Dispersion Copolymerization with Macromonomers	163
6.3	Emulsion Polymerization	167
6.4	Chain Conformation of Grafted Polymer Chains at Interfaces	171

| 7 | **Conclusions and Future** | 173 |

| 8 | **References** | 174 |

List of Symbols and Abbreviations

a	exponent in Mark-Houwink-Sakurada equation
AIBN	2,2'-azobisisobutyronitrile
a'	bead spacing
a_s	area occupied by a surfactant molecule
α_s	expansion factor
ATR	attenuated total reflection
B	excluded-volume strength
b	bond length
β	binary cluster integral
BMA	n-butyl methacrylate
C_s	surfactant concentration
[cmc]	critical micelle concentration
d_m	density of monomer
d_p	density of polymer
DP	degree of polymerization
DP_n	number-average DP
DP_n^o	DP_n without chain transfer
DP_w	weight-average DP
DV	differential viscosity
ESCA	electron spectroscopy for chemical analysis
ESR	electron spin resonance
$[\eta]$	limiting viscosity
f	initiator efficiency
f'	number of branches
FTIR	Fourier transform infrared spectroscopy
g	shrinking factor
γ	ratio of molecular weights of branch and backbone

GTP	group transfer polymerization
HEMA	2-hydroxyethyl methacrylate
[I]	initiator concentration
k_d	decomposition rate constant
KP	Kratky-Porod
k_p	propagation rate constant
k_t	termination rate constant
k_2	diffusion-controlled rate constant for coalescence between similar-sized particles
L	contour length
LALLS	low-angle laser light scattering
λ^{-1}	Kuhn segment length
M	molecular weight
M_o	molecular weight of monomeric unit
[M]	monomer concentration
MALLS	multiangle laser light scattering
M_D	molecular weight of macromonomer
μ	rate of particle volume growth
M_L	shift factor
MMA	methyl methacrylate
mp	melting point
$[M]_p$	equilibrium concentration of monomer swelling particle
MW	molecular weight
M_w	weight average molecular weight
n	number of bonds
N	number of particles
$\bar{n}$	average number of radicals per particle
n_K	Kuhn segment number
n'	number of grafted chains onto surface
v	kinetic chain length
N_A	Avogadro's number
NAD	nonaqueous dispersion
NMR	nuclear magnetic resonance
PAA	poly(acrylic acid)
PBMA	poly(*n*-butyl methacrylate)
PCL	poly(ε-caprolactone)
PDMS	poly(dimethylsiloxane)
PE	polyethylene
PEO	poly(ethylene oxide)
PHBd	hydrogenated poly(1,3-butadiene)
PHEMA	poly(2-hydroxyethyl methacrylate)
ϕ_m	volume fraction of monomers swelling particles
PHSA	poly(12-hydroxystearic acid)
PIB	polyisobutylene
PIp	polyisoprene

PLMA	poly(lauryl methacrylate)
PMA	poly(methacrylic acid)
PMMA	poly(methyl methacrylate)
PNIPAM	poly(N-isopropylacrylamide)
POXZ	polyoxazolines
PP	polypropylene
PSt	polystyrene
PTBA	poly(t-butyl acrylate)
PTBMA	poly(t-butyl methacrylate)
$P(\theta)$	particle scattering factor
PVA	poly(vinyl alcohol)
PVAcA	poly(N-vinylacetamide)
PVC	poly(vinyl chloride)
PVP	poly(vinylpyrrolidone)
P4VP	poly(4-vinylpyridine)
q	persistence length
r_i	reactivity ratio of i species
R	radius of particle
R_{crit}	radius of particle at critical point
ρ	density
ρ'	rate of radical generation
ROMP	ring-opening methathesis polymerization
R_p	rate of polymerization
S	surface area occupied by a macromonomer chain
$<S^2>$	mean square radius of gyration
SAXS	small-angle X-ray scattering
SANS	small-angle neutron scattering
S_{crit}	surface area occupied by a macromonomer chain at critical point
SEC	size exclusion chromatography
STM	scanning tunneling electron microscopy
TBA	t-butyl acrylate
TEMPO	2,2,6,6-tetramethylpiperidinyloxy
θ	fractional conversion of monomer
θ_{crit}	fractional conversion of monomer at critical point
θ_D	fractional conversion of macromonomer
θ_{Dcrit}	fractional conversion of macromonomer at critical point
T_g	glass transition temperature
T_g^{∞}	glass transition temperature of polymer with infinite molecular weight
v	excess free volume at a chain end
v_m	free volume per monomeric unit
W_D	weight of macromonomer polymerized
W_{do}	initial weight of macromonomer
W_M	weight of monomer polymerized

W_{Mo} initial weight of monomer
x   fraction of disproportionation in termination
z   excluded-volume parameter
$\tilde{z}$ scaled excluded-volume parameter

1
Introduction

A macromonomer is any polymer or an oligomer with a polymerizable functionality as an end group. Formally, the macromonomer homopolymerizes to afford a star- or comb-shaped polymer and copolymerizes with a conventional monomer to give a graft copolymer. Thus the macromonomer serves as a convenient building block to constitute arms or branches of known structure in the resulting polymer. A large number of macromonomers, differing in the type of the repeating monomer and the end-group, have so far been prepared, thereby offering the possibility of construction of an enormous number of branched polymers in a variety of architectures, combinations, and compositions. Polymerization and copolymerization of macromonomers have also been studied in great detail in order to understand their unique behavior in comparison with that of conventional monomers. Their useful application in design of polymeric microspheres has also been appreciated recently. Some interesting properties of poly(macromonomers) have also been explored very recently as a simple model of brush polymers which are of increasing interest. Comparatively, however, the characterization and properties of graft copolymers with randomly distributed branches have not been investigated to the same extent in spite of their theoretical and practical importance.

The present article is intended to discuss the state-of-the-art of the design and characterization of the branched polymers obtained by the macromonomer technique, with particular stress on the characterization and the properties of the brush polymers obtained by the homopolymerization of macromonomer. The synthetic aspects of the macromonomer technique, including preparation of various kinds of macromonomers, have been recently reviewed by one of the authors [1]. Therefore, we intend here to outline briefly the macromonomer technique and describe only the very recent important developments in syntheses. Preparation and characterization of the polymeric microspheres by use of macromonomers as reactive (copolymerizable) emulsifiers or dispersants will be described in some detail to represent one of their unique applications.

Some comprehensive reviews covering earlier references include those by Kawakami [2], Meijs and Rizzard [3], Velichkova and Christova [4], and those in books edited by Yamashita [5] and by Mishra [6] among others.

2
Survey of Macromonomer Techniques

A macromonomer is usually defined as a polymeric or an oligomeric monomer with a polymerizable or copolymerizable functional group at one end. They afford a comb-shaped polymer with regularly and densely attached branches by homopolymerization, and a graft copolymer with randomly and loosely distributed branches by copolymerization with a conventional, low molecular weight (MW) comonomer, as illustrated in Fig. 1a,b, respectively. A formally comb-shaped poly(macromonomer) may actually be forced to take a conformation that looks like a star as in Fig. 1c or a brush as in Fig. 1e, depending on the relative lengths of the macromonomer branch vs the poly(macromonomer) backbone. A graft copolymer with a relatively short backbone as compared to the branches may also look like a star as in Fig. 1d in a solvent which is selective for the branches, while that with a long backbone with few but long branches may take a flower-like conformation as in Fig. 1f with some of their backbone segments looped outside in a selective solvent for the backbone. These isolated conformations favored in dilute solutions are expected to coalesce to some organ-

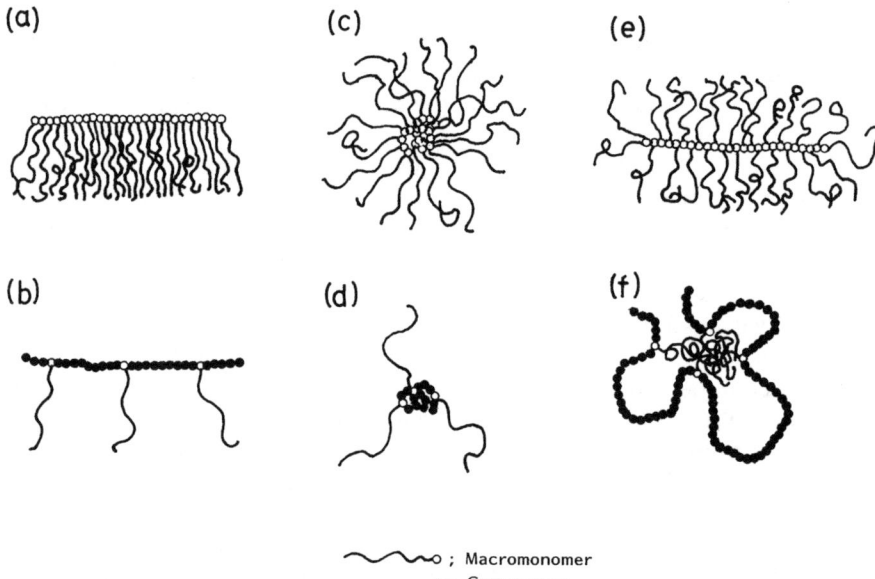

Fig. 1a–f. Various branched architectures obtained by the macromonomer technique: **a,b** comb-like; **c,d** star-like; **e** brush; **f** flower-like. **a c**, and **e** are poly(macromonomers) obtained by homopolymerization, while **b, d**, and **f** are graft copolymers obtained by copolymerization

ized structure or morphology in concentrated solutions or in solids. In fact, a number of possible conformations or morphologies that can be expected from self-organization of the branched polymers has been a matter of increasing study for the macromonomer technique.

The variety of branched architectures that can be constructed by the macromonomer technique is even larger. Copolymerization involving different kinds of macromonomers may afford a branched copolymer with multiple kinds of branches. Macromonomer main chain itself can be a block or a random copolymer. Furthermore, a macromonomer with an already branched or dendritic structure may polymerize or copolymerize to a hyper-branched structure. A block copolymer with a polymerizable function just on the block junction may homopolymerize to a double comb or double-haired star polymer.

If we extend the definition of the macromonomer to include all polymers or oligomers with a multiple number of (co)polymerizable functional groups at any positions, then we can design an even larger number of branched polymers by their polymerization and copolymerization. For example, a "telechelic macromonomer" with two (co)polymerizable functional groups, each on one end, may be useful to design a network structure in copolymerization with control over the inter-crosslink length and/or crosslink density. A "multifunctional macromonomer" with a multiple number of (co)polymerizable functional groups along their chain may include already well-known resins such as unsaturated polyesters used in thermosetting. Although these "macromonomers" are no doubt practically important in applications, the scope becomes too broad and complicated and the authors prefer to adhere to the original, simpler definition of the macromonomer as that with a single (co)polymerizable end group that affords star- and comb-shaped polymers and/or graft copolymers with their branches (side chains) of known structure as in Fig. 1.

So far, a great number of well-defined macromonomers as branch candidates have been prepared as will be described in Sect. 3. Then a problem is how to control their polymerization and copolymerization, that is how to design the backbone length, the backbone/branch composition, and their distribution. This will be discussed in Sect. 4. In brief, radical homopolymerization and copolymerization of macromonomers to poly(macromonomers) and statistical graft copolymers, respectively, have been fairly well understood in comparison with those of conventional monomers. However, a more precise control over the backbone length and distribution by, e.g., a living (co)polymerization is still an unsolved challenge.

Needless to say, the best established architecture which can be designed by the macromonomer technique has been that of graft copolymers. With this technique we now have easy access to a variety of multiphased or microphase-separated copolymer systems. This expanded their applications into a wide area including polymer alloys, surface modification, membranes, coatings, etc. [5].

One of the most unique and promising applications of the technique may be found in the design of polymeric microspheres. In this technique macromonomers are reactive emulsifiers or dispersants in emulsion or dispersion systems, respectively. Since the macromonomers are already polymers, they serve as effective steric stabilizers of the resulting microspheres. They are surface grafts after copolymerization with the substrate comonomer. A number of hydrophilic or polar macromonomers have been designed for aqueous emulsion or alcoholic dispersion systems. They are the counterpart of the nonpolar macromonomers which were indeed the first "macromonomers" developed for the well-known nonaqueous (petroleum) dispersion polymerization (NAD) by ICI [7].

3
Syntheses of Macromonomers

Macromonomers are synthesized by introducing an appropriate (co)polymerizable end-group, generally by one of the following methods: (a) end-capping of a living polymer (termination method), (b) initiation of living polymerization (initiation method), (c) transformation of any functional end-group, and (d) polyaddition. Methods (a) and (b) are simple and usually afford most well-defined macromonomers of a controlled degree of polymerization with a narrow MW distribution, but depend on proper combination of any living polymerization with an effective terminator or initiator carrying a polymerizing group or its protected one. Method (c) utilizes any end-functionalized polymers such as those obtained from chain-transfer-controlled radical polymerization and polycondensation. Method (d) involves the polyaddition reactions between vinyl and silane groups (hydrosilylation), for an example. Since more than one hundred macromonomers have been reviewed previously [1], including polyolefins, polystyrenes, polydienes, polyvinylpyridines, poly(meth)acrylates and their derivatives, poly(vinyl ethers), poly(vinyl acetate) and derivatives, halogenated vinyl polymers, poly(alkylene oxides), poly(dimethylsiloxanes), poly(tetrahydrofuran) and polyacetals, polyoxazolines and poly(ethyleneimines), polylactones and polylactide, polylactams and poly(amino acids), and macromonomers prepared by polycondensation and polyaddition, only very recent developments will be described here in a way to supplement them.

3.1
Polyolefins

End-functionalized polyethylene (PE) [8, 9], polypropylene (PP) [10], and polyisobutylene (PIB) [11] have been transformed to their corresponding macromonomers carrying (meth)acrylate, oxazoline, and methacrylate end groups, 1, 2, and 3, respectively. Polybutadienyl lithium was terminated with chlorodimethylsilane, followed by hydrogenation to saturated polyolefin (PHBd) [12]. Hydrosilylation of the end silane with allyl glycidyl ether afforded an epoxidized macromonomer, 4, and subsequent hydrolysis gave a dihydroxy-ended macromonomer, 5, to be used for polycondensation to polyester-g-PHBd.

$$PE-\underset{\underset{O}{\|}}{O}CCH=CH_2 \tag{1a}$$

$$PE-\underset{\underset{O}{\|}}{O}C\overset{CH_3}{\underset{}{C}}=CH_2 \tag{1b}$$

$$PP-\underset{O}{\overset{N}{\diagup}}\rceil \tag{2}$$

$$PIB-CH_2\underset{\underset{CH_3}{|}}{\overset{CH_3}{\underset{|}{C}}}-O\underset{\underset{O}{\|}}{\overset{CH_3}{\underset{|}{C}}}C=CH_2 \tag{3}$$

$$PHBd-\underset{\underset{CH_3}{|}}{\overset{CH_3}{\underset{|}{Si}}}(CH_2)_3OCH_2CH-CH_2 \atop \diagdown O \diagup \tag{4}$$

$$PHBd-\underset{\underset{CH_3}{|}}{\overset{CH_3}{\underset{|}{Si}}}(CH_2)_3OCH_2\underset{\underset{OH}{|}}{C}H-\underset{\underset{OH}{|}}{C}H_2 \tag{5}$$

3.2
Polystyrenes

Polystyrene (PSt) macromonomers, 6, almost quantitatively functionalized with p-styrylalkyl end groups have been prepared by termination of living polystyryllithium with corresponding p-styrylalkyl bromide or iodide [13]. Termination of PSt-Li with epichlorohydrin, in benzene plus tetrahydrofuran, was successful after end-capping with 1,1-diphenylethylene to afford epoxide-ended PSt macromonomer, 7 [14]. Living polystyryllithium was end-capped with ethylene oxide, followed by reaction with 5-norbornene-2-carbonyl chloride to afford ω-norbornenyl PSt macromonomer, 8, which was also successfully subjected to living, ring-opening methathesis polymerization (ROMP) to afford regular comb PSt, 9, with both the branch and the backbone well-controlled with regard to

MW and MW distribution [15]. ω-Norbornenyl macromonomers of poly(styrene-*b*-ethylene oxide) have similarly been prepared, as will be described later in Sect. 3.4.

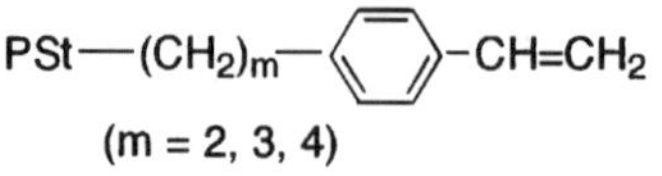
PSt—(CH2)m—
-CH=CH2
(m = 2, 3, 4)

(6)

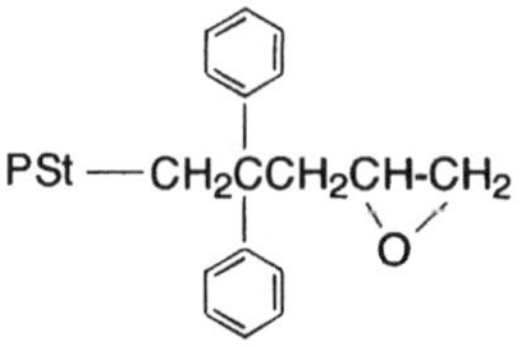
PSt —CH2CCH2CH-CH2
O

(7)

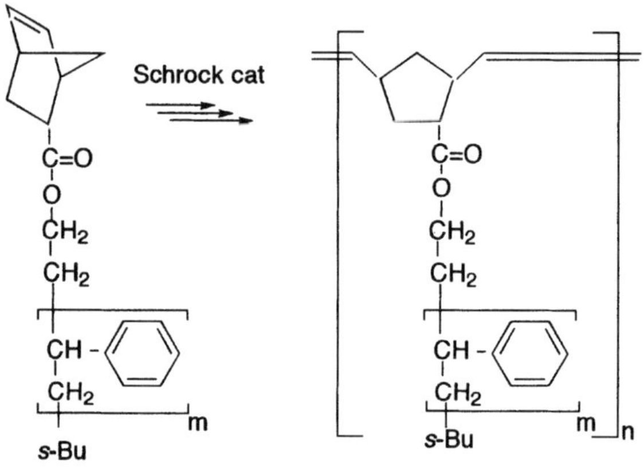
Schrock cat
C=O
O
CH2
CH2
CH -
CH2
m
s-Bu
n

(8) **(9)**

Very recently, a multifunctional, "orthogonal" initiator, **10**, has been developed by Puts and Sogah [16]. Living free radical polymerization of styrene, initiated with the styryl-TEMPO moiety as an active site, afforded ω-oxazolinyl PSt macromonomer, which was in turn polymerized through cationic ring-opening of the oxazoline end groups by methyl trifluoromethanesulfonate, to give a regular comb PSt with poly(oxazoline) as a backbone, **11**.

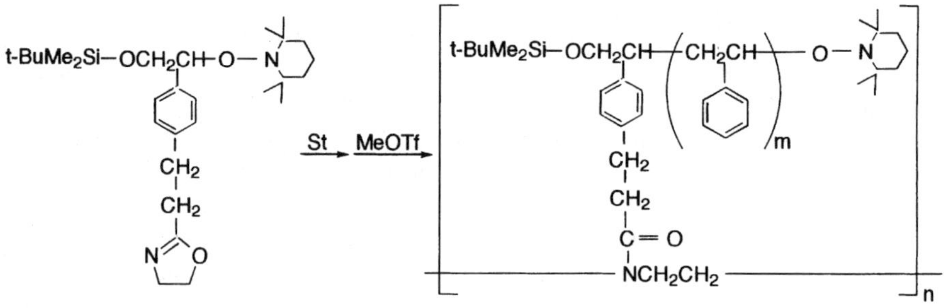
t-BuMe2Si—OCH2CH-O—N
CH2
CH2
N
O
St
MeOTf
t-BuMe2Si-OCH2CH
CH2CH
O—N
m
CH2
CH2
C= O
NCH2CH2
n

(10) **(11)**

3.3
Polyacrylates

1,3-Pentadienyl-terminated poly(methyl methacrylate) (PMMA) as well as PSt, **12**, have been prepared by radical polymerization via addition-fragmentation chain transfer mechanism, and radically copolymerized with St and MMA, respectively, to give PSt-*g*-PMMA and PMMA-*g*-PSt [17, 18]. Metal-free anionic polymerization of *tert*-butyl acrylate (TBA) initiated with a carbanion from diethyl 2-vinyloxyethylmalonate produced vinyl ether-functionalized PTBA macromonomer, **13** [19].

$$\text{PMMA or PSt} -CH_2-CH=CH-CH=CH_2 \tag{12}$$

$$CH_2=CHOCH_2CH_2-\underset{\underset{COOC_2H_5}{|}}{\overset{\overset{COOC_2H_5}{|}}{C}}-PTBA \tag{13}$$

Highly stereoregular PMMA macromonomers, **14**, prepared by Hatada and coworkers, have recently been fractionated by supercritical fluid chromatography into completely uniform fractions with no structural distribution [20, 21]. They have been oligomerized with a radical (AIBN) or an anionic initiator (3,3-dimethyl-1,1-diphenylbutyllithium). After a new fractionation by SEC comb or star polymers of completely uniform architecture are obtained. No doubt, these samples will be most promising to investigate the branched structure-property relationship.

$$t\text{-Bu}-\left[CH_2-\underset{\underset{COOCH_3}{|}}{\overset{\overset{CH_3}{|}}{C}}\right]_n-(CH_2)_3-O\underset{\underset{O}{\|}}{C}\overset{CH_3}{\underset{|}{C}}=CH_2 \tag{14}$$

3.4
Poly(ethylene oxide)

Norbornenyl-ended macromonomers from poly(ethylene oxide) (PEO), **15**, as well as from PEO-*b*-PSt or PSt-*b*-PEO block copolymers, **16a**, **16b**, have been prepared by the initiation or termination method of living anionic polymerization [22, 23]. The ROMP of **16** afforded various types of controlled, core-shell type star polymers, and block copolymerization of **8** and **15** produced a Janus-type or two-faced star polymer.

$$\text{norbornenyl}-CH_2O\underset{n}{+}CH_2CH_2O\underset{}{+}CH_2Ph \tag{15}$$

$$\text{norbornenyl}-CH_2\underset{m}{+}CH_2CH(Ph)\underset{}{+}\underset{n}{+}CH_2CH_2O\underset{}{+}CH_2Ph \tag{16a}$$

$$\text{norbornenyl}-\underset{O}{\overset{\|}{C}}\underset{n}{+}OCH_2CH_2\underset{}{+}\underset{m}{+}CHCH_2(Ph)\underset{}{+}\text{s-Bu} \tag{16b}$$

Polymerization of ethylene oxide with an acetal-protected alkoxide afforded α-aldehyde-ω-methacryloyl PEO macromonomer, **17**, after termination with methacrylic anhydride followed by acid hydrolysis [24].

$$O=CH\text{-}CH_2CH_2O\underset{n}{+}CH_2CH_2O\underset{}{+}\underset{O}{\overset{CH_3}{\underset{\|}{CC}}}=CH_2 \tag{17}$$

Epoxide-terminated PEO macromonomer, **18** [25], and a mesogen-substituted PEO macromonomer, **19** [26], have been prepared and polymerized to liquid crystalline comb polymers.

$$RO\underset{n}{+}CH_2CH_2O\underset{}{+}CH_2\text{-}CH\text{-}CH_2\overset{O}{\diagup} \tag{18}$$

(R = $C_{10}H_{21}$, $C_{12}H_{25}$)

$$CH_2=\underset{O}{\overset{CH_3}{\underset{\|}{C}}}COCH_2CH_2O\underset{n}{+}CH_2CHO\underset{}{+}H \tag{19}$$

(with pendant –CH$_2$–O–C$_6$H$_4$–C(=O)–O–C$_6$H$_4$–OR)

(R = CH_3, C_8H_{17})

3.5
Some Other New Macromonomers

Polymerization of hexamethylcyclotrisiloxane with 3-butadienyllithium afforded butadienyl-ended polysiloxane macromonomer, 20 [27]. Polycondensation of a chiral methyl β-hydroxyisobutyrate at a temperature higher than 150 °C with Ti(O-nBu)$_4$ afforded directly a biodegradable polyester macromonomer, 21 [28].

$$\begin{array}{c} CH_2 \\ \parallel \\ CH \\ | \\ C \\ \parallel \\ CH_2 \end{array} - \left[\begin{array}{c} CH_3 \\ | \\ Si-O \\ | \\ CH_3 \end{array} \right]_n - Si(CH_3)_3 \tag{20}$$

$$CH_2=CC \begin{array}{c} CH_3 \\ | \\ \\ \parallel \\ O \end{array} - \left[\begin{array}{c} CH_3 \\ | \\ OCH_2CHC \\ \parallel \\ O \end{array} \right]_n - OCH_3 \tag{21}$$

Glycopeptide macromonomers, 22, were prepared from p-vinylbenzylamine-initiated ring-opening polymerization of sugar-substituted α-amino acid N-carboxyanhydrides [29]. They have been copolymerized with acrylamide to afford the corresponding sugar-grafts with molecular recognition ability.

$$\text{(22)}$$

(X = OH, NHAc; Y = H, Ac)

4
Homopolymerization and Copolymerization of Macromonomers

Since macromonomers are already polymers with MW between 10^3 and 10^4, their polymerization and copolymerization involves polymer-polymer reactions. Thus a question of continuing concern has been how and why a macromonomer is different in its reactivity from a corresponding conventional monomer of low MW.

4.1
Homopolymerization

Radical homopolymerization kinetics of some typical macromonomers, such as those from PSt, **23, 24** [30, 31], and PMMA, **25** [32, 33], have been studied in detail by means of ESR methods.

$$s\text{-Bu}\!-\!\!\left[\text{CH}_2\text{CH}(\text{C}_6\text{H}_5)\right]_n\!\!-\!\text{CH}_2\!-\!\text{C}_6\text{H}_4\!-\!\text{CH}=\text{CH}_2 \quad (23)$$

$$s\text{-Bu}\!-\!\!\left[\text{CH}_2\text{CH}(\text{C}_6\text{H}_5)\right]_n\!\!-\!\text{CH}_2\text{CH}_2\text{OC}(=\!O)\text{C}(\text{CH}_3)\!=\!\text{CH}_2 \quad (24)$$

$$t\text{-Bu}\!-\!\!\left[\text{CH}_2\text{C}(\text{CH}_3)(\text{COOCH}_3)\right]_n\!\!-\!\text{CH}_2\!-\!\text{C}_6\text{H}_4\!-\!\text{CH}=\text{CH}_2 \quad (25)$$

(highly isot. or syndiot.)

The kinetics apparently follow the conventional square-root equation for the overall rate of polymerization, R_p:

$$R_p = k_p \left(\frac{2k_d f}{k_t}\right)^{1/2} [I]^{1/2} [M] \quad (1)$$

where k_p and k_t are the rate constants of propagation and termination, respectively, k_d and f are rate constants of initiator decomposition and initiation efficiency, respectively, and [I] and [M] are the concentrations of initiator and monomer, respectively. Therefore, we also have the conventional expression for the kinetic chain length, ν:

$$\nu = \frac{k_p[M]}{(2k_d f k_t)^{1/2}[I]^{1/2}} = \frac{(1+x)}{2} DP_n^\circ \quad (2)$$

where DP_n° is an instantaneous number-average degree of polymerization assuming no chain transfer and x is the fraction of disproportionation in the termination step.

Table 1 [1] summarizes the relevant kinetic parameters. Clearly, the polymerization of macromonomers, **23–25**, is characterized by very low k_t values and by less reduced k_p values, compared to those of the corresponding conventional monomers such as styrene and MMA. This means that the propagation involving the macromonomer and the multibranched radical is slightly less favored

Table 1. Kinetic parameters of some macromonomers in radical polymerization as compared with conventional monomers

Monomer	MW (g/mol)	Solvent/initiator[a]	Temp. (°C)	k_p (l/mol) s	k_t^b (l/mol s)	f	Ref.
PSt-VB 23	4980	benz/AIBN	60	4	1.3×10^{3c}	0.05^c	30
					2.6×10^{3d}	0.03^d	
PSt-MA 24	400	benz/AIBN	60	18	8.4×10^{3c}	0.2^c	30
					16.8×10^{3d}	0.1^d	
PMMA-VB 25							
(isotactic)	2900	tol/AIBN	60	50	1.4×10^{5b}	0.28	32
(syndiotactic)	2720	tol/AIBN	60	4.7	1.3×10^{3b}	0.22	32
PEO-VB26 (m=1)	2260	benz/tBPO	20	40	1.8×10^3	0.15	34
		water/AVA	20	1100	5.4×10^3	0.9	34
Styrene	104	–	60	176	7.2×10^7	0.7	35
		–	20	60	3.5×10^7	–	35
MMA	100	–	60	515	2.6×10^7	–	35

[a] Solvent: benz=benzene, tol=toluene; initiator: AIBN=2,2'-azobisisobutyronitrile, tBPO=*tert*-butyl peroxide, AVA=4,4'-azobis(4-cyanovaleric acid)
[b] The values are doubled from those in [30] and [32], where a convention of $2k_t$ instead of k_t in Eqs. (1) and (2) was used in evaluation of k_t
[c] Assumed recombination for termination
[d] Assumed disproportionation for termination

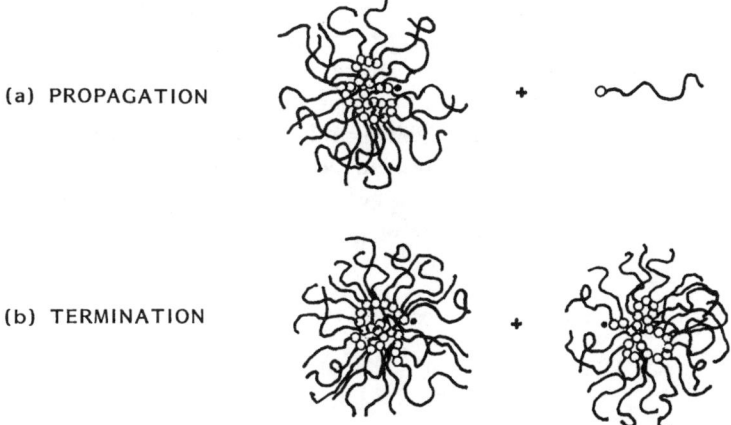

(a) PROPAGATION

(b) TERMINATION

Fig. 2a,b. Models of **a** propagation reaction of a poly(macromonomer) radical with a macromonomer; **b** bimolecular termination between poly(macromonomer) radicals

while the diffusion-controlled termination between two multibranched radicals is severely restricted, as expected from the steric requirements involved as illustrated in Fig. 2.

As a result, by virtue of Eqs. (1) and (2), the macromonomers may polymerize even more rapidly and to a higher degree of polymerization than the corresponding small monomers, provided the polymerization is conducted at the same molar concentration of [M] and [I]. Unfortunately, because the MW of the macromonomers is very high, solutions with high [M] and [I] are impractical.

In any case, we can use Eqs. (1) and (2) as a basis for the designed preparation of comb-like poly(macromonomers). Indeed, Eq. (2) predicts that their backbone length, DPn, may be controlled by changing the ratio [M]/[I]. Some deviation from the simple rate expressions, however, have been observed since a macromonomer solution is already viscous from the beginning of polymerization. Presumably this makes the diffusion-controlled termination constant, k_t, a decreasing function with respect to [M]. For example, the exponents of the [M] dependence of R_p or DP were found to be 1.5 or even higher compared to unity as required from Eqs. (1) or (2) [34, 36]. The very low initiator efficiency, f, around 0.2 or even smaller, as shown in Table 1, found in the solution polymerization of macromonomers also appears to come from the initiator decomposition in the high viscosity medium, resulting in an enhanced probability of recombination or disproportionation of the primary radicals generated.

Polymerization of p-styrylalkyl-ended poly(ethylene oxide) (PEO) macromonomers, **26**, in benzene followed the similar trend in k_p and k_t as discussed above [34]. Most interestingly, however, these amphiphilic macromonomers polymerize unusually rapidly in water to very high DPs, apparently because they organized into micelles with their hydrophobic, polymerizing end groups locally concentrated in the cores. Furthermore, the true k_p and k_t values in the hypothetically isolated micellar organization, estimated from the apparent values given in Table 1 by just multiplying by the macromonomer weight fraction (0.11), appears to be enhanced and reduced, respectively, compared to those in benzene. The initiator efficiency, f, was also high in that case. ω-Methacryloyloxyalkyl PEO macromonomers, **27a** (m=6, 11), also polymerize very rapidly in water [37, 38]. The results, therefore, suggest that, by taking advantage of the polymeric nature of the macromonomers, the control of their organization in solution will lead to unique and useful applications.

$$CH_3O\text{-}[CH_2CH_2O]_n\text{-}(CH_2)_m\text{-}C_6H_4\text{-}CH=CH_2 \quad (m = 1, 4, 7) \tag{26}$$

$$CH_3O\text{-}[CH_2CH_2O]_n\text{-}(CH_2)_m\text{-}OC(=O)C(CH_3)=CH_2 \quad (m = 0, 6, 10, 11) \tag{27a}$$

$$CH_3O\text{-}[CH_2CH_2O]_n\text{-}C(=O)C(CH_3)=CH_2 \tag{27b}$$

Conventional radical polymerization usually produces polymers with a broad distribution in DP. The polymers are mixtures of the instantaneous polymers with DP_w/DP_n of at least 1.5 for the termination by recombination or 2.0 either for the termination by disproportionation or for the chain transfer to small molecules. In this respect, any living polymerization with rapid initiation will afford polymers with a narrow DP distribution of the Poisson type. Ring-opening methathesis polymerization of norbornenyl-terminated macromonomers, **8**, **15**, and **16**, appears promising in this regard [22, 23].

4.2
Copolymerization

A number of copolymerizations involving macromonomer(s) have been studied and almost invariably treated according to the terminal model, Mayo-Lewis equation, or its simplified model [39]. The Mayo-Lewis equation relates the instantaneous compositions of the monomer mixture to the copolymer composition:

$$\frac{d[A]}{d[B]} = \frac{1 + r_A[A]/[B]}{1 + r_B[B]/[A]} \tag{3}$$

where $d[A]/d[B]$ is the molar ratio of the monomers A to B incorporated into the copolymers instantaneously formed from the monomer mixture with the molar ratio $[A]/[B]$, and r_A and r_B are the respective monomer reactivity ratios.

Copolymerization between a conventional comonomer (A) and a macromonomer (B) affords a so-called graft copolymer with A as a backbone and B as statistically distributed branches, as in Fig. 1b,d. Since usually $[A]/[B] \gg 1$ in order to obtain a balanced composition (in weight) of backbone and branches, Eq. (3) is approximated to a simplified form:

$$\frac{d[A]}{d[B]} = r_A \frac{[A]}{[B]} \tag{4}$$

Therefore, the copolymer composition or the frequency of the branches is essentially determined by the monomer composition and the monomer reactivity ratio of the comonomer.

The relative reactivity of the macromonomer in copolymerization with a common comonomer, A, can be assessed by $1/r_A = k_{AB}/k_{AA}$, i.e., the rate constant of propagation of macromonomer B relative to that of the monomer A toward a common poly-A radical. In summarizing a number of monomer reactivity ratios in solution copolymerization systems reported so far [3, 31, 40], it appears reasonable to say that the reactivities of macromonomers are similar to those of the corresponding small monomers, i.e., they are largely determined by the nature of their polymerizing end-group, i.e., essentially by their chemical reactivity.

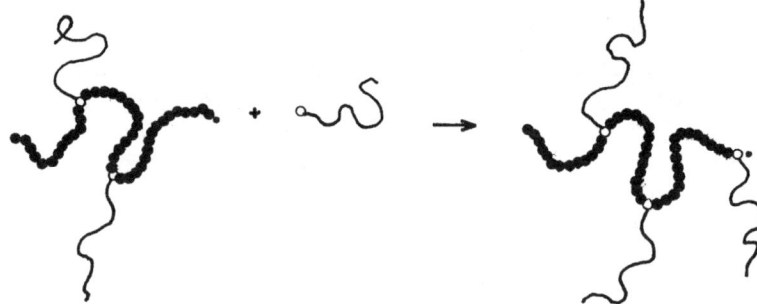

Fig. 3. Model of cross-propagation between a grafted poly(comonomer) radical and a macromonomer

In some but not so rare cases, however, reactivity of macromonomers was found to be apparently reduced by the nature of their polymer chains. For example, *p*-vinylbenzyl- or methacrylate-ended PEO macromonomers, **26** (m=1) or **27b**, were found to copolymerize with styrene (as A) in tetrahydrofuran with increasing difficulty ($1/r_A$ is reduced to one half) with increasing chain length of the PEO [41]. Since we are concerned with polymer-polymer reactions, as shown in Fig. 3, the results suggest that any thermodynamically repulsive interaction, which is usually observed between different, incompatible polymer chains, in this case PEO and PSt chains, may retard their approach and hence the reaction between their end groups, polystyryl radical and *p*-vinylbenzyl or methacrylate group. Such an incompatibility effect was discussed in terms of the degree of interpenetration and the interaction parameters between unlike polymers to support the observed reduction in the macromonomers copolymerization reactivity [31, 40]. Similar observations of reduction of the copolymerization reactivity of macromonomers have recently been reported for the PEO macromonomers, **27a** (m=11) with styrene in benzene [42], **27b** with acrylamide in water [43], and for poly(L-lactide), **28**, with dimethyl acrylamide or *N*-vinylpyrrolidone in dioxane [44].

$$CH_2=\underset{O}{\overset{CH_3}{\underset{\|}{C}}}COCH_2CH_2O\left[CH_2\overset{CH_3}{\underset{|}{C}}HO\right]_n H \tag{28}$$

The composition distribution of the graft copolymers obtained by the macromonomer method has been shown theoretically to be statistically broader than in the corresponding conventional linear copolymer, due to the high MW of the macromonomer branches [45, 46]. This has been experimentally confirmed by Teramachi et al. with PSt macromonomers, **23** or **24**, copolymerized with MMA [47–49]. The chemical composition distribution was found to broaden with increasing MW and decreasing frequency of the macromonomer branches, as well as with increasing conversion, as expected.

Narrow distribution in the backbone length as well as in the chemical composition or the branch frequency may be expected from a living-type copolymerization between a macromonomer and a comonomer provided the reactivity ratios are close to unity. This appears to have been accomplished to some extent with anionic copolymerizations with MMA of methacrylate-ended PMMA, **29**, and poly(dimethylsiloxane) macromonomers, **30**, which were prepared by living GTP and anionic polymerization, respectively [50, 51]. Recent application [8] of nitroxide (TEMPO)-mediated living free radical process to copolymerizations of styrene with some macromonomers such as PE-acrylate, **1a**, PEO-methacrylate, **27b**, polylactide-methacrylate, **28**, and poly(ε-caprolactone)-methacrylate, **31**, may be a promising approach to this end.

$$CH_2=\overset{CH_3}{\underset{\underset{O}{\|}}{C}}COCH_2CH_2O\overset{CH_3}{\underset{\underset{O\ CH_3}{\|\ |}}{C}}C\text{---}\left[CH_2\overset{CH_3}{\underset{COOCH_3}{C}}\right]_n\text{---}H \qquad (29)$$

$$n\text{-Bu}\text{---}\left[\overset{CH_3}{\underset{CH_3}{Si}}\text{-O}\right]_n\text{---}\overset{CH_3}{\underset{CH_3}{Si}}(CH_2)_3\text{-O}\overset{CH_3}{\underset{\underset{O}{\|}}{C}}C=CH_2 \qquad (30)$$

$$CH_2=\overset{CH_3}{\underset{\underset{O}{\|}}{C}}COCH_2CH_2O\text{---}\left[\underset{\underset{O}{\|}}{C}(CH_2)_5O\right]_n\text{---}H \qquad (31)$$

In an apparently homogeneous solution, macromonomers, possibly together with the resulting graft copolymers, may lead to some structure formation such as micelle or looser association, which may in turn change the apparent reactivities due to some specific solvation or partition of the monomers around the active sites. Such a "bootstrap" effect [52] may be responsible for some complicated dependency of the apparent reactivities on the monomer concentration and composition in radical copolymerization of **29** with n-butyl acrylate [53].

Use of macromonomers as reactive (copolymerizable) surfactants in heterogeneous systems such as emulsion and dispersion constitutes an increasingly important application in the design of polymeric microspheres, as will be discussed later in Sect. 6. Here the macromonomers copolymerize in situ with some of the substrate comonomers to afford the graft copolymers, the grafts (branches) of which serve as effective steric stabilizers by anchoring their backbone onto the surfaces of the particles. In general, however, the copolymerization reactivities of macromonomers in such systems are not well understood yet.

Copolymerization involving two or more kinds of macromonomers appears interesting in providing comb polymers with multiple kinds of branches, but also in reflecting some explicit polymer effects involved in the polymer-polymer reactions. Two macromonomers with different polymer chains but with the same polymerizing end groups may copolymerize azeotropically if the reaction is solely chemically controlled and there are no polymer effects. PSt and polyisoprene (PIp) macromonomers, both with a *p*-vinylbenzyl end group, have been copolymerized in benzene with a free radical (AIBN) and an anionic initiator (*n*-butyllithium) [54, 55]. The results show a nearly azeotropic polymerization when the macromonomers have similar DPs but with some preference for incorporation of higher MW PIp macromonomer, suggesting some polymer effect caused by the morphology of the double comb copolymers formed.

We copolymerized PSt and PEO macromonomers carrying the same methacrylate end groups, **24** (n=27) and **27b** (n=16, 48), with AIBN in benzene, and found the latter more reactive [56]. In contrast, copolymerization between the macromonomers with the same polymer chain but with different polymerizing groups, PEOs with *p*-vinylbenzyl and methacrylate, **26** (m=1, n=48) and **27b** (n=48), was nearly azeotropic, i.e., $r_A \approx r_B \approx 1$, in benzene or in methanol. Therefore, the PEO chains appear to make the intrinsic reactivity difference of their end groups almost insignificant. In water with 4,4'-azobis(4-cyanovaleric acid), however, PEO macromonomers with more hydrophobic polymerizing end groups are apparently more reactive in copolymerization in the order of **26** (m=4)>**26** (m=1)>**27b**. This clearly supports the micellar copolymerization mechanism which favors an amphiphilic monomer with a more hydrophobic polymerizing moiety to participate more readily in the reaction sites (micelles).

To summarize, macromonomers in polymerization and copolymerization are only fairly well understood compared to the conventional monomers. Effects, such as conformational, morphological, or due to incompatibility caused by the macromonomer chains, remain to be further investigated. As a result, the macromonomer technique is expected to lead to other unique applications including construction of novel branched architectures.

5
Characterization of Star and Comb Polymers

Homopolymerization of macromonomer provides regular star- or comb-shaped polymers with a very high branch density as shown in Fig. 1a,c,e. Such polymacromonomers, therefore, are considered to be one of the best models for understanding of branched architecture-property relationships. Their properties are expected to be very different from the corresponding linear polymers of the same MW both in solution and the bulk state. Indeed, during the past decade, remarkable progress has been accomplished in the field of static, dynamic, and hydrodynamic properties of the polymacromonomers in dilute and concentrated solutions, as well as by direct observation of the polymers in bulk.

On the other hand, copolymerization of a conventional monomer with a macromonomer also affords well-defined graft copolymers at least in the sense that the chain length of the macromonomer which forms the branches is predetermined, as shown in Fig. 1b,d,f. Nevertheless, both the branched structure and heterogeneities in MW and composition make the relevant characterization techniques, such as SEC and light scattering, greatly inefficient in a strict sense. In spite of the fact that characterization of graft copolymers prepared by the macromonomer method is of essential importance, their characterization is still not fully realized. Only a very few literature references are available in which a precise characterization of the MW and compositional distributions of the graft copolymers is described.

Two papers to be noted here have been reported by Ward's group [51] and more recently, by Müller's group [53] for the synthesis and characterization of model PMMA-g-PMMA. The former group synthesized model PMMA-g-PMMA with narrow MW and compositional distributions by anionic copolymerization of MMA and a methacrylate-terminated PMMA macromonomer, 29. The graft polymers were characterized by several methods including membrane osmometry, static light scattering, and hyphenated techniques such as SEC-LALLS and SEC-differential viscosity (DV). The combination of these techniques demonstrated that the PMMA-g-PMMA containing up to 40 mass % of long-chain branching obeyed the universal calibration in SEC. The small characteristic ratio values were also determined for the graft copolymers by applying Stockmayer-Fixman (S-F) plot, though the application of S-F plot to the branched polymer system is questionable. The shrinking factor, $g=<S^2>_b/<S^2>_l$ (see below), determined by SEC-DV, was found to increase with increasing MW; that is, the apparent branching density decreased with M_w.

Müller et al. [53] prepared similar PMMA-g-PMMA by radical copolymerization of MMA with methacrylate-terminated PMMA macromonomer, 29, and characterized the samples by SEC-multiangle laser light scattering (MALLS). The power law exponent, **a**, in the equation, $<S^2>^{1/2} \propto M^a$, was found to be 0.36. In remarkable contrast to the result of Ward et al. [51], the shrinking factor decreased with increase of MW. This may imply that the difference in graft copolymerization method, anionic or radical, results in the graft copolymers with very different branch distribution.

The characterization and solution properties of graft copolymers in which the backbone polymers are chemically different from the branches require many difficulties to be overcome, from the viewpoints of the determination of MW, the branching rate, and their distributions.

In the next section, therefore, we review recent studies of simpler cases, i.e., homopoly(macromonomers), star- and comb-shaped polymers, followed by some interesting properties of the graft copolymers to be used as polymeric surfactants, surface modifiers, and compatibilizers for blends.

5.1
Characterization and Solution Properties of Poly(macromonomers)

Polymacromonomers can be geometrically classified into two types of regular branched forms, i.e., stars and combs, depending on the degree of polymerization of the backbone and side chains. The poly(macromonomers) are probably better treated as star polymers when the number of arms is small. The "bottlebrush" conformation, characteristic of poly(macromonomers), develops as the number of branches increases. Crossover between stars and bottlebrushes, therefore, would be expected to appear at a certain degree of polymerization of macromonomer, as will be described later.

The effect of branching on the solution properties is usually discussed in terms of the comparison with those of corresponding linear polymers. The mean-square radius of gyration of branched polymers, $<S^2>_b$, is characterized using a dimensionless parameter, the shrinking factor, g, which is defined as

$$g = \frac{\langle S^2 \rangle_b}{\langle S^2 \rangle_l} \tag{5}$$

where $<S^2>_l$ is the mean-square radius of gyration of the linear polymer of the same MW. For Gaussian chains, the value of g for star- (g_s) and comb-shaped (g_c) polymers is theoretically given as [57, 58]

$$g_s = \frac{3f'-2}{f'^2} \tag{6}$$

$$g_c = \frac{1 + 2f'\gamma + (2f'+f'^2)\gamma^2 + (3f'^2-2f')\gamma^3}{(1+f'\gamma)^3} \tag{7}$$

where f' is the number of branches and γ is the ratio of the MWs of a branch and the backbone. When excluded-volume effects exist, the value of g_s for the star-shaped branched polymer near the θ-temperature may be modified to [59, 60]

$$g_s = \frac{3f'-2}{f'^2} \frac{(1+K_b z + \cdots)}{(1+1.276 z + \cdots)} \tag{8}$$

where z is the excluded-volume parameter which is defined as $z=(3/(2\pi b^2))^{3/2}\beta n^{1/2}$ with the bond length b, the number of the bond n, and the binary cluster integral β, and K_b is given by

$$K_b = \frac{3}{f'^{1/2}(3f'-2)} \left[\frac{67 \times 2^{7/2}}{315}(f'-1) - \frac{134}{315}(f'-2) + \frac{4}{45}(101 \times 2^{0.5}-138)(f'-1)(f'-2) \right] \tag{9}$$

The g factors of some star-shaped polymacromonomers with relatively limited number of arms have been investigated and compared with the theory mentioned above. Tsukahara et al. [61] estimated the g factors of PSt polymacromonomers from 24 by SEC-LALLS measurement and compared with Eqs. (6) and (8). The results suggest that these poly(macromonomers) behave like star polymer. The experimental value of g is larger than the theoretical one based on Eq. (6) in agreement with results of studies on model star polymers [62].

Gnanou and coworkers [15, 22, 23] prepared several types of regular star- and comb-shaped polymers by living, ROMP of ω-norbornenyl macromonomers, **8**, **15** and **16**. Some of them were characterized by means of the universal calibration in SEC to discuss the chain density, radius of gyration, and shrinking factor [63].

Hatada and coworkers [64] have prepared a series of uniform oligo(PMMA macromonomer)s prepared by a radical or anionic polymerization of uniform PMMA macromonomer, **14**, followed by fractionation by SEC. The MW dependence of the limiting viscosity, [η], was investigated with monomer to tetramer of the PMMA macromonomer using SEC-DV.

Ito et al. [65] investigated the MW dependence of the limiting viscosity for a series of regular polymacromonomers from PEO macromonomers, **26** (m=1) and demonstrated that the universal SEC calibration holds for these polymers. The exponent, **a**, in the Mark-Houwink-Sakurada equation defined by

$$[\eta] = KM^a \tag{10}$$

decreased with increase of the chain length of the branch. It was found at least in the MW range investigated that the value of **a** steeply approaches zero when n is above 44. Very low values of **a** clearly suggest that polymacromonomers behave hydrodynamically as a non-draining rigid sphere and/or constant segment density particles. The similar results were reported by Tsukahara et al. [66] in which they studied [η] of PSt polymacromonomers from **24**. They also observed that in high M_w region, [η] goes up and increases with M_w. This rising of [η] with M_w was suggestive of the change in the molecular conformation of polymacromonomers from starlike to bottlebrush.

Schmidt et al. [67, 68] have first demonstrated that polymacromonomers from a series of PSt macromonomers, **24**, behave as semi-flexible polymer chains in dilute toluene solution by the measurements of SEC-MALLS, SEC-DV, and dynamic light scattering (DLS). The Kuhn statistical segment length was reported to increase monotonously up to ca. 300 nm with branch chain length. They termed them, therefore, as "molecular bottlebrushes" (see Fig. 1e). Subsequently, the diffusion and sedimentation experiments studied by Nemoto et al. [69] clearly showed that their MW dependence is quantitatively described by the prolate ellipsoid model with the values of the main and the minor axes calculated from a planar zigzag PMMA backbone and Gaussian PS branched chains, respectively. Small-angle X-ray scattering (SAXS) experiments also supported that the side chains assume a random coil conformation [68].

Figure 4 shows a double logarithmic plot of radius of gyration, $<S^2>_z$, (open circles) of a polymacromonomer from PEO macromonomer, **26** (m=4, n=50) (C_1-PEO-C_4-S-50) in 0.05 N NaCl solution against M_w [70]. The $<S^2>_z$ data for M_w lower than 1×10^5 (filled circles) were calculated using Eq. (6) and the equation, $<S^2>_z = 4.08\times10^{-4}M_w^{1.16}$ for a PEO chain in water at 25 °C [71]. Both data are seen to superimpose in the region of $M_w \sim 1\times10^5$. It can be expected from Fig. 4 that the plot of $\log<S^2>_z$ vs $\log M_w$ of the polymacromonomers assumes a re-

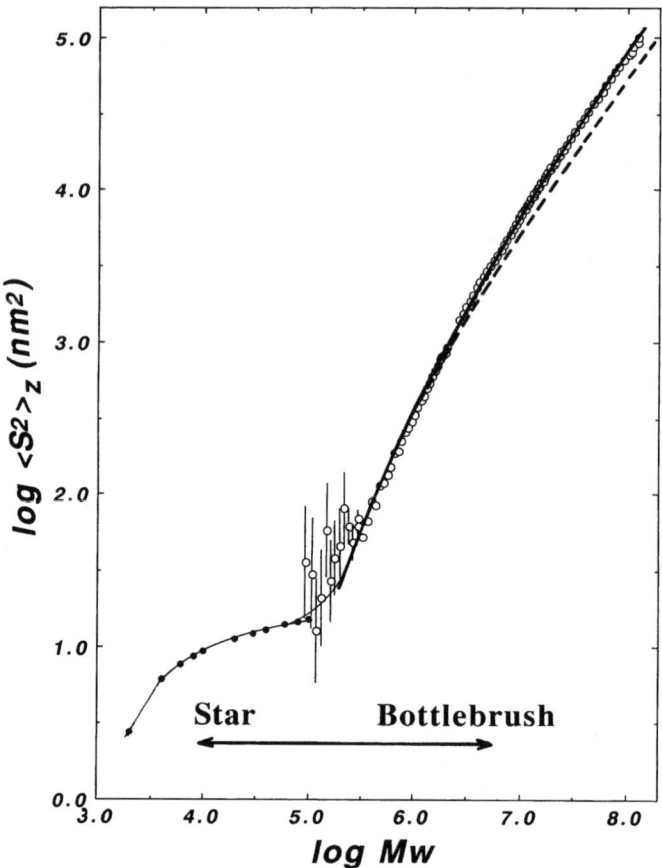

Fig. 4. Double logarithmic plots of $<S^2>_z^{1/2}$ vs M_w of poly(macromonomer) of **26** (m=4, n= 50) in 0.05 N NaCl H$_2$O at 25 °C. *Open symbols* (o) are experimental results. The *closed symbols* (l) are calculated values by Eqs. (5) and (6). The *thick solid line* is theoretical, calculated from Eqs. (11) and (14–17) for the perturbed KP chain with q=17 nm, M_L=1.03×10^4 nm^{-1}, and B=5.78 nm; the *dashed line* is theoretical, calculated for the unperturbed KP chain (B=0)

verse S-shaped curve. That is, in the M_w lower than ca. 1×10^5, the polymacromonomer assumes a star-shaped conformation and above it the characteristic bottlebrush conformation of the polymacromonomer appears.

With M_w higher than 3×10^5, the experimental points could be fitted by a smooth convex curve. The slope of the convex curve was about 1.54 for M_w between 5×10^5 and 1×10^6 and about 1.15 for M_w between 3×10^6 and 3×10^7. This change in slope implies that the molecule is rodlike at lower MW and approaches a spherical coil as M_w increases, which is the characteristic behavior of semiflexible polymers.

According to Benoit and Doty [72], the unperturbed $\langle S^2 \rangle_o$ of a monodisperse wormlike chain is expressed by

$$\langle S^2 \rangle_o = \frac{qM}{3M_L} - q^2 + \frac{2q^3 M_L}{M}\left[1 - \frac{qM_L}{M}\left(1 - e^{-\frac{M}{qM_L}}\right)\right] \quad (11)$$

with the molecular weight M, persistence length q, and shift factor $M_L = M/L$, where L is the contour length of a wormlike chain. Murakami et al. [73] showed that when $M/(2qM_L) > 2$, Eq. (11) can be approximated by

$$\left(\frac{M}{\langle S^2 \rangle_o}\right)^{1/2} = \left(\frac{3M_L}{q}\right)^{1/2}\left[1 + \frac{3qM_L}{2M}\right] \quad (12)$$

Zhang et al. [74] showed that for $M/(2qM_L) < 2$, Eq. (11) may be replaced by

$$\left(\frac{M^2}{12\langle S^2 \rangle_o}\right)^{2/3} = M_L^{4/3} + \frac{2M_L^{1/3}}{15q} M \quad (13)$$

When the Zhang plot is applied to the present data for $5 \times 10^5 < M_w < 1 \times 10^6$, a straight line was obtained to give q=17 nm and $M_L = 1.03 \times 10^4$ nm^{-1}. The value of M_L is in good agreement with that ($M_L = 1.05 \times 10^4$) calculated from the molecular structure. The theoretical curve (broken line) in Fig. 4 was computed from Eq. (11) with q=17 nm and $M_L = 1.03 \times 10^4$ nm^{-1}. It can be seen that the theoretical curve describes the chain length dependence of the dimension of polymacromonomers for $M_w < 1.5 \times 10^6$ but above it deviates from the experimental data. This is most likely due to excluded-volume effects. The Kuhn segment number, n_K (= $M_w/(2qM_L)$, $M_w = 1.5 \times 10^6$) at which onset of excluded-volume effects appear is calculated to be 4.28, in agreement with Yamakawa-Stockmayer perturbation theory [75].

Yamakawa-Stockmayer-Shimada (YSS) theory [75–77] predicts that the radius expansion factor α_s (=$(\langle S^2 \rangle / \langle S^2 \rangle_o)^{1/2}$) is a universal function of the scaled excluded-volume parameter $\tilde{z}$ defined by

$$\tilde{z} = \left(\frac{3}{4}\right) K(\lambda L) z \quad (14)$$

with

$$z = \left(\frac{3}{2\pi}\right)^{3/2} (\lambda B)(\lambda L)^{1/2} \quad (15)$$

and

$$K(\lambda L) = \frac{4}{3} - 2.711(\lambda L)^{-1/2} + \frac{7}{6}(\lambda L)^{-1} \quad \text{for } \lambda L > 6 \quad (16)$$

or

$$K(\lambda L) = (\lambda L)^{-1/2} \exp\left[-6.611(\lambda L)^{-1} + 0.9198 + 0.03516(\lambda L)\right] \quad \text{for } \lambda L \leq 6 \quad (17)$$

where λ^{-1} is the Kuhn segment length (=2q), z is the conventional excluded-volume parameter for the Kratky-Porod (KP) chain, and B is the excluded-volume strength (=β/a'^2) for the KP chain with a' being the bead spacing. The YSS theory with the Domb-Barrett function [78] for the α_s^2 is applied to the experimental data of $<S^2>_z$ and M_w in Fig. 4. It is seen that the experimental values in $3\times10^5<M_w<1\times10^8$ are quantitatively described in terms of the YSS theory (solid line) with B=5.78 nm, q=17 nm, and M_L=1.03×10^4 nm^{-1} within experimental error. It is concluded, therefore, that the polymacromonomer of **26** (m=4, n=50), (C_1-PEO-C_4-S-50) behaves in water as an unperturbed semi-flexible polymer in the region n_K<4.3 and as a perturbed semi-flexible polymer when n_K>4.3. The experimental particle scattering function P(θ) of this polymer was also found to be described accurately by a theoretical curve for the wormlike chain with the same q and M_L parameters. In addition, fluorescence studies supported that the conformational mobility of the hydrophobic polystyrene backbone of the bottlebrush is strongly restricted in water. All these results support that the PEO polymacromonomer with M_w higher than 1×10^5 behaves as a bottlebrush in water, in accord with the results by Schmidt et al. [67, 68], but somewhat different from those by Richtering et al. [79], who carried out light scattering studies for amphiphilic polymacromonomer with oligo(ethylene oxide) chains as the branch.

The crossover from stars to bottlebrushes is expected to take place at M_w= 1×10^5 which corresponds to n_K=0.29 and L=9.7 nm in the polymacromonomers. The diameter of the bottlebrush may be calculated to be 7.8 nm from the $<S^2>_z^{1/2}$ of a PEO chain. Thus, the crossover may take place when the chain length of the backbone becomes more or less comparable to the diameter of the bottlebrush.

Recent Monte Carlo simulation studies for the polymacromonomers using the bond fluctuation model by Shiokawa et al. [80, 81] showed that the shape of the main chain varies gradually from a self-avoiding coil-like form to an extended rod-like form with increasing side chain length. On the other hand, the side chain conformation was shown to be independent of the side chain length and contour length. The SAXS profiles of polystyrene polymacromonomers in a concentrated solution was reported to show a sharp peak, which is also characteristic of the ordering of the semi-flexible polymers [82].

On the other hand, while the MW dependence of $<S^2>$, diffusion, and sedimentation coefficients of these bottlebrushes were quantitatively described by the wormlike chain model, a significant disagreement between that of [η] and the theory was also reported [65–67]. This is an unsolved subject to be studied further.

5.2
Bulk Properties

Polymacromonomers with polymeric branches on every second carbon atom of the backbone have an extremely high branched chain density. In other words, one polymacromonomer chain of degree of polymerization, DP has DP+2 ends.

Table 2. Glass transition temperatures of PSt polymacromonomers

MW of macromonomer		T_g of polymacromonomer/°C	
$10^{-3}M_n$	$10^{-3}M_w$	obs.	cal.[a]
0.81	0.9	56–68	44
2.9	3.1	84–90	84
12.4	13.1	98	96
14.0	14.6	100	97
27.0	28.1	–	98

[a] Calculated by Eq. (18), based on $T_g = T_g^\infty - 1.0 \times 10^5 2 M_w$ with $T_g^\infty = 100$

Therefore, their bulk properties are expected to be significantly different from those of the corresponding linear polymer. Indeed, detailed studies by Tsukahara et al. [31, 83] revealed that the glass transition temperature, T_g, of PSt polymacromonomers from **24** is predominantly determined by the excess free volume effect of end group per unit MW, as summarized in Table 2. Generally, the T_g value was found to increase with increase of the MW of macromonomer and also that of polymacromonomers. The chain end effect on the glass transition temperature is described by the relation [84]

$$T_g = T_g^\infty - \frac{2T_g^\infty v M_0}{v_m M} \tag{18}$$

where T_g^∞ is T_g for the polymer of infinite MW, M_0 and M are the MWs of monomeric unit and the macromonomer, v_m is the free volume per monomeric unit, and v is the excess free volume at a chain end. The value of T_g of the polymacromonomers calculated by Eq. (18) is given in Table 2. Hatada and Kitayama [33] also reported that the value of T_g of it-PMMA polymacromonomer from **25** (n=30) is 44 °C.

Tsukahara et al. [83] have reported that when PSt polymacromonomers are cast onto Teflon or glass plates, many cracks are generally created during the solvent evaporation and the resulting films are too brittle to handle. They ascribed the result to a lack of chain entanglement networks in the polymacromonomers [85]. In fact the SAXS profiles of the polymacromonomers in the bulk state demonstrated that the polymacromonomer molecules exist independently of each other [86, 87]. Such behavior was confirmed directly by STM [88] and tapping scanning force microscopy [89]. The incompatibility was also observed between high MW linear polystyrene and PSt polymacromonomers with high degree of polymerization, in spite of the fact that they are athermally mixing [85, 90, 91]. The PSt polymacromonomers have been reported to form lyotropic main chain liquid crystals in toluene solution and in the bulk state [82, 90, 91]. The crystallization behavior of PEO comb-shaped polymers from **27b** was reported by Wesslén et al. [92]. The polymacromonomers from n=9 and 23 were amorphous material with T_g=−55~−60 °C but those from n=45 were crystalline with mp= 38–44 °C.

5.3
Some Properties of Graft Copolymers

Amphiphilic graft copolymers are conveniently synthesized by copolymerization of a hydrophobic monomer with a hydrophilic macromonomer and vice versa. The resulting copolymers are of great interest from the point of view of their surface active properties.

Amphiphilic graft copolymers synthesized by HEMA with PSt macromonomer were found to form a micellar aggregate in methanol by ^{1}H NMR analysis [93, 94]; the spectrum of HEMA-rich graft copolymer (85% of HEMA content) in CD_3OD showed only peaks due to PHEMA segments, expected from their conformations like that shown in Fig. 1f. On addition of $CDCl_3$ to the solution of the graft copolymer, peaks ascribed to polystyrene segments also appeared, corresponding to the expected conformation with both segments expanded or coiled like in Fig. 1b. These data clearly indicate that the PSt segments, insoluble in methanol, constitute a micelle core which behaves as a solid on the NMR time scale. Consequently, their resonance peaks were too broad to be observed. The reverse phenomenon was observed in the case of polystyrene-rich graft copolymer (72% of PSt content); in the NMR spectrum measured in $CDCl_3$, only the peaks due to the PSt segments were observed, indicating the formation of a micelle with a PHEMA core surrounded by PSt segments. The conformation of the micelle is shown in Fig. 1d. In parallel to this line, some papers about surfactant properties of various kinds of graft copolymers have been reported [95, 96].

Control of surface properties of polymers is very important in technical fields such as coatings, adhesives, films, and fibers. Among various surface modification techniques, surface accumulation of graft copolymers is a convenient and promising method for the surface control. Yamashita et al. [97] investigated the surface activity of graft copolymers prepared from perfluoroalkylethyl acrylate and PMMA macromonomer in PMMA films, prepared by the solvent cast method. A very small amount of the graft copolymer was sufficient to improve the anti-wettability of PMMA films, as evaluated by contact angle of a water or dodecane droplet. Fluorine- [98], silicon- [99], PMMA-[100], and poly(2-methyl-2-oxazoline)-containing [96] graft copolymers have been prepared and studied with respect to their properties as a surface modifier.

The polymer composites have been investigated in order to achieve various functions and properties enhancements. In blending polymer materials, compatibility between the respective polymers becomes very important to affect the properties of the resulting polymer blend. In immiscible cases, the graft copolymers are often used as a compatibilizer. Poly(styrene-g-MMA) prepared by a macromonomer technique was reported to be an effective compatibilizer for blending between PSt and PVC; elongation and tensile strength of the composite containing the graft copolymer are superior to those without the compatibilizer [101].

Recently, much attention has been paid to selective gas permeable membranes. The basic requirements for these membrane are a high permeability coefficient, high selectivity, and self-supporting properties. Poly(dimethylsiloxane) is known to have a high permeability coefficient for oxygen although its selectivity for permeants and mechanical strength are low. Kawakami et al. [102] have prepared various polymers having oligosiloxane side chains by the homopolymerization of the macromonomers. The polymer with short siloxane chains was found to be better than that with long branches in selectivity for permeation of oxygen and in mechanical strength.

6
Design of Polymeric Microspheres Using Macromonomers

Polymerizations are usually carried out in a good solvent for both a monomer and its resulting polymer. When in the presence of suitable dispersants one carries out the polymerization in non-solvents, however, polymeric microspheres are produced. Much attention has been paid recently to this type of heterogeneous polymerization to afford monodisperse polymeric microspheres because of various applications in technical and biomedical fields.

One of the fascinating applications of macromonomers is in the field of heterogeneous polymerization such as emulsion and dispersion systems. This was first reported for nonaqueous dispersions (NAD) by the ICI group [7]. The heterogeneous polymerization in the presence of suitable stabilizers affords submicron- to micron-sized polymeric microspheres, often of excellent monodispersity. Among a variety of methods of preparing polymeric microspheres, the macromonomer technique is unique and flexible in that the macromonomers themselves act as reactive emulsifiers or dispersants without any conventional surfactants. The macromonomers are graft-copolymerized during copolymerization and accumulate on the particle surface, so that the resulting emulsions and dispersions are very effectively sterically stabilized against flocculation. As the counterpart of NAD, a number of emulsion or dispersion systems in water or in alcoholic media have been recently developed using hydrophilic macromonomers to meet increasing concerns for environmentally friendly systems.

6.1
Dispersion Polymerization

Dispersion polymerization is defined as a type of precipitation polymerization by which polymeric microspheres are formed in the presence of a suitable steric stabilizer from an initially homogeneous reaction mixture. Under favorable circumstances, this polymerization can yield, in a batch process, monodisperse, or nearly monodisperse, latex particles with a relatively large diameter (up to 15 μm) [103]. The solvent selected as the reaction medium is a good solvent for both the monomer and the steric stabilizer, but a non-solvent for the polymer being formed and therefore a selective solvent for the graft copolymer. This restriction on the choice of solvent means that these reactions can be carried out

Table 3. Examples of dispersion copolymerization with macromonomers

Macromonomer	Monomer	Medium	Ref.
PHSA 32	MMA	Hydrocarbon	7
PLMA 33	MMA	Hydrocarbon	7
PE 1b	MMA	Dodecane, PE	9
POXZ 34	MMA	MeOH-H_2O	104
POXZ 34	Styrene	EtOH-H_2O	105
POXZ 34	CH_2=CHNHCHO	MeOH	106
POXZ 35	MMA	MeOH-H_2O	107
PEO 26 (m=1)	MMA, Styrene	EtOH-H_2O	108, 109
PEO 27b	Styrene	EtOH-H_2O	110
PEO 36, 2b(m=1)	Styrene	EtOH-H_2O	111
PEO 26 (m=1,4,7)	Styrene, MMA, BMA	MeOH-H_2O	112, 113, 114
PEO 27a(m=11)	Styrene	EtOH-H_2O	115
PEO 27a(m=6,10), 27b	Styrene, MMA	MeOH-H_2O	38
PEO 37	Styrene	EtOH-H_2O	116
PVP 38	Styrene, MMA	EtOH	117
PVAcA 39	Styrene	EtOH	118
PVA 40	MMA	EtOH-Water	119
P4VP 41	Styrene	EtOH	120
PNIPAM 42	Styrene	EtOH	121
PTBMA 43	Styrene	EtOH	120
PAA 44	MMA	EtOH-H_2O	122
PDMS 45	MMA, Styrene	CO_2	123
PCL 46	L,L-Lactide	Heptane-dioxane	124, 125
PMA 47	MMA	EtOH-H_2O	126
PMA 48	Styrene	MeOH-H_2O	127

with solvents with extremely low or high solubility parameters. The examples of dispersion polymerization using macromonomers are summarized in Table 3. Historically, non-aqueous dispersion (NAD) polymerization of polar monomers was first carried out in aliphatic hydrocarbon media with hydrophobic macromonomers, **32** and **33** [7]. These are copolymerized with MMA or other polar monomers to produce comb-graft copolymers which have limited solubility in pure aliphatic hydrocarbons but have adequate solubility in hydrocarbon-monomer mixtures. It is particularly effective in stabilizing PMMA NAD particles. PE macromonomers **1b** have been used for the dispersion copolymerization of MMA in dodecane and in PE melts to produce stable PMMA dispersion at a high temperature [9]. In the latter case, nanocomposite materials in which submicron-sized PMMA fine particles are uniformly dispersed in the PE bulk can be prepared during the copolymerization.

$$CH_2=\underset{O}{\overset{CH_3}{\underset{|}{C}}}COCH_2\underset{OH}{\overset{|}{CH}}CH_2O\left[\underset{O}{\overset{C_6H_{13}}{\underset{|}{C}}}(CH_2)_{10}CHO\right]_n\underset{O}{\overset{}{C}}C_{17}H_{35} \quad (32)$$

$$CH_2=\underset{O}{\overset{CH_3}{\underset{|}{C}}}COCH_2\underset{OH}{\overset{|}{CH}}CH_2O\underset{O}{\overset{}{C}}CH_2S\left[CH_2\underset{COOC_{12}H_{25}}{\overset{CH_3}{\underset{|}{C}}}\right]_n H \quad (33)$$

The technique has been recently extended to polar media, especially alcohols and their mixtures with water as a continuous phase. Kobayashi et al. [104–107] have reported that poly(2-oxazoline) macromonomers such as 34 and 35 are very effective for the dispersion copolymerization with styrene, MMA, and N-vinyl-formamide in methanol, ethanol, and mixtures of these alcohols with water. They reported that the particle size decreased with increasing initial macromonomer concentration and that poly(2-oxazoline) macromonomers graft-copolymerized are concentrated on the particle surface to act as steric stabilizers.

$$CH_2=CH-\underset{}{\underset{}{\bigcirc}}-CH_2\left[NCH_2CH_2\right]_n \quad R=Me, Et \quad (34)$$
$$\qquad\qquad\qquad C=O$$
$$\qquad\qquad\qquad R$$

$$HO\left[CH_2CH_2N\right]_n\underset{}{\overset{}{\diagup\!\!\!\diagdown}}\left[NCH_2CH_2\right]_n OH \quad (35)$$
$$\qquad C=O \qquad\qquad C=O$$
$$\qquad Me \qquad\qquad Me$$

Dispersion copolymerizations using poly(ethylene oxide) (PEO) macromonomers 26, 27, 36 and 37 in alcoholic media have been intensively studied by many researchers [38, 108–116]. They afford nearly monodisperse polymeric microspheres of submicron to micron size, covered with PEO chains on their surface. Several factors which affect the particle size and polymerization kinetics have been systematically studied. The theoretical model for particle nucleation in these systems has also been developed and compared with the experimental observations, as will be presented in Sect. 6.2.

$$HO\left[CH_2CH_2O\right]_n\underset{O}{\overset{CH_3}{\underset{|}{C}}}C=CH_2 \quad (36)$$

$$ROC\underset{O}{\overset{}{\diagup\!\!\!\diagdown}}CO\left[CH_2CH_2O\right]_nR' \quad R=H, C_{12}H_{25} \quad (37)$$
$$\qquad\qquad\qquad\qquad\qquad\qquad R'=H, CH_3$$

Several other hydrophilic macromonomers including **38–44** have been successfully applied to the dispersion polymerization [117–122]. These macromonomers were synthesized by radical polymerization in the presence of appropriate chain transfer agents, followed by transformation of the end group, as was previously summarized [1]. Akashi et al. [121] used PNIPAM macromonomer **42** in ethanol and have prepared thermosensitive microspheres 0.4–1.2 µm in diameter consisting of a PSt core and PNIPAM branches on their surface. The particles are particularly useful for many biomedical applications. Indeed, the particles have been reported to flocculate with increasing temperature together with the change in the light transmittance.

$$CH_2=CH-\langle\bigcirc\rangle-CH_2O\underset{O}{\overset{\|}{C}}CH_2CH_2S\left[CH_2\underset{COOtBu}{\overset{CH_3}{\underset{|}{C}}}\right]_n H \qquad (43)$$

$$\begin{array}{c} H_2C=HC-\langle\bigcirc\rangle-CH_2 \\ \\ H_2C=HC-\langle\bigcirc\rangle-CH_2 \end{array} \overset{Cl^-}{\underset{N}{\overset{+}{\underset{|}{N}}}}\left[CH_2-\underset{COOH}{\overset{|}{C}H}\right]_n H \qquad (44)$$

DeSimone and his co-workers have intensively studied polymerization reactions in an environmentally friendly solvent, CO_2. In the presence of CO_2-philic silicone-based macromonomer, 45, relatively monodisperse micron-sized polymer particles were obtained by the polymerization of MMA and styrene in supercritical CO_2 [123].

$$CH_2=\underset{O}{\overset{CH_3}{\underset{\|}{C}}}COC_3H_6-\underset{CH_3}{\overset{CH_3}{\underset{|}{Si}}}-\left[\underset{CH_3}{\overset{CH_3}{\underset{|}{OSi}}}\right]_n-C_4H_9 \qquad (45)$$

Sosnowski et al. [124, 125] have reported that uniform biodegradable polymeric particles with diameters of less than 5 µm can be prepared by ring-opening dispersion polymerization of L,L-lactide in heptane-dioxane mixed solvent in the presence of poly(dodecyl acrylate)-g-poly(ε-caprolactone), which were synthesized by copolymerization of dodecyl acrylate with poly(ε-caprolactone) macromonomers, 46. It is noted that the polymer particles consist of well-defined poly(L,L-lactide) polymers with $M_n \approx 1 \times 10^4$ and $M_w/M_n \approx 1.06$.

$$CH_2=\underset{O}{\overset{CH_3}{\underset{\|}{C}}}COCH_2CH_2O-\left[CH_2CH_2CH_2CH_2CH_2\overset{O}{\overset{\|}{C}}O\right]_n CH_2CH_3 \qquad (46)$$

Polyelectrolyte macromonomers 41, 44, 47 and 48 [120, 122, 126, 127] have been also prepared and applied to the dispersion copolymerizations to produce polymeric particles covered with polyelectrolyte chains. Evidently, the dependence of the conformational properties of polyelectrolyte brush chains attached to the latex surface on pH, degree of neutralization, and salt concentration have been the subject of a growing experimental and theoretical effort.

$$H_2C=HC-\langle\rangle-CH_2-\left[\begin{array}{c}CH_3\\|\\C-CH_2\\|\\COOH\end{array}\right]_n-C(Ph)_3 \tag{47}$$

$$H_2C=HC-\langle\rangle-CH_2-\left[\begin{array}{c}CH_3\\|\\C-CH_2\\|\\COOH\end{array}\right]_n-R \quad R=H, CH_3CH_2CH(CH_3)CH_2C(Ph)_2 \tag{48}$$

In all instances of the dispersion polymerization, amphiphilic graft copolymers produced in a selective solvent for the branches play a crucial role. They act as a steric stabilizer to provide colloidal stability to the system by adsorbing or becoming incorporated into the surface of the newly formed, precipitated polymers. Schematically, a microsphere thus obtained by copolymerization with a small amount of macromonomer has a core-shell structure as given in Fig. 5, with the core occupied by the insoluble substrate polymer chains and the shell by the soluble, graft-copolymerized macromonomer chains. The backbone chains of the graft copolymers, which must be insoluble in the medium, serve as the anchors into the core. The following section presents a general criterion for the size control of polymeric microspheres by the dispersion copolymerization using macromonomers.

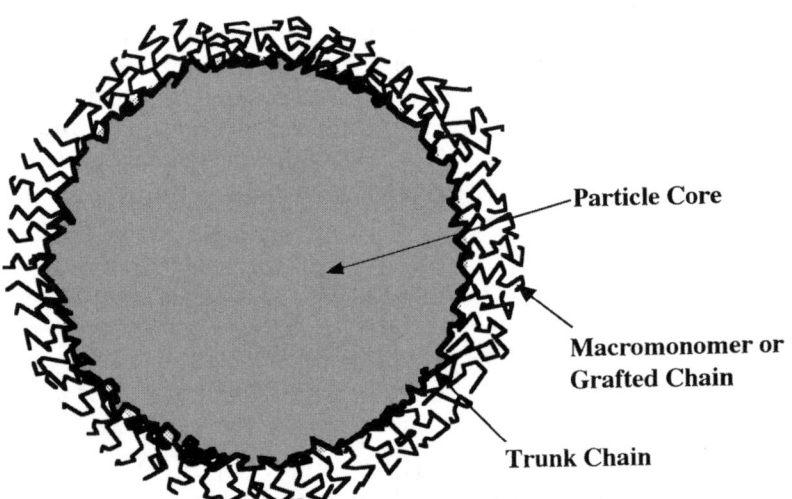

Fig. 5. Schematic picture of a microsphere obtained in emulsion and dispersion copolymerization using macromonomer technique. The grafted chains are exaggerated in size

6.2
Mechanistic Model of Dispersion Copolymerization with Macromonomers

According to the aggregative and coagulative nucleation mechanisms which have been all derived originally from the homogeneous nucleation theory of Fitch and Tsai [128], the most important point in the reaction is the instant at which colloidally stabilized particles form. After this point, coagulation between similar-sized particles no longer occurs, and the number of particles present in the reaction is constant. As shown in Fig. 6, the dispersion copolymerization with macromonomers is considered to proceed as follows. (1) Before polymerization, the monomer, macromonomer, and initiator dissolve completely into the

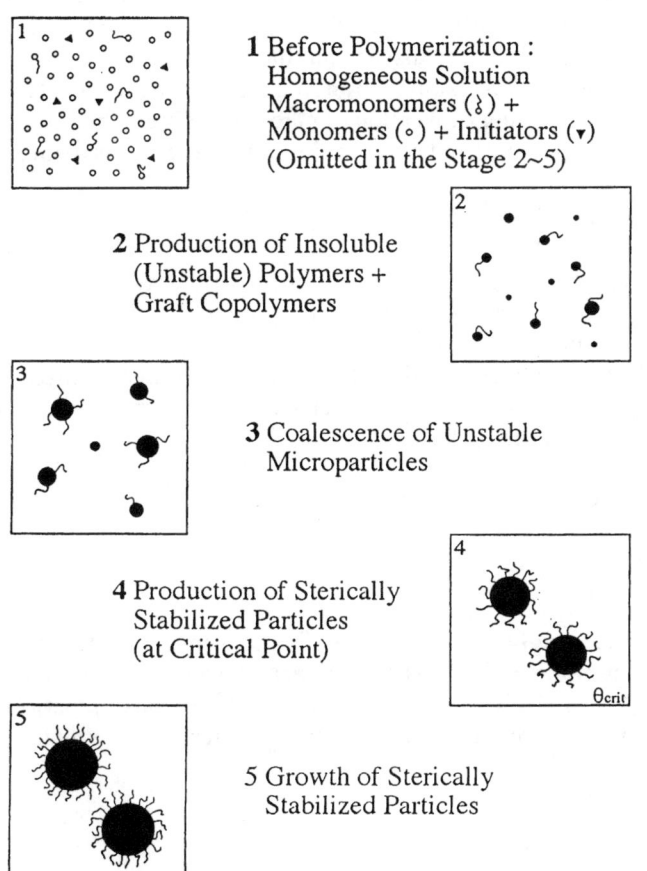

Fig. 6. Schematic model for the particle nucleation and growth of sterically stabilized particles in dispersion polymerization using macromonomer

solvent. (2) Accompanied by the decomposition of the initiator, linear oligomers, polymers, and graft copolymers are all produced by polymerization in the continuous phase. The solubility of these polymers is a function of their MW and the composition of the graft copolymer. Polymers with a MW larger than a certain critical value precipitate and begin to coagulate to form unstable particles. (3) These particles coagulate on contact, and the coagulation among them continues until sterically stabilized particles form. (4) This point is referred to as the critical point, and it occurs when all of the particles of interest contain sufficient stabilizer polymer chains on the surface to provide colloidal stability [112–114].

After this point, particles grow both by diffusive capture of oligomers and coagulation of very small yet unstable particles (nuclei, precursors) produced in the continuous phase and by polymerization of the monomer occluded within the particle. The total number of such sterically-stabilized particles remains constant so that their size is only a function of amount of polymers produced.

The particle size is determined at the critical point by the amount of polymers produced at that point. In the discussion that follows, one uses the term θ to describe the fractional conversion of monomer to polymer ($0 \leq \theta \leq 1$), and θ_D to describe the corresponding conversion of macromonomer. The weight (W_M in g/l) of the monomer polymerized at any point in the reaction is defined as

$$W_M = W_{Mo}\theta = \frac{4}{3}\pi R^3 \rho N \tag{19}$$

Here, W_{Mo} is the weight of monomer in the reactants; R is the radius (cm) of the particle occupied by the polymer chains only, N is the total number of particles per liter, and ρ is their density.

The surface area per sterically-stabilized particle is determined by the surface area (S) occupied by a macromonomer chain times the number (n') of these chains grafted onto the surface.

$$n' S = 4\pi R^2 \tag{20}$$

$$n' = \frac{W_{Do}\theta_D N_A}{N M_D} \tag{21}$$

W_{Do} is the weight (in g/l) of macromonomer in the reactants, N_A is Avogadro's number, and M_D is the molecular weight of the macromonomer. From Eqs. (19)–(21), one can obtain a universal relationship between the particle radius and the extent of polymerization for sterically stabilized particles:

$$R S = \frac{3 M_D W_{Mo} \theta}{\rho N_A W_{Do} \theta_D} \tag{22}$$

At the critical point, sterically stabilized particles are formed, and coalescence between similar-sized particles is terminated. At this point one has R=

R_{crit}, $S=S_{crit}$, $\theta=\theta_{crit}$, and $\theta_D=\theta_{Dcrit}$. Since the particle number N remains constant after this point, both R and S, at any subsequent conversion, can be described by the expressions

$$R = R_{crit}\left(\frac{\theta}{\theta_{crit}}\right)^{1/3} \tag{23a}$$

$$S = S_{crit}\left(\frac{\theta_{crit}}{\theta}\right)^{1/3}\left(\frac{\theta}{r_1\theta_D}\right) \tag{23b}$$

where r_1 is the reactivity ratio in copolymerization of monomer (M_1) with macromonomer (M_2). At low conversion, r_1 in this system is defined as

$$r_1 = \theta_{crit}/\theta_{D\,crit} \tag{24}$$

According to Paine [129], computer simulations using the multibin kinetic model for the coalescence between the unstable moieties indicate that the particle number (N) at the critical point is given by

$$N = \frac{N_A k_p}{0.386 k_2 \theta}\left(\frac{2fk_d[I]}{k_t}\right)^{1/2} \tag{25}$$

where k_p is the propagation rate constant ($M^{-1}s^{-1}$), k_t is the termination rate constant ($M^{-1}\,s^{-1}$), and k_2 is the diffusion-controlled rate constant for coalescence between similar-sized particles ($M^{-1}s^{-1}$). [I] is the initiator concentration (mol/l), and $f\,k_d$ is the product of initiator efficiency and the decomposition rate constant (s^{-1}) of the initiator. From Eqs. (19) and (25), the θ_{crit} can be written as

$$\theta_{crit} = (R_{crit})^{3/2}\left(\frac{(4/3)\pi\rho N_A k_p}{0.386 k_2 W_{Mo}}\right)^{1/2}\left(\frac{2fk_d[I]}{k_t}\right)^{1/4} \tag{26}$$

Substituting Eqs. (19) and (26) into Eq. (23) yields the equations

$$R = \theta^{1/3}\left(\frac{3W_{Mo}}{\rho N_A}\right)^{2/3}\left(\frac{M_D r_1}{W_{Do}S_{crit}}\right)^{1/2}\left(\frac{0.386 k_2}{4\pi k_p}\right)^{1/6}\left(\frac{k_t}{2fk_d[I]}\right)^{1/12} \tag{27}$$

$$S = \theta^{-1/3}\left(\frac{3W_{Mo}}{\rho N_A}\right)^{1/3}\left(\frac{M_D S_{crit}}{r_1 W_{Do}}\right)^{1/2}\left(\frac{4\pi k_p}{0.386 k_2}\right)^{1/6}\left(\frac{2fk_d[I]}{k_t}\right)^{1/12}\left(\frac{\theta}{\theta_D}\right) \tag{28}$$

In Eq. (27) one sees that the radius of latex particle follows simple scaling relationships with the key parameters in the system being $\theta^{1/3}$, $[\text{monomer}]_o^{2/3}$, $[\text{macromonomer}]_o^{-1/2}$, $[\text{initiator}]_o^{-1/12}$, where $[\ ]_o$ means initial concentration. These equations predict that the particle size and stabilization are determined by the magnitude of r_1. In addition, one sees in Eq. (28) that the surface area occupied by a stabilizer chain follows $\theta^{-1/3}$ in the case of azeotropic copolymerization, θ=θ$_D$. This means that the chain conformation for the grafts on the latex particle will change with grafting density, as will be presented in Sect. 6.4. The S value is closely related to the conformation of a single polymer chain as a stabilizer grafted onto the surface of a latex particle. According to de Gennes' "mushroom" model [130] for a polymer grafted to a noninteracting surface, the polymer chain occupies a volume determined by its mean-squared radius of gyration $<S^2>$. When the surface becomes crowded with chains, additional energy is needed to deform the polymer mushrooms into brushes. When the particle surfaces are covered completely with random coils of the polymer, they are also sterically stabilized against coagulation with other particles. One therefore defines S_{crit} as the maximum surface area occupied by a single polymer chain in the continuous phase. In this approximation one may treat the polymer chain as a rigid sphere composed of solvent and a random polymer chain, affixed on the surface of latex particle. In this case, S_{crit} is a cross section of the sphere and may be written in terms of the square radius of gyration as

$$S_{crit} = (5/3)\pi \langle S^2 \rangle \tag{29}$$

Figure 7 shows a comparison of Eq. (27) with the particle radius obtained by dispersion copolymerization of styrene with PEO macromonomer **26** (m=4, n=45) in methanol-water medium (9/1 v/v). One sees that the experimental particle

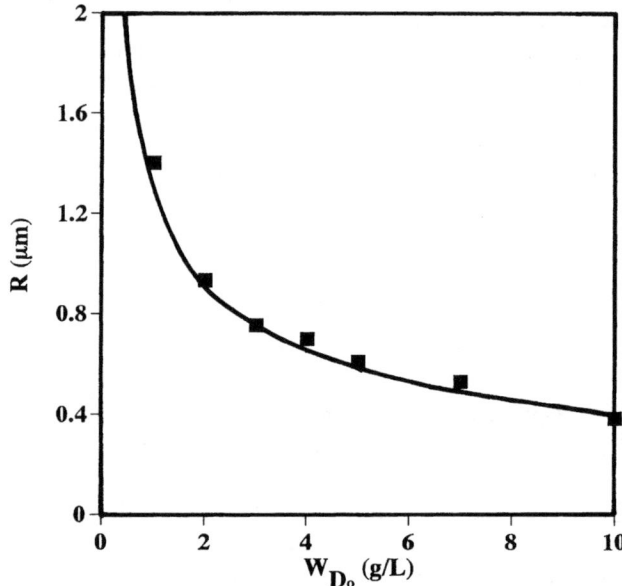

Fig. 7. Change of average particle radius (R) as a function of initial concentration (W_{D_o} in g/l) of PEO macromonomer, **26** (m=4 and n=45). W_{Mo}=100 g/l, $[I]_0$=0.0122 mol/l, θ=1, at 60 °C. A *solid line* is a theoretical curve calculated from Eq. (27) with S_{crit}/r_1=10 nm^2

Table 4. The power law exponents in dispersion copolymerization with PEO macromonomers, $R=K[\text{Monomer}]^a[\text{Macromonomer}]^b[\text{Initiator}]^c$

Macromonomer	Monomer	Medium	a	b	c	Ref.
theory, Eq. (27)			0.67	–0.50	–0.083	
26, m=4, n=45	Styrene	MeOH+H$_2$O(9:1)	0.6_3	-0.5_2	-0.06_8	112
26, m=1, 4, 7, n=53, 110	BMA	MeOH+H$_2$O(8:2)	0.8_2	-0.5_4	-0.1_0	113
26, m=1, n=45	MMA	MeOH+H$_2$O(8:2)	–	-1.1_7	–	114
26, m=1, n=45	MMA	MeOH+H$_2$O(7:3)	0.8_5	-1.1_5	-0.03_0	114
26, m=1, n=45	MMA	MeOH+H$_2$O(6:4)	–	-0.5_1	–	114
26, m=1, n=45	MMA	MeOH+H$_2$O(5:5)	–	-0.5_2	–	114
27a, m=11, n=40	Styrene	EtOH+H$_2$O(9:1)	1.0_2	-0.6_0	-0.09_0	115

radius is quantitatively described by the model with reasonable constants, $\theta = 1$, $\rho = 1.05$ g cm^{-3}, $N_A = 6.02 \times 10^{23}$, $k_2 = 10^9$ l mol^{-1} s^{-1}, $k_p = 352$ l mol^{-1} s^{-1}, $k_t = 6.1 \times 10^7$ l mol^{-1} s^{-1}, $k_d = 3.2 \times 10^{-7}$ s^{-1}, f=1, and $S_{crit}/r_1 = 10$ nm^2, $r_1 = 1$, where S_{crit} was calculated from Eq. (29) using the value of $<S^2>$ for PEO in methanol at 25 °C [131].

In the dispersion copolymerization with PEO macromonomers, the power law exponents in Eq. (27) have been experimentally determined and compared, as summarized in Table 4. Initial monomer concentration has a major influence on the final particle radius. The experimental power law exponents (0.82–1.02) is usually significantly larger than that in Eq. (27), except for 0.63 for styrene as a monomer with **26** (m=4, n=45). This is likely to be due to a solvency effect of the monomer. The values of the exponent of macromonomer and initiator concentration dependence in the polymerization of hydrophobic monomer, styrene and n-butyl methacrylate are in good agreement with those from Eq. (27). In remarkable contrast, unusually high exponent values (ca. 1.2) have been obtained in the dispersion copolymerization of a polar monomer, MMA in methanol-water (8:2 and 7:3 v/v) media. The exponent value decreases down to 0.51, when the water content is increased to higher than 40%. This significant change in the exponent value with the polarity of the continuous phase cannot be simply explained by the current theory and further refinement is needed.

6.3
Emulsion Polymerization

Emulsion polymerization is a free radical initiated chain polymerization in which a monomer or a mixture of monomers is polymerized in aqueous solution of a surfactant to form a product, known as a latex. The most important feature of emulsion polymerization is its heterogeneity from the beginning to the end of the polymerization, to yield in a batch process submicron-sized polymeric particles, often of excellent monodispersity. The main ingredients for conducting the polymerization include monomer, water, surfactants, and initiators. Upon mixing, the surfactant molecules cluster into micelles and their hydrophobic cores are swollen with monomer. Their diameters are typically of the order of 50–150 Å, and their number density is of the order of 10^{17}–10^{18} dm^{-3}. The bulk of the monomer, however, exists in the form of large-size droplets with surfactant molecules adsorbed on their surfaces; their diameters are typically in the range of 1–10 µm, and their number density of the order of 10^9–10^{11} dm^{-3}. Therefore, the polymerization is believed to take place in the micellar phase, the aqueous phase, the monomer droplet phase, and the particle phase. Three major mechanisms have been proposed for particle formation in emulsion polymerization – micellar, homogeneous, and droplet nucleation – depending on the surfactant concentration, the monomer solubility in the aqueous phase, and the extent of subdivision of the monomer droplets. This has been a central subject in the emulsion polymerization for a long time and is still a controversial issue.

Instead of conventional surfactant molecules, amphiphilic water soluble macromonomers, especially PEO macromonomers, have been used extensively as a reactive emulsifier and as steric stabilizer polymer, as summarized in Table 5. Generally speaking, however, the mechanism for the particle nucleation in the emulsion polymerization systems using macromonomers has been poorly established when compared to the dispersion copolymerizations with macromonomers as mentioned earlier.

The first work on the emulsion copolymerization with a macromonomer has been reported by Ito et al. [132] for styrene using PEO macromonomer 49. Subsequently, several research groups have reported the synthesis and some properties of latex particles containing PEO chains from 26, 27, 36, 50–52, attached to their particles surface [132–140]. Ottewill and Satgurunathan [134] and Westby [135] reported the preparation of such particles in a multistage emulsion po-

Table 5. Examples of emulsion copolymerization with macromonomers

Macromonomer	Monomer	Ref.
PEO, 49	Styrene	132
PEO, 36	Styrene	133
PEO 27b	Styrene	134
PEO 27b	Butyl acrylate	135
PEO 50	Styrene	136
PEO 26 (m=7)	Styrene	137
PEO 26(m=1, 11), 27a(m=10), 27b, 51, 52	Styrene	138
PEO 27a (m=11)	Styrene	139
PEO 26(m=1, 4, 7)	Styrene	140
POXZ 34	Styrene	143
POXZ 53	Styrene	144
POXZ 54	Styrene	145
POXZ 55	Vinyl acetate	146
PMA 48	Styrene	127

lymerization process in which PEO macromonomer **27b** was introduced in the final stage. These latexes were extremely stable to the addition of electrolyte and to the freeze-thaw [141]. Winnik et al. [142] have examined the influence of such surface grafted PEO chains on the latex film formation of PBMA latex particles by using nonradiative energy transfer method.

$$C_{12}H_{25}O\text{-}[CH_2CH_2O]_n\text{-}\underset{\underset{O}{\parallel}}{C}\underset{|}{\overset{CH_3}{C}}=CH_2 \tag{49}$$

$$CH_3O\text{-}[CH_2CH_2O]_n\text{-}\underset{\underset{O}{\parallel}}{C}NHCH_2CH_2O\underset{\underset{O}{\parallel}}{C}\underset{|}{\overset{CH_3}{C}}=CH_2 \tag{50}$$

$$\begin{array}{l}\text{-CO-[CH}_2\text{CH}_2\text{O-]}_n\text{-CH}_3\\ \text{-CO-[CH}_2\text{CH}_2\text{O-]}_n\text{-CH}_3\end{array} \tag{51}$$

$$CH_3\text{-}[OCH_2CH_2]_n\text{-}O(CH_2)_{10}OCH_2\text{-}\langle\!\!\!\bigcirc\!\!\!\rangle\text{-}CH=CH_2 \tag{52}$$

Kobayashi et al. [143–146] have synthesized several types of amphiphilic poly(2-oxazoline), **34** and its block cooligomers, **53–55**, and applied them to soap-free emulsion copolymerization of styrene and vinyl acetate to produce monodisperse, submicron-sized latex particles. They found that the particle size significantly depended on the type of macromonomer used and generally decreased with increasing the macromonomer concentration.

$$CH_2=CH\text{-}\langle\!\!\!\bigcirc\!\!\!\rangle\text{-}CH_2\text{-}[NCH_2CH_2]_m\text{-}[NCH_2CH_2]_n\text{-}OH \quad (C=O, Bu)(C=O, Me) \tag{53}$$

$$CH_2=CH\text{-}\langle\!\!\!\bigcirc\!\!\!\rangle\text{-}CH_2\text{-}[NCH_2CH_2]_m\text{-}[NCH_2CH_2]_n\text{-}N^+Et_3TsO^- \quad (C=O, Bu)(C=O, Me) \tag{54}$$

$$CH_2=CHOCCH_2\text{-}[NCH_2CH_2]_m\text{-}[NCH_2CH_2]_n\text{-}I \quad (C=O, Bu)(C=O, Me) \tag{55}$$

The latex particle diameter produced in the emulsion copolymerization of styrene with partially neutralized poly(methacrylic acid) macromonomers, **48**, was studied as a function of degree of neutralization [127]. The latex particle size was found to increase with increasing the neutralization of the polyelectrolyte macromonomers.

The rate of polymerization, R_p, in the emulsion polymerization is generally given by the equation

$$R_p = \bar{n} N k_p[M]_p/N_A \tag{30}$$

where $\bar{n}$ is the average number of radicals per particle, N the number of latex particles per unit volume, and $[M]_p$, the equilibrium concentration of the monomer swelling the latex particle. According to the Smith-Ewart theory [147], in which the main locus for particle nucleation is assumed to take place in the surfactant micelles, the number of particles is given as

$$N = k(\rho'/\mu)^{0.4}(a_s C_s)^{0.6} \tag{31}$$

where ρ' is the rate of radical generation, μ is the rate of particle volume growth, a_s is the area occupied by a surfactant molecule, and C_s is the total amount of surfactant in the micelles. Gardon [148, 149] thoroughly reviewed and confirmed the assumptions of Smith-Ewart theory and derived the more convenient equation for numerical calculation:

$$N = 0.208 \, (\rho'/K)^{0.4}(a_s C_s)^{0.6} \tag{32}$$

where $\rho'=2N_A k_d f[I]$, $C_s=N_A([C_s]-[cmc])$, and $K=[(3/4\pi)(k_p/N_A)(d_m/d_p)\phi_m]/(1-\phi_m)$ with the monomer and polymer density, d_m, d_p, and the volume fraction of monomers swelling the particles at equilibrium, ϕ_m. These equations have been well established to hold for the conventional emulsion polymerization of hydrophobic, water insoluble monomers such as styrene. Therefore, when one assumes that micellar entry dominates the particle nucleation in the emulsion copolymerization of styrene with a macromonomer, N may be written by the following power law relationship:

$$N \propto [I]^{0.4}[\text{Macromonomer}]^{0.6} \tag{33}$$

Sauer and coworkers [138] obtained $N \propto [I]^{0.82}[\text{Macromonomer}]^{-0.2 \sim +0.82}$ in the styrene emulsion copolymerization with various types of PEO macromonomers, 26, 27, 51, and 52. On the other hand, it has been found that using a PEO macromonomer, 26 (m=1, 4, 7), with a relatively short PEO chain (n=16), $N \propto [\text{Macromonomer}]^{0.6}$ while using a macromonomer with a moderately long PEO chain (n=45), $N \propto [\text{Macromonomer}]^{1.8}$ [140]. A mechanistic model for the emulsion copolymerization using amphiphilic macromonomers is under study [140] in which both micellar and homogeneous nucleations by homo- and copolymerization with macromonomer in the continuous phase are considered.

Another interesting heterogeneous polymerization using macromonomers is a microemulsion copolymerization to produce particles 10–100 nm in diameter. Gan and coworkers [150] have prepared transparent nanostructured polymeric materials by direct polymerization of bicontinuous microemulsions consisting of MMA, HEMA, water, ethylene glycol dimethacrylate, and PEO macromonomers, **27a** (m=11, n=40). The transparent polymeric materials with pore sizes from about 1 to 10 nm in diameter have been obtained.

6.4
Chain Conformation of Grafted Polymer Chains at Interfaces

Polymer adsorption has been a subject of both theoretical and experimental interest, because the adsorption behavior of polymer at solid-liquid interface is strongly connected with many technologically important processes such as flocculation, adhesion, coating, and lubrication in addition to the colloidal stabilization already discussed above. This subject has been recently reviewed by Cosgrove and Griffiths [151], Fleer et al. [152], and Kawaguchi and Takahashi [153]. Among a variety of adsorption forms of the polymers, two are of interest with macromonomers. One is the adsorption of comb and graft copolymers with highly grafted chain density onto the solid surface, which is referred to as "brush adsorption", as illustrated in Fig. 8a. Another is the attachment by the chemical reaction of the double bond of the macromonomer with the solid surface, which is referred to as "terminally-attached adsorption", as illustrated in Fig. 8b. The conformational properties of these grafted polymer chains have been a subject of growing attention, from the point of view of the "mushroom-brush" transition proposed by de Gennes [130, 154] and Alexander [155], as shown in Fig. 8c.

While there have been many studies on the conformational properties of terminally-attached polymer chains, prepared either by adsorption of block copolymers in a selective solvent onto a solid surface [156] or by a reaction of a solid surface with reactive groups of polymer [157], little has been reported for the graft copolymer chains prepared from macromonomers. Cairns et al. [158] carried out a SANS study of a non-aqueous dispersion system comprised of deuterated PMMA latex grafted with poly(12-hydroxystearic acid), **32** The thickness of the layer was found to correspond to about 2/3 of the extended chain length. Comb-like PEO polymers were grafted onto low density PE sheets by corona discharge treatment followed by homopolymerization of PEO macromonomers, **27b** (n=1, 5, and 10) and the gradient PEO concentration at the surface was characterized by measurement of the water contact angle, FTIR-ATR, and ESCA [159]. The gradient surfaces can be used to investigate the interactions between biological species and the surface PEO chains. Hadziioannou and coworkers [160] prepared terminally attached cationic polyelectrolyte brushes on a gold-coated Si-wafer by end-grafting styryl-terminated poly(vinylpyridine) macromonomer, followed by quanternization with methyl iodide. The surface was characterized by means of scanning force microscopy, ellipsometry, and FTIR-ATR.

Wu et al. [161] studied the surface properties of PS and PMMA microspheres stabilized by PEO macromonomer, **26** (m=1) and **27b** using dynamic light scattering and claimed that for PMMA microspheres the surface area occupied by a PEO molecule is nearly twice as large as that for PS microspheres, assuming that

(a) Brush adsorption

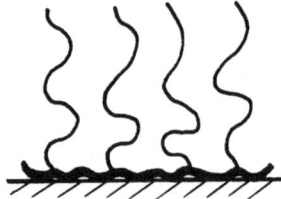

(b) Terminally-attached adsorption

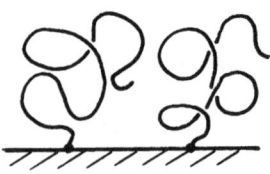

(c) Mushroom-brush transition

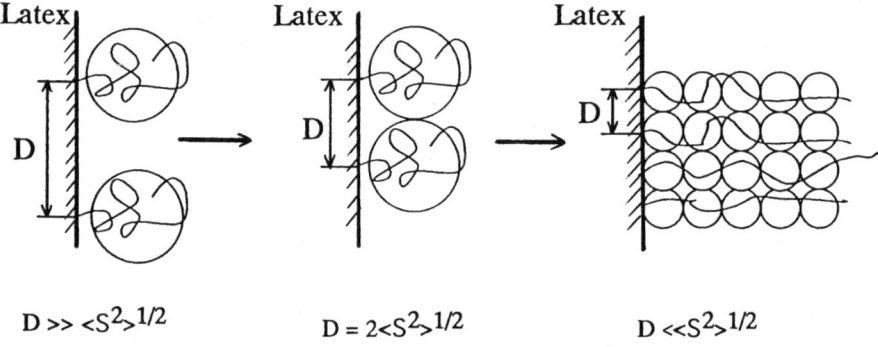

Fig. 8. Schematic representation of the possible conformations of adsorbed (co)polymers prepared using macromonomer technique: **a** brush adsorption of graft copolymer; **b** terminally-attached adsorption; **c** the mushroom-brush transition for strongly overlapping chains proposed by de Gennes [130] and Alexander [155]

100% macromonomer is copolymerized to attach onto their latex surface. However, this is not the case for styrene copolymerization with PEO macromonomers in which only 10% PEO macromonomer was copolymerized [112]. In contrast, it was confirmed that 100% of PEO macromonomers were copolymerized for the MMA and BMA dispersion copolymerization [113, 114].

^{1}H NMR studies have been carried out for the dispersion copolymerization of BMA with PEO macromonomer, **26** (m=7), in a deuterated methanol-water medium [162]. The fractional composition and surface grafted PEO concentration were monitored as the function of conversion and particle size. In Fig. 9, the mobile fraction f_M of PEO chains incorporated into the particles is plotted against interchain spacing D as shown in Fig. 8c, which can be calculated using the values of particle size and conversions. One sees that the values of f_M increase sharply with decreasing D in the region of D below 1.6 nm and become constant

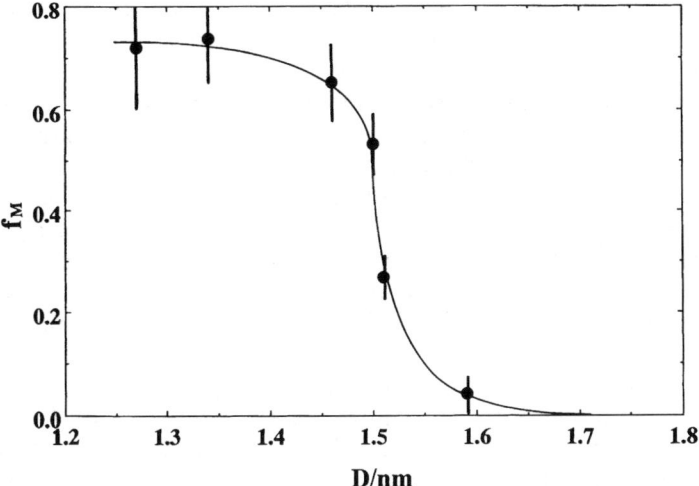

Fig. 9. Plots of mobile fraction(f_M) of surface anchored PEO chains against the estimated mean separation D between PEO anchor points on the surface of the particles. The D values were calculated from the particle size and number, assuming that all PEO chains were located at the surface

below D=1.4 nm. The radius of gyration of the PEO chain coil in methanol is calculated to be 1.6 nm, which corresponds to the D value at which the sharp increase in f_M occurs. This result may suggest that the onset of a pancake-to-brush transition of grafted chains at the interface occurs when $D \approx <S^2>^{1/2}$, as was expected from theory [130, 154, 155]. This kind of subject is of importance for a better understanding of the true nature of the steric stabilization existing in many dispersion systems. With these guidelines, the macromonomer technique will be used extensively to construct model colloidal systems.

7
Conclusions and Future

A number of well-defined macromonomers differing in the types of the monomer and the end functionality have been made available in these two decades. Their polymerization and copolymerization have provided a relatively easy access to a variety of branched polymers and copolymers, including comb-, star-, brush-, and graft-structures. Progress will no doubt continue to disclose further different types of macromonomers and branched polymers.

Radical homopolymerization and copolymerization of macromonomers are fairly well understood and reveal their characteristic behaviors that have to be compared with those of conventional monomers. A detailed mechanism of the polymer-polymer reactions involved, however, appears still to be an issue. Ionic or, desirably, living polymerization and copolymerization are still an important subject because these methods can provide branched polymers with backbones that are also well-defined in terms of the length, the composition, and their distribution.

Characterization of the poly(macromonomers) prepared by homopolymerization has proved that they provide a useful probe for discussing the structural characteristics of the star and brush polymers. Graft copolymers have been and will be a most important area of application of the macromonomer technique since a variety of multi-phased and microphase-separated systems can easily be designed just by an appropriate combination of a macromonomer and a conventional monomer. In general, however, characterization of their absolute MW, branch/backbone composition as well as their distributions remain to be studied in more detail.

Amphiphilic macromonomers work effectively as reactive (copolymerizable) emulsifiers and dispersants for heterogeneous polymerizations. They are successfully applied to the design of the polymeric microspheres of submicron to micron size. Introduction of additional functions, such as environment-sensitivity, will be a matter of interest in the potential application of the macromonomer technique.

Acknowledgments. Support of the authors' macromonomer studies by Ministry of Education, Science and Culture, Japan, as well as by Kansai Paint Co., Ltd., NOF Corporation, and Toagosei Co., Ltd. are greatly appreciated. The authors would like to acknowledge a number of valuable comments to the manuscript by the editor, Dr. J. Roovers of National Research Council, Ottawa, Canada.

8
References

1. Ito K (1998) Prog Polym Sci 23:581
2. Kawakami Y (1994) Prog Polym Sci 19:203
3. Meijs GF, Rizzard E (1990) J Macromol Sci Rev C30:305
4. Velichkova RS, Christova DC (1995) Prog Polym Sci 20:819
5. Yamashita Y (ed) (1993) Chemistry and industry of macromonomers. Hüthig & Wepf Verlag, Basel
6. Mishra MK (ed) (1994) Macromolecular design: concept and practice. Polymer Frontiers International, New York
7. Barrett KEJ (ed) (1975) Dispersion polymerization in organic media. Wiley, London
8. Hawker CJ, Mecerveyes D, Elce E, Dao J, Hedrick JL, Barakat I, Dubois P, Jérome R, Volksen W (1997) Macromol Chem Phys 198:155
9. Okada T, Kawaguchi S, Ito K (1997) Polym Prepr Jpn 46:113
10. Wörner C, Rösch J, Höhn A, Mülhaupt R (1996) Polym Bull 36:303
11. Haenz K, Stadermann D (1996) Angew Makromol Chem 242:183
12. Jayaraman RB, Facinelli JV, Riffle, JS, George SE (1996) J Polym Sci:A:Polym Chem 34:1543
13. Hirao A, Hayashi M, Nakahama S (1996) Macromolecules 29:3353
14. Quirk RP, Zhuo Q (1997) Macromolecules 30:1531
15. Heroguez V, Gnanou Y, Fontanille M (1996) Macromol Rapid Commun 17:137
16. Puts RD, Sogah DY (1997) Macromolecules 30:7050
17. Zink HO, Colombani D, Chaumont P (1997) Eur Polym J 33:1433
18. Jiang S, Viehe HG, Oger N, Chaumont D (1996) Am Chem Soc Polym Prepr 37(2):523

19. Hizal G (1996) Polymer 37:541
20. Hatada K, Nishiura T, Kitayama T, Hirotani S (1996) Macromol Rapid Commun 18:37
21. Hatada K, Kitayama T, Ute K, Nishiura T (1997) Kobunshi Ronbunshu 54:661
22. Heroguez V, Breunig S, Gnanou Y, Fontanille M (1996) Macromolecules 29:4459
23. Heroguez V, Gnanou Y, Fontanille M (1997) Macromolecules 30:4791
24. Nagasaki Y, Ogawa, R, Yamamoto S, Kato M, Kataoka K (1997) Macromolecules 30:6489
25. Schwarzwälder C, Meier W (1997) Macromolecules 30:4601
26. Libiszowski J, Goethals EJ, Mijs WJ (1996) Polym Bull 37:7
27. Kawakami Y, Ajima K, Nomura M, Hishida T, Mori A (1977) Polym J 29:95
28. Mitra M, O'Hagan D (1996) Polym Bull 36:311
29. Aoi K, Tsutsumiuchi K, Aoki E, Okada M (1996) Macromolecules 29:4456
30. Tsukahara Y, Tsutsumi K, Yamashita Y, Shimada S (1990) Macromolecules 23:5201
31. Tsukahara Y (1994) In: Mishra MK (ed) Macromolecular design: concept and practice. Polymer Frontiers International, New York, chap 5
32. Masuda E, Kishiro S, Kitayama T, Hatada K (1991) Polym J 23:847
33. Hatada K, Kitayama T (1994) In: Mishra MK (ed) Macromolecular design: concept and practice. Polymer Frontiers International, New York, chap 3
34. Nomura E, Ito K, Kajiwara A, Kamachi (1997) Macromolecules 30:2811
35. Brandrup J, Immergut EH (eds) (1975) Polymer handbook, 2nd edn. Wiley, New York
36. Tsutsumi K, Okamoto Y, Tsukahara Y (1990) Polymer 35:2205
37. Liu J, Chew CH, Gan LM (1996) J Macromol Sci-Pure Appl Chem A33:337
38. Furuhashi H, Kawaguchi S, Itsuno S, Ito K (1997) Colloid Polym Sci 275:227
39. Ito K (1993) In: Yamashita Y (ed) Chemistry and industry of macromonomers. Hüthig & Wepf Verlag, Basel, chap 3
40. Tsukahara Y (1993) In: Yamashita Y (ed) Chemistry and industry of macromonomers. Hüthig & Wepf Verlag, Basel, chap 4
41. Ito K, Tsuchida H, Hayashi A, Kitano, T, Yamada E (1985) Polym J 17:827
42. Liu J, Chew, CH, Wong SY, Gan LM (1996) J Macromol Sci-Pure Appl Chem A33:1181
43. Xiao H, Pelton R, Hamielec A (1996) Polymer 37:1201
44. Equiburu JL, Fernandez-Berridi MJ, Román JS (1996) Polymer 37:3615
45. Stejskal J, Kratochvil P, Jenkins AD (1987) Macromolecules 20:181
46. Stejskal J, Kratochvil P (1987) Macromolecules 20:2624
47. Teramachi S, Hasegawa A, Matsumoto T, Kitahara K, Tsukahara Y, Yamashita Y (1992) Macromolecules 25:4025
48. Teramachi S, Sato S, Shimura H, Watanabe S, Tsukahara Y (1995) Macromolecules 28:6183
49. Tanaka S, Uno M, Teramachi S, Tsukahara Y (1995) Polymer 36:2219
50. DeSimone JM, Hellstern AH, Siochi EJ, Smith SD, Ward TC, Gallagher PM, Krukonis VJ, McGrath JE (1990) Macromol Chem, Macromol Symp 32:21
51. Siochi EJ, DeSimone JM, Hellstern AM, McGrath JE, Ward TC (1990) Macromolecules 23:4696
52. Harwood J (1987) Macromol Symp 10/11:331
53. Radke W, Roos S, Stein HM, Müller AHE (1996) Macromol Symp 101:19
54. Ishizu K, Kuwahara K (1994) Polymer 35:4907
55. Ishizu K, Kuwahara K (1995) Polymer 36:4163
56. Nishimura H, Hadama T, Ito K (1995) Polym Prepr Jpn 44:1013; Nishimura H (1996) Master thesis. Toyohashi University of Technology; Nishimura H, Hadama T, Kawaguchi S, Ito K (to be published)
57. Zimm BH, Stockmayer WH (1949) J Chem Phys 17:1301
58. Berry GC (1968) J. Polym. Sci. A-2 6:1551
59. Berry GC, Orofino TA (1964) J Chem Phys 40:1614
60. Yamakawa H (1971) In: Modern theory of polymer solutions. Haper &Row, NY
61. Tsukahara Y, Mizuno K, Segawa A, Yamashita Y (1989) Macromolecules 23:1546

62. Roovers J, Zhou LL, Toporowski PM, Zwan MVZ, Iatrou H, Hadjichristidis N (1993) Macromolecules 26:4324
63. Brennig B, Heroguez V, Gnanou Y, Fontanille M (1995) Macromol Symp 95:151
64. Hatada K, Nishiura T, Kitayama T, Hirotani S (1997) Macromol Rapid Commun 18:37
65. Ito K, Tomi Y, Kawaguchi S (1992) Macromolecules 25:1534
66. Tsukahara Y, Kohjiya S, Tsutsumi K, Okamoto Y (1994) Macromolecules 27:1662
67. Wintermantel M, Schmidt M, Tsukahara Y, Kajiwara K, Kohjiya S (1994) Macromol Rapid Commun 15:279
68. Wintermantel M, Gerle M, Fisher K, Schmidt M, Wataoka I, Kajiwara K, Tsukahara Y (1996) Macromolecules 29:978
69. Nemoto N, Nagai M, Koike A, Okada S (1995) Macromolecules 28:3854
70. Kawaguchi S, Akaike K, Zhang ZM, Matsumoto H, Ito K (1998) Polym J 30:1004
71. Kawaguchi S, Imai G, Suzuki J, Miyahara A, Kitano T, Ito K (1997) Polymer 38:2885
72. Benoit H, Doty P (1953) J Phys Chem 57:958
73. Murakami H, Norisuye T, Fujita H (1980) Macromolecules 13:345
74. Zhang L, Liu W, Norisuye T, Fujita H (1987) Biopolymers 26:333
75. Yamakawa H, Stockmayer WH (1972) J Chem Phys 57:2843
76. Yamakawa H, Shimada J (1985) J Chem Phys 83:2607
77. Shimada J, Yamakawa H (1985) J Chem Phys 85:591
78. Domb C, Barrett AJ (1976) Polymer 17:179
79. Richtering W, Loffler R, Burchard W (1992) Macromolecules 25:3642
80. Shiokawa K (1995) Polym J 27:871
81. Shiokawa K, Itoh K, Nemoto N (1998) Macromolecules (in press)
82. Wintermantel M, Fisher K, Gerle M, Ries R, Schmidt M, Kajiwara K, Urakawa H, Wataoka I (1995) Angew Chem 107:1606
83. Tsukahara Y, Tsutsumi K, Okamoto Y (1992) Macromol Chem Rapid 13:409
84. Fox TG, Flory PJ (1950) J. Appl Phys 21:581
85. Tsukahara Y (1997) Kobunshi Ronbunshu 54:661
86. Tsukahara Y, Inoue J, Ohta Y, Kojiya S (1994) Polymer 35:5785
87. Tsukahara Y (1997) Kobunshi 46:738
88. Unayama S, Nakajima K, Ikehara T, Nishi T, Tsukahara Y (1996) Jpn J Appl Phys 35:2280
89. Sheiko SS, Gerle M, Fisher K, Schmidt M, Möller M (1997) Langmuir 13:5368
90. Tsukahara Y, Ohta Y, Senoo K (1995) Polymer 36:3413
91. Tsukahara Y (1995) Macromolecular Reports A32:821
92. Bo G, Wesslén B, Wesslén KB (1992) J Polym Sci Part A Polym Chem 30:1799
93. Ito K, Masuda Y, Shintani T, Kitano T, Yamashita Y (1983) Polym J 15:443
94. Yoshida H, Istuno S, Ito K (1995) Can J Chem 73:1757
95. Chujo Y, Shishino T, Tsukahara Y, Yamashita Y (1985) Polym J 15:443
96. Shoda S, Masuda E, Furukawa M, Kobayashi S (1992) J Polym Sci Polym Chem Ed 30:1489
97. Yamashita Y, Tsukahara Y, Ito K, Okada K, Tajima Y (1981) Polym Bull 5:335
98. Chujo Y, Hiraiwa A, Kobayashi H, Yamashita Y (1988) J Polym. Sci. Polym Chem Ed 26:2991
99. Kawakami Y, Aswatha R, Murthy N, Yamashita Y (1984) Makromol Chem 185:9
100. Yamashita Y, Tsukahara Y, Ito K (1982) Polym Bull 7:289
101. Azuma K, Tsuda T (1987) Kino Zairyo 10:5
102. Kawakami Y, Karasawa H, Aoki T, Yamamura Y, Hisada H, Yamashita Y (1985) Polym J 17:1159
103. Candau F, Ottewill R (eds) (1990) An introduction to polymer colloids. Kluwer Academic Press, Dordrecht
104. Kobayashi S, Uyama H, Choi JH, Matsumoto Y (1991) Proc Jpn Acad Ser. B 67:140
105. Kobayashi S, Uyama H, Lee SW, Matsumoto Y (1993) J Polym Sci Part A Polym Chem 31:3133

106. Uyama H, Kato H, Kobayashi S (1993) Chem Lett 261
107. Kobayashi S, Uyama H, Narita Y (1992) Makromol Chem Rapid 13:337
108. Akashi M, Chao D, Yashima E, Miyauchi N (1990) Appl Polym Sci 39:2027
109. Capek I, Riza M, Akashi M (1992) Polym J 24:959
110. Prestige C, Tadros ThF (1988) J Coll Interf Sci 124:660
111. Capek I, Riza M, Akashi M (1992) Makromol Chem 193:2843; Riza M, Capek I, Kishida A, Akashi M (1993) Angew Makromol Chem 206:69
112. Nugroho M, Kawaguchi S, Ito K (1995) Macromolecular Reports A32:593
113. Kawaguchi S, Winnik MA, Ito K (1995) Macromolecules 28:1159
114. Ito K (1994) Kobunshi Kako 30:510; Hattori T, Nugroho MB, Kawaguchi S, Ito K (1996) Polym Prepr Jpn 45:157
115. Liu J, Gan LM, Chew CH, Quek CH, Gan LH (1997) J Polym Sci Part A Polym Chem 35:3575; Liu J, Chew CH, Wong SY, Gan LM, Lin J, Tan KL (1998) Polymer 39:283
116. Lacroix-Desmages P, Guyot A (1996) Macromolecules 29:4508
117. Akashi M, Yanagi T, Yahsima E, Miyauchi N (1989) J Polym Sci Part A Polym Chem 27:3521
118. Iwasaki I, Yashima E, Akashi M, Miyauchi N, Marumo K (1991) Polym Prepr Jpn 40:2597
119. Ishizu K, Tahara N (1996) Polymer 37:1729
120. Riza M, Tokura S, Kishida A, Akashi M (1993) Polym Prepr Jpn 42:4617
121. Chen MQ, Kishida A, Akashi M (1996) J Polym Sci Appl Polym Chem 34:2213
122. Ishizu K, Yamashita M, Ichimura A (1997) Polymer 38:5471
123. Shaffer KA, Jones TA, Canelas DA, Desimone JM, Wilkinson SP (1996) Macromolecules 29:2704
124. Sosnowski S, Gadzinowski M, Slomkowski S, Penczek S (1994) J Bioactive & Compatible Polymers 9:345
125. Sosnowski S, Gadzinowski M, Slomkowski S (1996) Macromolecules 29:4556
126. Ishizu K, Tahara N (1996) Polymer 37:2853
127. Obayashi N, Kawaguchi S, Ito K (1997) Polym Prepr Jpn 46:162
128. Fitch RM (ed) (1971) In: Polymer colloids. Plenum Press, New York
129. Paine AJ (1990) Macromolecules 23:3109
130. de Gennes PG (1980) Macromolecules 13:1069
131. Zhou P, Brown W (1990) Macromolecules 23:1131
132. Ito K, Yokoyama S, Arakawa F (1986) Polym Bull 16:345
133. Hoshino F, Sakai M, Kawaguchi H (1987) Polym J 19:383
134. Ottewill RH, Satgurunathan R (1987) Coll Polym Sci 265:845
135. Westby MJ (1988) Coll Polym Sci 266:46
136. Levesque G, Moitie V, Bacle B, Deraetere P (1988) Polymer 29:2271
137. Chao D, Ito K (1992) 34th IUPAC Symp Macromol, Prepr 1P-19 Prague
138. Brown R, Stutzel B, Sauer T (1995) Macromol Chem Phys 196:2047
139. Liu J, Chew CH, Gan LM, Teo WK, Gan LH (1997) Langmuir 13:4988
140. Tano K, Kawaguchi S, Ito K (1997) Polym Prepr Jpn 46:1187
141. Ottewill RH, Satgurunathan R (1988) Coll Polym Sci 266:265
142. Farinha JPS, Martinho JMG, Kawaguchi S, Yekta A, Winnik MA (1996) J Phys Chem 100:12,552
143. Kobayashi S, Uyama H, Yamamoto I (1990) Makromol Chem Rapid 13:3115
144. Uyama H, Matsumoto Y, Kobayashi S (1992) Chem Lett 2401
145. Uyama H, Sato M, Matsumoto Y, Kobayashi S (1993) Bull Chem Soc Jpn 66:3124
146. Uyama H, Honda M, Kobayashi S (1993) J Polym Sci Part A Polym Chem 31:123
147. Smith WV, Ewart RH (1948) J Chem Phys 16:592
148. Gardon JL (1968) J Polym Sci A-1 6:623
149. Gardon JL (1968) J. Polym. Sci. A-1 6:643; 665; 687; 2853; 2859; Gardon JL (1971) J. Polym. Sci. A-1 9:2763
150. Gan LM, Lin J, Poon LP, Chew CH, Gan LH (1997) Polymer 38:5339

151. Cosgrove T, Griffiths PC (1992) Adv Colloid Interface Sci 42:175
152. Fleer GJ, Cohen Stuart MA, Scheutjens JMHM, Cosgrove T, Vincent B (1993) In: Polymers at interfaces. Chapman & Hall London
153. Kawaguchi M, Takahashi A (1992) Adv Colloid Interface Sci 37:219
154. de Gennes PG (1987) Adv Colloid Interface Sci 278:189
155. Alexander S (1977) J. Phys (Paris) 38:983
156. Auroy P, Auvray L, Leger LL (1991) Physica A172:269
157. Taunton HJ, Toprakcioglu C, Fetters LJ, Klein J (1990) Macromolecules 23:571
158. Cairns RJR, Ottewill RH, Osmond DWJ, Wagstaff I (1976) J Coll Interface Sci 54:45
159. Jeong BJ, Lee JH, Lee HB (1996) J Coll Interface Sci 178:757
160. Werts MPL, van der Vegte EW, Hadziioannou G (1997) Langmuir 13:4939
161. Wu C, Akashi M, Chen MQ (1997) Macromolecules 30:2187
162. Kawaguchi S, Winnik MA, Ito K (1996) Macromolecules 29:4465

Received: May 1998

Dendrimers and Dendrimer-Polymer Hybrids

Jacques Roovers[1], Bogdan Comanita
Institute for Chemical Process and Environmental Technology, National Research Council, Ottawa, Ontario CANADA, K1A 0R6; [1] e-mail: jacques.roovers@nrc.ca

Abstract. The synthesis and study of dendrimers has been truly dramatic in the last ten years. This review gives a brief introduction to some of the key concepts and main synthetic strategies in dendrimer chemistry. The focus of the chapter is a survey of modern analytical techniques and physical characterization of dendrimers. Results of model calculations and experiments probing the dimensions and conformation of dendrimers are reviewed. In the final sections the experimental work on dendrimer-polymer hybrids is highlighted. The dense spherical conformation of dendrimers has been combined with the loose random-coil conformation of ordinary polymers to form new hybrids with potentially interesting new properties.

Keywords. Dendrimer, Dendrimer-polymer hybrid, Conformation, Branched polymers

List of Abbreviations and Symbols 180

1 Introduction 181

1.1 Dendritic Architecture 181
1.2 Synthetic Highlights 182

2 Analysis of Dendrimers 187

2.1 NMR Analysis 187
2.2 Mass Spectrometry 187
2.3 Size Exclusion Chromatography 193

3 Conformation of Dendrimers 194

3.1 Theoretical Models 194
3.2 Experimental Dimensions of Dendrimers 195
3.2.1 Radius of Gyration 195
3.2.2 Hydrodynamic Radii 197
3.2.3 Effect of Solvent on Dendrimer Dimensions 199

4 Dendrimer-Polymer Hybrids 200

4.1 Polymeric Dendrimers 200
4.2 Dendrimers on Polymers 206

4.2.1 Dendrimers on Flexible Polymers 206
4.2.2 Dendrimers on Stiff Backbones 208
4.3   Linear Polymers on Dendrimers 211
4.3.1 Single Polymer-Dendrimer Hybrids 211
4.3.2 Multiple Polymer-Dendrimer Hybrids 216

5   Hybrids of Dendrimers and Biological Polymers 219

5.1   Dendrimer-Peptide Hybrids 219
5.2   Dendrimer-DNA Complexes 221
5.3   Dendrimer-Antibody Conjugates 222

6   References 224

List of Abbreviations and Symbols

ATRP	atom transfer radical polymerization
D	Dalton
DAB	diaminobutane based poly(propylene imine) dendrimer
DNA	deoxyribonucleic acid
D_o	translational diffusion coefficient at zero concentration
DP	degree of polymerization
DSM	Dutch State Mines
ESI-MS	electron spray ionization mass spectrometry
FAB-MS	fast ion bombardment mass spectroscopy
IUPAC	International Union of Pure and Applied Chemistry
MALDI-TOF	matrix-assisted laser desorption/ionization time-of-flight spectroscopy
MALLS	multiple angle laser light scattering
Mn	number-average molecular weight
mRNA	messenger ribonucleic acid
Mw	weight-average molecular weight
MW	molecular weight
MWD	molecular weight distribution
n	number of segments
NMR	nuclear magnetic resonance
Pam	phenylacetamidomethyl
PAMAM	poly(amidoamine) dendrimer
PBd	polybutadiene
PEG	poly(ethylene glycol)
PEO	poly(ethylene oxide)
PI	polyisoprene
PPE	poly(2,6-dimethylphenylene ether)
PS	polystyrene
q	scattering vector

R_g	radius of gyration
R_η	hydrodynamic radius from intrinsic viscosity
R_h	hydrodynanic radius from translational diffusion coefficient
SEC	size exclusion chromatography
TEMPO	2,2,6,6-tetramethylpiperidine oxide
V_e	elution volume
$[\eta]$	intrinsic viscosity
Θ	scattering angle
λ	wavelength of radiation
ν	scaling exponent
σ_c	chromatographic dispersion due to the instrument
σ_d	chromatographic dispersion due to sample polydispersity
σ_T	total chromatographic dispersion

1
Introduction

1.1
Dendritic Architecture

Dendrimers are molecules with regularly placed branched repeat units. They are also known as Starburst, Cascade or Arborols. These names describe aspects of their molecular architecture. Dendrimers consist of different parts (see Fig. 1). Each dendrimer has a core or focal point. The core is the central unit of the dendrimer and can formally be regarded as the center of symmetry for the entire molecule. The core has its characteristic branching functionality, i.e. the number of chemical bond by which it is connected to the rest of the molecule (Fig. 1a). The focal point plays the same role as the core. Moreover, it has a chemical functional group not found elsewhere in the dendrimer.

Attached to the core or focal point is a first layer of branched repeat units or monomers (Fig. 1b). This layer is alternatively considered to be the zeroed or first generation of the dendrimer. Each successive generation is end-standingly placed onto the previous generation (Fig. 1c). Each generation usually but not necessarily contains the same branched repeat units. The process of growth is extendable to several more generations. Because of the multifunctionality of each repeat unit, the number of segments in each generation grows exponentially. The end-standing groups of the outermost generation are called peripheral or terminal groups.

The description of dendrimers as outlined suggests that there is a fixed spacial arrangement in dendrimers whereby the core or focal group forms the center, successive generations radiate outwardly, and end-groups of the outermost generation form an outer surface. This is only partly true. A dendrimer is indeed a framework of chemical bonds and bond angles between atoms that vary little; however, the torsion angles about the σ bonds allow for a wide range of conformations and numerous dynamic transitions between them. Therefore, the core

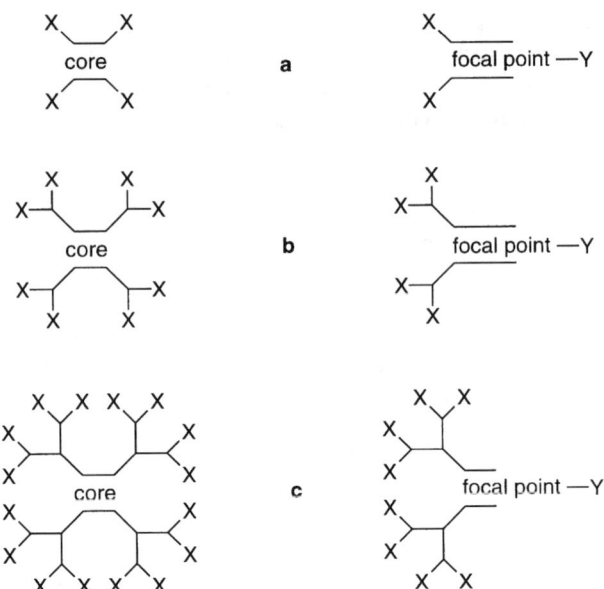

Fig. 1a–c. Schematic representation of the different parts of a dendrimer; —< stands for the repeat branching unit (monomer); X are end-standing (terminal) functional groups; Y is the functional group of the focal point: **a** core or focal point; **b** generation one dendrimer or dendron; **c** generation two homologues. The branching functionality of the core is four while the branching functionality of the monomer is three

is not necessarily the physical center of the dendrimer nor are the end groups necessarily permanently located at the periphery of the dendrimer.

The steady branching pattern of the dendrimer architecture is paralleled by an exponential increase of the molecular mass with each successively added generation. Dendrimers with more than a few generations have molecular weights that resemble those of step-growth polymers (10^4–10^5 D). For that reason and for the presence of an identifiable (branched) repeat unit, higher generation dendrimers are considered polymeric molecules.

1.2
Synthetic Highlights

Retrosynthetic analysis [1] of the generic dendritic structure (**1**) suggests two possible solutions for the synthesis of dendrimers (see Fig. 2). A first disconnection along Path A leads to generation n-1 dendrimer (**2**) and the branching monomer synthon (**3**). This rationale can be successively applied until the problem is reduced to the reaction of a core synthetic equivalent (**4**) with the branching monomer (**3**). Alternatively, along Path B, synthon (**4**) can react with the branched dendron (**5**) to provide the target dendrimer (**1**). After further iterative

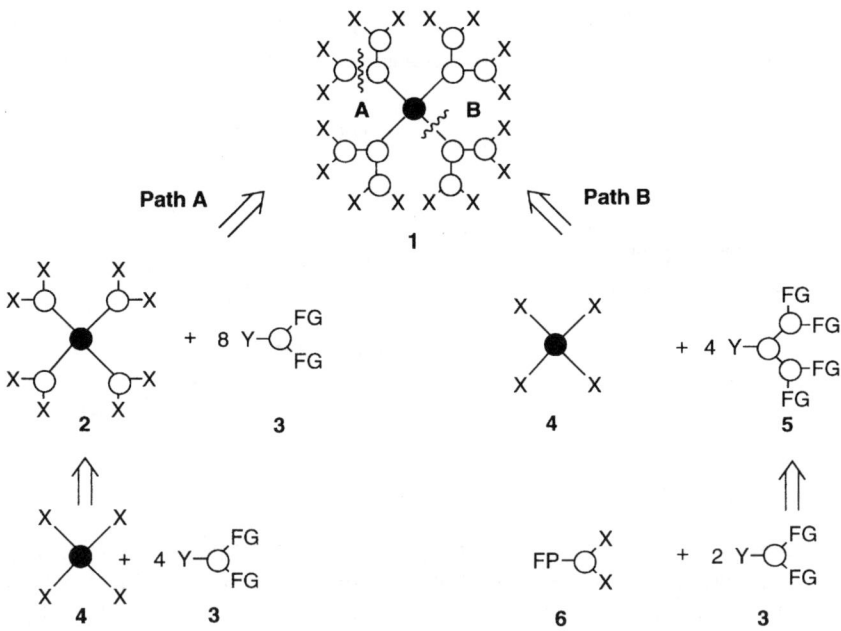

Fig. 2. Retrosynthetic analysis for the dendritic structures; *FG*, *FP* and *X*, *Y* are respectively interconvertible functional groups; *Path A* is the divergent synthesis; *Path B* is the convergent synthesis; ● stands for the core structure; ○ stands for the branching repeat unit

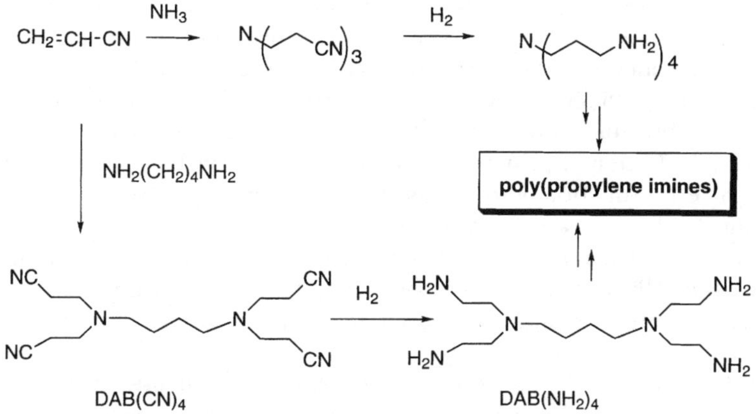

Scheme 1

disconnection of (**5**), the problem is reduced in this case to the reaction of the focal point synthon (**6**) and the branching monomer (**3**).

Path A, the divergent method, was introduced by Vögtle et al. [2] and extensively applied by Tomalia and his coworkers at Dow [3]. Working on an inside-

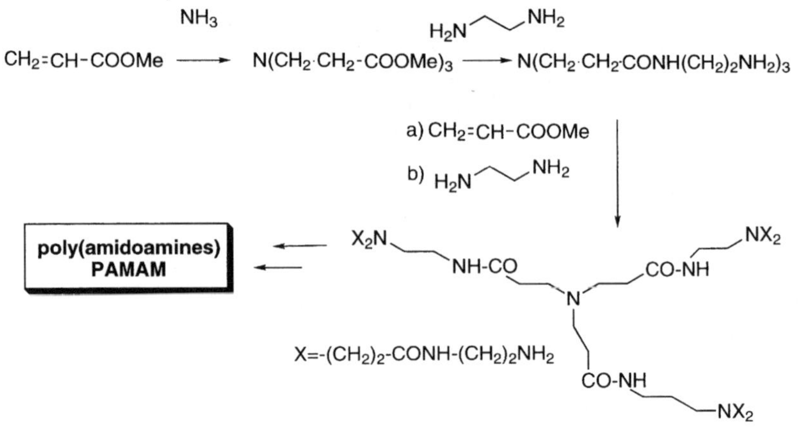

Scheme 2

out scheme starting from the core and proceeding to the periphery, Vögtle synthesized poly(alkylene imine)s by means of two alternating reactions [2]: (1) the Michael addition of a primary amino group to acrylonitrile, (2) the hydrogenation of the nitrile group to regenerate the amino group (Scheme 1). The primary amines are now available for a new cycle of Michael addition and hydrogenation. The overall yield was originally limited by the poor yield of the hydrogenation step. Improved hydrogenation methods have been found later independently by Wörner and Mülhaupt [4] and workers at DSM [5, 6]. This made the large scale synthesis of poly(propylene imine)s possible. The DSM dendrimers AS-TRAMOL are based on the 1,4-diaminobutane core and are available to generation 5 which contains 64 primary amine groups (see Scheme 1).

Other commercially available dendrimers containing nitrogen branching points were introduced by Tomalia at Dow and Dendritech. They are based on the Michael addition of primary amines to methyl acrylate followed by aminolysis of the ester function with excess ethylene diamine [3] (see Scheme 2). The resulting dendrimers are poly(amidoamine)s (PAMAM) and have been prepared to the 10th generation. Details of the reaction conditions and limitations brought about by side reactions have been given [7]. Dendrimers with carbon branch points are more difficult to prepare. They have been synthesized and are known as "Arborols" [8].

Path B in Fig. 2 is the convergent method. It is the outside-inward method, proposed independently by Miller and Neenan [9] and by Hawker and Fréchet [10]. This method is well suited when the branch point is an aromatic ring. As an example of the convergent process we show in Scheme 3 the preparation of poly(benzyl ether) dendrimers. The phenol functionality of 2,5-dihydroxybenzyl alcohol is first protected by Williamson reaction with benzyl bromide to provide the first generation dendron [G-1]-OH. The benzyl alcohol in [G-1]-OH is then converted to the benzyl bromide form [G-1]-Br. This in turn reacts with

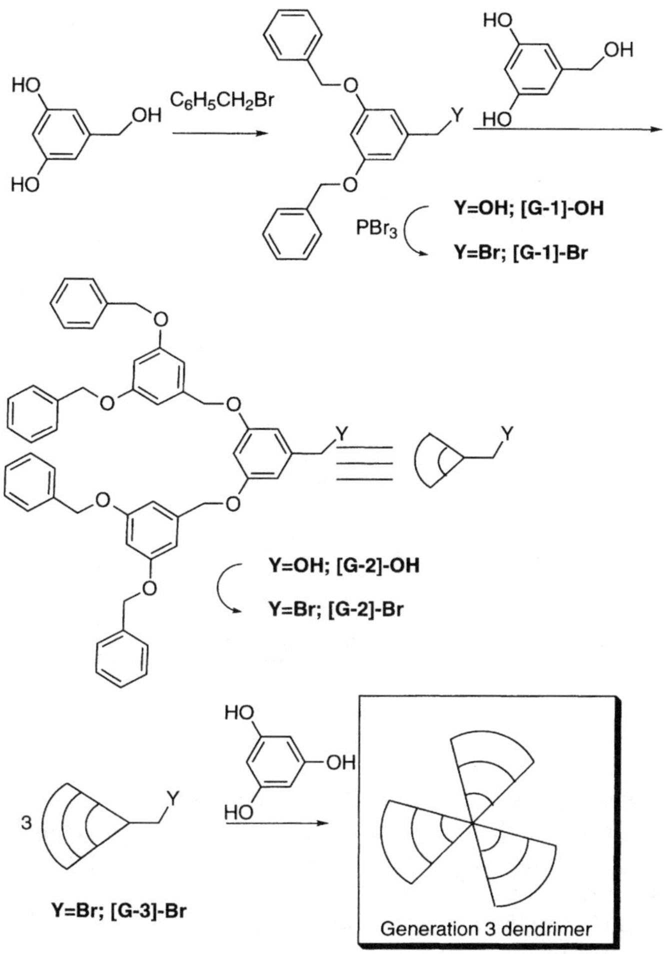

Scheme 3

2,5-dihydroxybenzyl alcohol to yield [G-2]-OH. Scheme 3 illustrates also the synthesis of the generation-3 dendrimer from a generation-3 dendron in a self-explanatory pictorial manner.

The advantages of the convergent method over the divergent method are that each generation requires limited (usually two) reactions per molecule. Furthermore, unreacted material is easily separable because it is substantially different in molecular weight from the product. As a consequence, organic reactions producing lower yields ($\geq$90%) can be tolerated in convergent synthesis. In contrast, the divergent synthesis involves an increasing number of identical reactions per molecule and requires high yield (>99%) reactions in order to minimize imperfect products that are practically unseparable. The main disadvantage of the

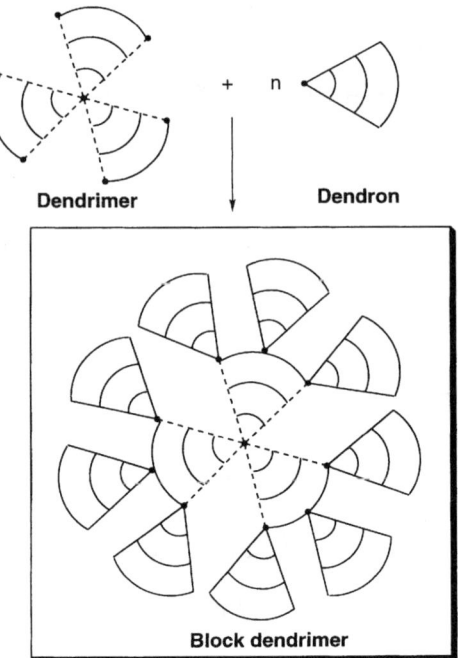

Scheme 4

convergent method lies in the decrease of the reactivity of the focal group which is present at decreasingly lower concentration for higher generation dendrons. Sixth generation dendrons have been prepared and coupled with a trifunctional core to generate a dendrimer of 40,000 D [10].

The convergent method lends itself to accelerated growth. Fréchet et al. have shown how a dendrimer with n end-standing functional groups can be used as a core for reaction with n convergent dendrons each containing a reactive focal group [11]. These are dendrimer-dendron reactions (Scheme 4). In this manner, intermediate generations can be bypassed for an overall gain in time and yield. Moreover, the double growth process allows the formation of radial block dendrimers in one step because the core dendrimer and the peripheral dendrimer can be of different chemical composition [11–13]. The limits of the double growth process have been explored for the poly(benzyl ether) dendrimers [14, 15]. The results suggest that growth is not affected by steric crowding up to the fifth generation. The double growth process has also been applied to the synthesis of chiral dendrimers [16] and poly(phenylacetylene) dendrimer [17], as will be discussed in Sect. 2.2.

This review does not attempt an exhaustive survey of the progress in the synthetic chemistry of dendrimers. A number of reviews have already been dedicated to this rapidly expanding subject [18–30].

The nomenclature according to IUPAC rules has proven particularly unwieldy in the case of dendrimers. Several proposals have been made [25]. The major classes of dendrimers described here are represented in Schemes 1–4 in short-hand form. Other particular dendritic structures will be characterized by reaction schemes or by their branch unit in the text.

2
Analysis of Dendrimers

2.1
NMR Analysis

All standard analytical techniques of organic chemistry are applicable to dendrimers. Of these, NMR spectroscopy is the most powerful for the analysis of low MW dendrimers. However, as with polymers in general, the high molecular weight of dendrimers and the similar composition of all dendritic segments often make it difficult to detect small quantities of irregularities in the dendrimer structure.

Small chemical shift differences between the core or focal group, the interior spacers and the terminal groups are usually observed in the ^{1}H and ^{13}C NMR spectra of dendrimers and these can be used for analysis of low generation dendrimers. As the number of generations increases, signals of the core or focal group become relatively weak and the ratio of signals from the interior spacers and terminal groups reaches an asymptotic limit (usually 1.0) that does not allow accurate quantification of structural imperfections [7, 10, 31] In some cases, it has been possible to observe distinct NMR signals for similar atoms belonging to different generations. For example, four different ^{29}Si resonances are observed in some carbosilane dendrimers [32] and five distinct ^{15}N resonances are reported in DAB (CN)$_{32}$. See Scheme 1 for related structure. In the latter case, intensities of the resonances qualitatively match the theoretical ratio of the number of nitrogen atoms in each shell [33].

2.2
Mass Spectrometry

Various modern mass spectrometric methods have been applied to the analysis of dendrimers. The earliest publication describes fast ion bombardment mass spectrometry (FAB-MS) on polyether dendrimers derived from pentaerythrytol [31] (Scheme 5). The first generation dendrimer containing 12 hydroxyl groups yields the expected molecular ion (M+H$^+$=608 D) and a peak at 490 D identified as the parent dendrimer minus one -CH$_2$C(CH$_2$OH)$_3$ group. The impurity could not be quantified. The second generation dendrimer with 36 hydroxyl groups shows the 2025 D parent peak with no low MW impurities observable in the noise. However, SEC of this generation shows a high molecular weight shoulder

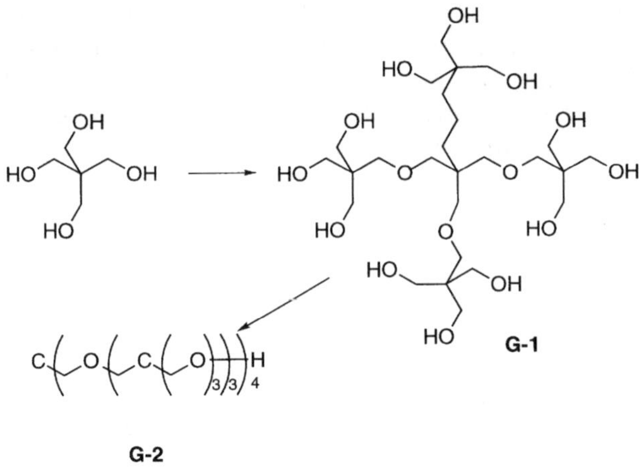

Scheme 5

that might not be detected by FAB-MS. The third generation dendrimer could not be analyzed by this method.

A PAMAM dendrimer of generation four was analyzed by electron spray ionization mass spectrometry (ESI-MS) with detection in the 600–1600 D range [34]. Multiple charged ions with +7 to +11 are observed. After deconvolution two groups of species are recognized. The first group consists of the expected parent peak at 10632 D and six species with MW fitting the relation 10,632–n×114 with n=1 to 6. These are recognized as compounds missing from 1 to 6 $CH_2=CHCONHCH_2CH_2NH_2$ groups out of a total 48 possible terminal groups due to incomplete Michael addition (see Scheme 6). Another series of compounds with MW=10,632–n×60 is due to the formation of cyclic structure in the amidation step leading to 1 to 6 cyclic groups in the outer shell. If it is assumed that the mass spectrometric intensities are proportional to the number of molecules then this fourth generation dendrimer sample contains only about 8% perfect dendrimers and 92% of the dendrimers are deficient in from one to ten terminal amine functional groups. The authors calculated that this molecular mass distribution is representative of an overall yield of 97.5% in the combined two-step reaction to form each generation. A more detailed study of the side reactions leading to these defects was made on lower generation dendrimers by chemical ionization mass spectrometry [7]. Schwartz et al. have advanced the analysis of PAMAM dendrimers by ESI-MS to the tenth generation [35]. Spectra up to generation four have clearly resolved multiple ion bands. For higher generation dendrimers species with different molecular weight and charge number form one envelope which cannot be deconvoluted. A comparison of m/z values with the theoretical MW indicates that the charge (z) on the dendrimer increases with the size of the dendrimer.

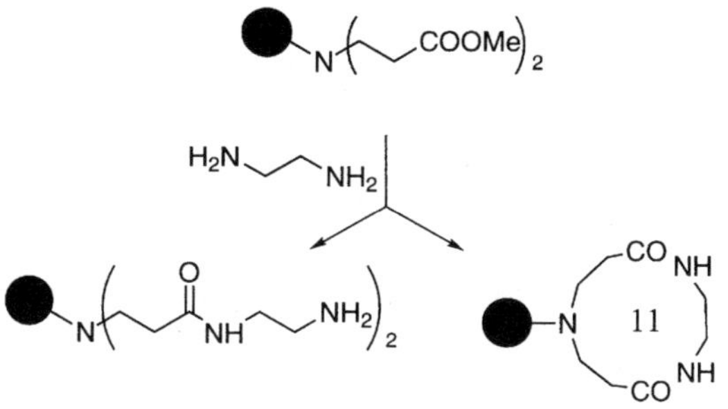

Scheme 6

A complete ESI-MS analysis of generations 1–5 has allowed one to determine quantitatively the extent of the two possible defect introducing reactions in DAB dendrimers [36] (see Scheme 1). The Michael addition is on average 99% complete. During hydrogenation 0.5% of end groups form rings. As a consequence, the fifth generation DAB(NH_2)$_{64}$ contains 23% pure compound. The remainder are known impurities lacking one or more end groups. It should be noted that all these impurities do not increase the polydispersity of the dendrimer beyond Mw/Mn=1.001. An ESI-MS has also been coupled to the outlet of a capillary electrophoresis instrument allowing the mass identification of different elution peaks. When applied to DAB(CN)$_8$ different isomers of defective compounds can be identified [37]. ESI-MS was also used to study a polystyrene-DAB (NH_2)$_8$ hybrid. The series of peaks corresponding to charge z=+4 displayed a spacing equal to 26, characteristic of the styrene unit. Other smaller peaks may be due to dendrimer imperfections or fragmentation [38].

Matrix-assisted laser desorption/ionization time-of-flight mass spectrometry MALDI-TOF appears the newest technique particularly suitable for the study of oligomers and dendrimers because, under appropriate conditions, the parent peak is obtained uncontaminated by fragmentation species. However, in some cases supramolecular clusters have been observed which could be misinterpretated as dimers and higher multiplets [13]. The molecular weight range available reaches 50,000 D with a potential resolution between 0.01 and 0.05%.

Polyesters dendrimers obtained by the divergent method by means of 1,3-dicyclohexylcarbodiimide esterification and catalytic hydrogenation to remove the benzylic protecting group gave essentially the parent molecule peak with a width at half height of 4–8 D [39]. The highest MW observed is 5147 D. SEC data corroborated the MALDI-TOF results with Mw/Mn values between 1.005 and 1.007 for dendrimers with benzyl terminal groups and 1.007 to 1.017 for dendrimers with hydroxyl terminal groups.

MALDI-TOF data have been obtained on a sixth generation poly(benzyl ether) dendrimer containing 64 perdeuterobenzyl terminal groups [40] (see Scheme 3). The correct molecular species (M+K$^+$=13,965 D) with a width at half

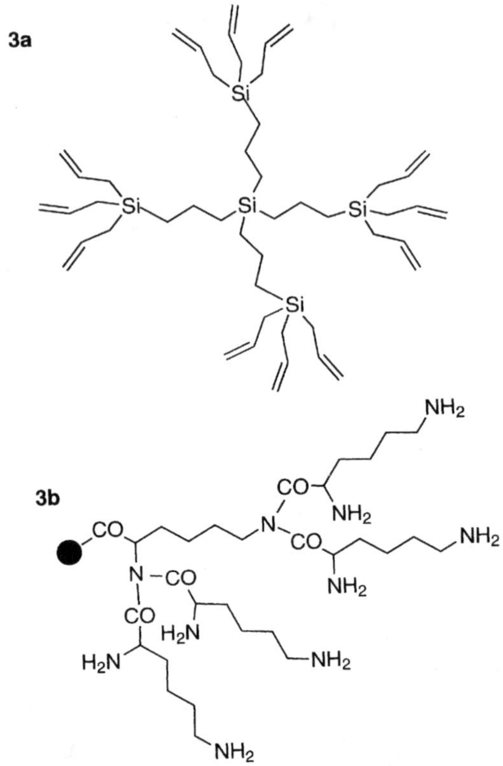

Fig. 3a,b. Other classes of dendrimers: **a** carbosilane; **b** poly(α,ϵ-L-lysine)

height of about 100 D is observed together with an impurity (20%) at about 13,100 D possibly due to acid hydrolysis of a benzyl ether linkage. MALDI-TOF spectra provide clear proof of the superior quality of dendrimers made by the convergent method.

The MALDI-TOF spectrum of [G-3] poly(benzyl ether) dendrimer-poly(ethylene glycol) triblock copolymer shows a broad band of peaks between 4300 and 6100 D with resolution of the individual ethyleneoxide (44 D) units. The MALDI-TOF spectrum of a [G-3] dendrimer with two polystyrene blocks (molecular peak=8073 D) shows material with 6000–11,000 D and a broad band corresponding to material with 2 M+Ag$^+$. SEC can be used to prove that the latter species is indeed an artifact of the mass spectroscopic method. The authors claim almost exact agreement between the polydispersities derived from MALDI-TOF and SEC [40]. This does, however, not leave any room for the unavoidable column spreading in the latter method. Furthermore, anionically prepared low MW polymers have a minimum polydispersity given by (1+1/DP) [41].

Frey and coworkers critically discuss MALDI-TOF spectra of carefully chromatographed G2 and G3 poly(carbosilane) dendrimers with 36 and 108 terminal allyl groups [42] (see Fig. 3a for related structure). They quoted these samples to have apparent MW distributions in the range of 1.05 to 1.06 by SEC. The G2 sample has two main peaks with MWs for dendrimers with 36 and 34 allyl end groups, respectively. The peak intensities indicate that 80% of the parent G1 dendrimer is completely hydrosilylated and that the remaining 20% is hydrosilylated 11 out of 12 times. This is equivalent to a 98.3% reaction yield. In the case of the G3 dendrimer a wider range of dendritic species is observed. The most abundant species has 106 allyl end groups (35 out of 36 allyl group are hydrosilylated) but species with 100 to 108 allyl groups are also present. A dendrimer with 100 allyl groups is the product of 32 out of 36 hydrosilylations and it contains 4 allyl groups belonging to the G2 shell. Nevertheless, from a polymer perspective, these samples are very monodisperse, Mw/Mn being of the order of 1.01. The allyl groups in the carbosilane dendrimers have also been converted to primary alcohols via hydroboration oxidation. In the MALDI-TOF spectra of the hydroxylated G2 compound traces of lower MW material together with the main 36 and 34 hydroxyl containing dendrimers are observed. In the case of the hydroxylated G3 dendrimer the MALDI-TOF spectrum almost completely reflects the MW pattern of the parent allylic G3 dendrimer.

Recently, MALDI-TOF results on poly(aryl ether) dendrimers allowed the detection of a small impurity characterized by an extra -C_6H_4S- group (108 D) and confirmed upon oxidation by a small extra species with a -$C_6H_4SO_2$- group (140 D) in the sulfone form [43]. Two impurities are detected in the third generation dendrimer with one and two extra -C_6H_4S- groups, respectively. However, it was not possible to quantify the amount of these defects. Fourth generation dendrimers (18,212 D) could not be analyzed by the MALDI-TOF method.

Finally, the MALDI technique was used for the characterization of poly(phenylacetylene) [17]. These dendrimers were synthesized through a double growth process (Scheme 7). The most advanced application involved the reaction of a third generation dendrimer with 16 reactive terminal groups (G-3-16) with a third generation dendron. The aim was to skip directly to the sixth generation dendrimer (G-6-256) with 256 end groups. The reaction occurred in 86% yield after optimization of the conditions. MALDI-TOF analysis revealed the major compound to be the desired dendrimer contaminated only with a small amount of a compound having 16 fewer end-groups. Chemical ionization, infra-red laser desorption proved the superior method for generation 0 to 2 dendrimers with MW up to 3631. MALDI-TOF is more successful with generation 3 and 4. In all MALDI spectra dendrimer dimers and sometimes dendrimer trimers are observed, despite very low sample to matrix ratios. The spectra are also somewhat complicated by partial fragmentation of these dendrimers during the analytical process [44, 45].

New reports indicate that MALDI-TOF is beginning to be used on a routine basis (like NMR) to monitor the synthesis and modification of each batch of den-

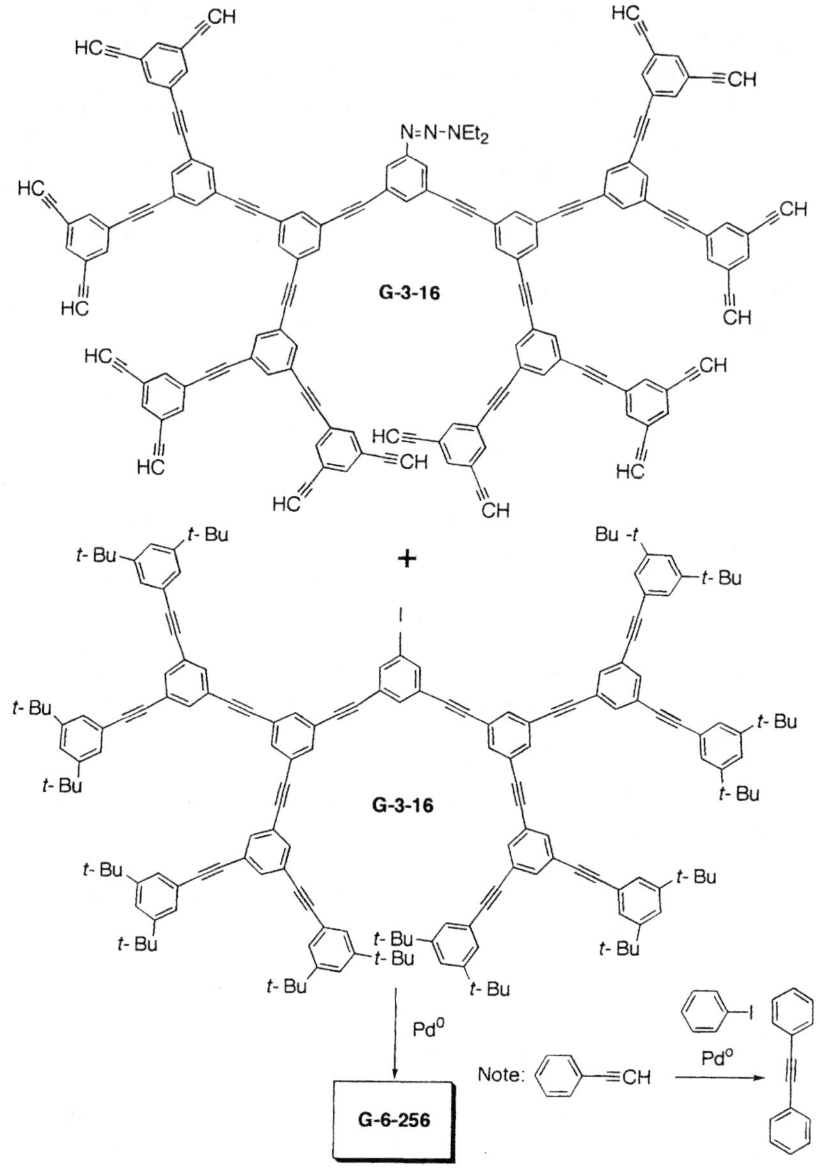

Scheme 7

drimers [13, 46]. Hopefully, a comparison of the analytical capabilities of mass spectrometric methods and SEC will be attempted. The least such a comparison will accomplish is to provide for an absolute method to evaluate the effect of column spreading in the determination of MW distributions (MWD) by SEC.

2.3
Size Exclusion Chromatography

SEC by itself is not an absolute MW determination method but the analysis of the elution peak has been used extensively for estimating the molecular purity of dendrimers. If the shape of the elution peak of a size exclusion chromatography experiment is Gaussian, the total dispersion, σ_T of the curve is given by the sum of squares [47]:

$$(\sigma_T)^2 = (\sigma_c)^2 + a(\sigma_d)^2 \tag{1}$$

where $(\sigma_c)^2$ is the sum of the squares of dispersions due to the injector, column spreading and detector, σ_d is the dispersion due to the polydispersity of the sample, and a is the experimental slope of the calibration of the elution volume, V_e, against MW

$$V_e = B - a \ln MW \tag{2}$$

$(\sigma_d)^2 = \ln(Mw/Mn)$. Mw/Mn is a measure of polydispersity of the sample. Since experimentally σ_T has almost always been substituted for σ_d, SEC has rarely given the true MW distribution, the deviation being larger the narrower the MWD of the sample. This is especially the case for near-monomolecular dendrimers. Methods have been described for obtaining the true MW distribution of narrow MWD polymers by SEC [47]. Furthermore, when the calibration is performed with a set of linear polymers, it can only be used directly with the same type of linear polymers. When other polymers are analyzed, the principle of universal calibration is invoked.

$$V_e = B' - a' \ln [\eta]M \tag{3}$$

Analysis of polydispersity can still be made but only when the exponent in the Mark-Houwink relation $[\eta] = KM^a$ is identical for the calibrating polymer and the new polymer. As will be shown in the next section this is clearly not the case for highly branched dendrimers. The validity of the universal calibration principle for dendrimers has been questioned [48]. SEC analysis with multiple detectors, especially with a low angle laser light scattering detector, obviously removes many of these objections. The results of such experiments [48] will be described in the next section. SEC does not have the capability for analyzing small defects in a dendrimer sample. However, it is a quick method for quantifying the monomer-dimer and higher multiple content.

3
Conformation of Dendrimers

3.1
Theoretical Models

The first model for the conformation of a dendrimer was proposed by de Gennes and Hervet [49]. It is based on a modified Edwards self-consistent field for dendrimers with large linear segments between consecutive branch points in a good solvent. It places the spacers of consecutive generations in concentric shells. The segment density is rather low in the center and increases parabolically (according to r^2, r being the distance from the center) to a generation for which the segment density approaches unity. As a consequence, the radius of dendrimers, R, increases according to $M^{0.2}$ up to the steric saturation generation. The radius of incomplete dendrimers grown beyond this generation increases according to $R \sim M^{1/3}$, i.e., like constant density objects. This model is related to the Maciejewski box [50] and may have merit when phase separation forces the end-standing groups to lie on the surface of the dendrimer or in case the spacers are stiff and cannot fold back.

Simulations and models developed more recently for dendrimers with flexible spacers provides a quite different picture. Generally, the segment density decreases from the center of the dendrimer to zero at the surface [51–54]. In particular, some studies suggest a small minimum in the segment density near the core [52], others show evidence for an extended plateau region of near constant segment density [52, 54]. The end-standing groups are found throughout the entire dendrimer volume [52, 54] as a result of backfolding of the spacers, although most of them are found in the large volume near the dendrimer surface. Segments of lower generation spacers are more localized in the interior of the dendrimer [54] and have a stretched conformation [53].

The dependence of the radius of gyration, R_g, on the mass of the dendrimer is complex. For small generation dendrimers the exponent ν in

$$R_g \sim M^\nu \tag{4}$$

is 0.5 [51], 0.4 [52], and 0.5 [53]. For high generation dendrimers the limiting value for ν is 0.22 [51], 0.24 [52], 0.20 [53], and 0.3 [54]. The latter exponents are comparable with ν=0.25 for randomly branched polymers [55, 56]. There is general agreement on the increasing sphericity of dendrimers as the number of generations increases [57] and on a limited overlap of different dendrons inside the dendrimer volume [54, 58]. Based on a calculation of the intrinsic viscosity of model dendrimers, Mansfield and Klushin concluded that the hydrodynamic radius, R_η, of low generation dendrimers is smaller than R_g but that the reverse is true for high generation dendrimers [59].

The variation of the size of the dendrimer with the number of segments, n, between two branch points, at constant architecture (generation), has been considered. Lescanec and Muthukumar found $R_g \sim n^{0.5}$ [51]. Others have established that the good solvent limit for dendrimers and linear polymers is the same, i.e., $R_g \sim n^{3/5}$ [60, 61].

Many of these conclusions on the conformation of dendrimers are in qualitative agreement with the known behavior of other types of branched polymers. For example, the preferred stretching of the lower generation segments in a dendrimer is comparable to the increased expansion of the interior segments of star polymers with many arms [62, 63], and with the preferential expansion of the backbone segments in comb polymers. The limited overlap of dendrons is comparable with the limited long range interaction of the two halves of a linear polymer in a good solvent or of the different blocks in a block copolymer. The ratio $R_\eta/R_g<1$ predicted for low generation dendrimers is the ratio found in linear polymers and star polymers with few arms [64–66]. The ratio $R_\eta/R_g>1$ is, however, observed in star polymers with many arms. It would be interesting to see whether for large dendrimers $R_\eta/R_g=1.29$, the value for equal density spheres, or approaches $R_\eta/R_g=1.00$, the asymptotic value for a hollow sphere.

The effect of the quality of the solvent on the dimensions of dendrimers has been considered [54]. The size of the dendrimers increases with an increased interaction with solvent. However, in contrast to linear polymers or regular star polymers, the exponent ν in Eq. (4) was found to be independent of the quality of the solvent for high MW dendrimers [54].

The structure factor for dendrimers has also been calculated. The structure factor provides a description of the relative scattering intensity from a collection of scatterers as a function of the scattering vector $q=(4\pi/\lambda)\sin(\theta/2)$. λ is the wavelength of the radiation in the medium and θ is the angle between incident and scattered radiation. The calculated structure factor is necessary for comparison with experimental scattering curves. The structure factor of dendritic polymers has been calculated on the assumption of Gaussian statistics, i.e., with "ghost" segments [67, 68]. Burchard also studied the dynamic structure factor and established the limiting value for $R_h/R_g=1.023$ for high generation dendritic polymers. Structure factors can also be computed from simulated segment density distributions [52, 53].

3.2
Experimental Dimensions of Dendrimers

3.2.1
Radius of Gyration

The earliest systematic study of the radii of gyration of dendrimers was performed on poly (α,ε-L-lysine) dendrimers up to generation 10 (see Fig. 3b for the branch unit). The polylysine dendrimers are atypical in so far as only one of each consecutive generation is placed end-standingly. Radii of gyration between 0.8 nm (generation 3, MW=1900) and 4.3 nm (generation 10, MW= 2.3×10^5) have been obtained [69]. These small dimensions attest to the com-

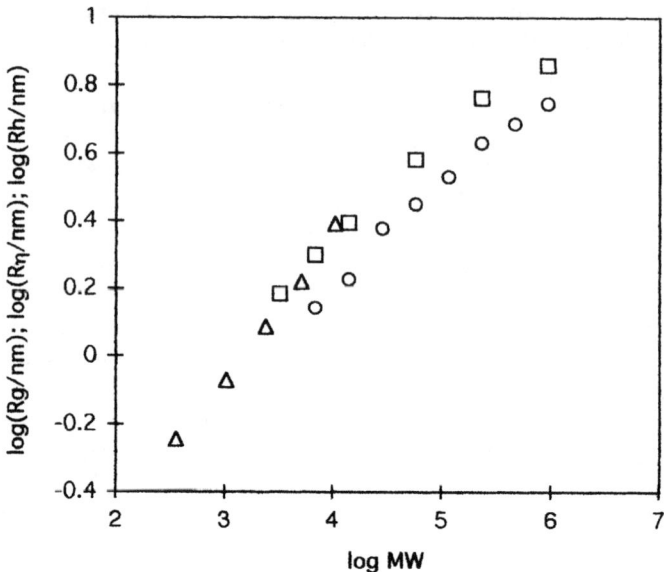

Fig. 4. Dependence of radii on the molecular weight of PAMAM dendrimers in methanol; ○ R_g; △ R_η; □ R_h

pactness of the dendrimers. The exponent ν in Eq. (4) is equal to 1/3, a value expected for equal density spheres.

No full account of a systematic study of dendrimer dimensions is at present available in spite of the great interest. Preliminary values of R_g, obtained by SANS on dilute PAMAM solutions in CD_3OH, have recently appeared [70]. The MW dependence is shown in Fig. 4. The dependence of R_g on MW is stronger for low MW dendrimers (ν=0.36) than for the high MW dendrimers (ν=0.20). This latter exponent is probably a minimum value because the nominal rather than the measured MWs have been used in Fig. 4 and synthetic problems suggest that the real MWs are lower than the nominal ones. The low exponent for the high MW dendrimers is in the range predicted by various models discussed in Sect. 3.1. The 1/3 slope predicted for high MW dendrimers synthesized under conditions of saturation substitution [49] is clearly not observed.

Some data have also appeared for the first five generations of the smaller $DAB(CN)_x$ and $DAB(NH_2)_x$ dendrimers (see Scheme 1 for structure). For $DAB(CN)_x$ in acetone-d_6 SANS yields ν=0.31 (generations 2–5). For $DAB(NH_2)_x$ in D_2O the low MW dendrimers follow $R_g \sim M^{0.30}$ [71]. In general, it is difficult to determine R_g of low MW dendrimers accurately [70]. Furthermore, all dendrimers studied have ionizable groups and may act like polyelectrolytes. Therefore solvent conditions need to be carefully controlled and specified.

3.2.2
Hydrodynamic Radii

More extensive data are available for the hydrodynamic radii of dendrimers than for the radii of gyration, because they can be derived from intrinsic viscosity measurements according to

$$R_\eta = (3[\eta]M)^{1/3}(10\pi N_A)^{-1/3} \tag{5}$$

or from the translational diffusion coefficient, D_o, according to

$$R_h = kT(6\pi\eta_s D_o)^{-1} \tag{6}$$

where N_A is Avogadro's number, k is the Boltzmann constant, and η_s the solvent viscosity. R_η and R_h are the radii of the equivalent sphere of constant density. Equations (5) and (6) may not apply to the smallest dendrimers whose dimensions are comparable to the dimensions of the solvent molecules. Furthermore, R_η and R_h cannot be compared directly to the predictions of the models because the ratio R_η/R_g and R_h/R_g are not independent of the mass and branching architecture of the dendrimers [52]. The earliest hydrodynamic radii have been obtained on poly (α,ε-L-lysine) dendrimers [72] (see Fig. 3b). For these dendrimers it was found that $[\eta]=2.5$ ml/g independent of generation. It follows from Eq. (5) that $R_\eta \sim M^{1/3}$, this exponent is expected for equal-density spheres and is unusual for dendrimers and ascribed to the asymmetric nature of the branching pattern in poly (α,ε,-L-lysine) [73]. The experimental ratio R_η/R_g for these dendrimers varies between 1.14 and 1.0.

Hydrodynamic radii of poly(benzyl ether) dendrimers are shown in Fig. 5. Data for monodendrons with a hydroxyl focal group and tridendrons fall on the same curve. The value of the exponent v in Eq. (4) is 0.46 of low MW. At high MW it is 0.26 [48]. Data on low MW linear polystyrene in benzene [74] have been included in Fig. 5 for comparison. They highlight the little difference in the actual values of the hydrodynamic radii of linear polystyrene and low MW poly(benzyl ether) dendrimers. Deviations are observed only when MW$>5\times10^3$. Furthermore, the MW dependence of the radii of polystyrene and poly(benzyl ether) dendrimers are the same at low MW. This indicates that it remains impossible to draw major conclusions about the conformation of the low MW dendrimers from their global properties. The low values of the hydrodynamic radii of the high MW dendrimers, on the other hand, attest to their compact conformation. A similar transition to more compact dendrimers has recently been shown in a direct comparison of linear and dendritic poly(benzyl ethers) [75].

Recently, Stechemesser and Eimer published hydrodynamic radii obtained from translational diffusion coefficients measured by means of recovery after photobleaching experiments on selected generations of PAMAM dendrimers [76]. The values of R_h obtained in methanol are compared with R_g data on the same dendrimers in Fig. 4. It can be calculated that the ratio $R_h/R_g \sim 1.4$ for all but the tenth generation dendrimer. This ratio is somewhat higher than expect-

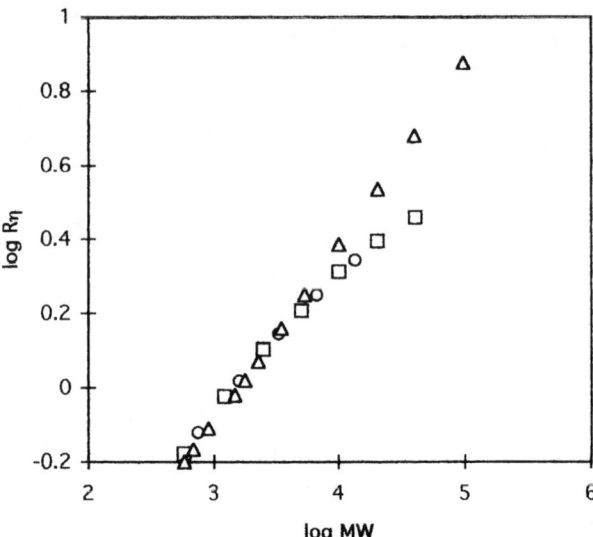

Fig. 5. Dependence of the hydrodynamic radius of poly(benzyl ether) dendrimers on molecular weight; ○ monodendron; ◻ tridendron; data on low MW polystyrene (△) in a good solvent are included for comparison

ed especially for the low MW dendrimers. The same study [76] reported results in three different solvents of decreasing polarity, water at pH 8.0 with 0.1 mol/l NaCl, methanol and n-butanol. At low MW up to the fourth generation, $R_\eta \sim M^{0.4}$ and the dimensions of the dendrimers vary little with the solvent quality. This behavior is comparable to that of low MW organic compounds and oligomers. For example, Yamakawa et al. showed that poor and good solvent distinctions disappear for polystyrene oligomers with MW<10^4 [74]. Although only three high MW generations (6, 8, and 10) have been studied it is clearly established that $R_\eta(H_2O) > R_\eta(MeOH) > R_\eta(n\text{-BuOH})$. Such behavior is typical of polymers and reflects the influence of solvent on long-range interactions. It is remarkable that hydrodynamic radii of PAMAM dendrimers in MeOH based on the ammonia core are in good agreement with the hydrodynamic radii of the dendrimers based on ethylenediamine core [3, 77]. These results have also been included in Fig. 4.

Intrinsic viscosity results on DAB(CN)$_x$ and DAB(NH$_2$)$_x$ have been quoted in various publications [5, 71, 78]. Unfortunately, the solvent and experimental conditions are not specified. R_η is found to increase by a constant increment for five generations. In a double logarithmic plot of R_η against MW it is found that the slope decreased from slightly larger than 0.4 at low MW to about 0.3 at high MW. Intrinsic viscosities of four generations of carbosilane dendrimers [79] (see Fig. 3a for structure) and four generations of poly(dimethylsiloxane) dendrimers [80] lead to $R_\eta - M^\nu$ dependencies with $\nu = 0.4$ for these low MW dendrimers.

The studies of R_g as well as R_η and R_h of a number of dendrimer systems establish a transition from $v=0.45+0.05$ to $v=0.25+0.05$ dependence on MW as shown in Figs. 4 and 5. This result is qualitatively consistent with the simulation results and has several consequences. Because $[\eta] \sim R_\eta^3/M$, the value of $[\eta]$ will increase with the MW of the dendrimer when $v>1/3$, be independent of MW when $v=1/3$, and decrease with MW when $v<1/3$. This last, unusual, behavior of high MW dendrimers has been pointed out [48]. The maximum in the value of $[\eta]$ at intermediate MW resembles the behavior of star polymers when values of $[\eta]$ are plotted against f, the arm functionality at constant arm MW [62, 81]. It is also observed in comb and graft copolymers with increasing grafting density, other variables being kept constant [82].

Furthermore, if the dependence of R_g on MW of dendrimers is correctly represented by the data of Figs. 4 and 5 then there cannot be constant incremental increase of the dimensions of consecutive generations of dendrimers but over a narrow MW range [48]. For example, in the case of PAMAM dendrimers the increase in the radius is about 0.6 nm per generation, a value that approaches the fully extended spacer length estimated at 0.87 nm. In the case of poly(benzyl ether) dendrimers the increment per generation is about 0.45 nm for an estimated fully extended spacer of 0.6 nm. The increase of the dimensions of a dendrimer over that of the previous generation cannot be assigned solely to the terminal generation because the interior spacers stretch and chains can backfold. It is therefore possible that the increment of dimension is larger than the fully extended added spacer length.

Another consequence of the size-mass relations of dendrimers shown in Figs. 4 and 5 is that the average segmental volume fraction decreases with increasing generation at low generation when $v>1/3$ but increases at high generation when $v<1/3$. This may be physically observable by means of partial specific volume measurements. Indirect indication is found in the minimum in the refractive index increment of dendrimers [18, 48].

A further consequence of the experimental size-mass relation of Figs. 4 and 5 is that, assuming that all terminal groups lie on the surface of a sphere, the surface area per terminal group is nearly constant for low MW dendrimer where $R_g \sim M^{1/2}$. However, even for moderately large dendrimers, R_g depends less strongly on MW and the area per end-group decreases rapidly with increasing MW. The area per end group can reach the van der Waals dimension unless alleviated by chain backfolding. The theoretical surface area per end group of a tenth generation PAMAM is of the order of 1.6–2.2 nm² [18].

3.2.3
Effect of Solvent on Dendrimer Dimensions

The decrease of PAMAM dimensions with decreasing polarity of solvent have been mentioned [76]. Newkome and coworkers have studied the dimensions of dendrimers with different terminal groups [83, 84]. Dendrimers with CH_2OH end groups have the same dimension in acidic, neutral, and basic aqueous solutions. However, dendrimers with COOH end groups have a 35% larger hydrodynamic radius in neutral or basic than in acidic solution. Dendrimers terminated with CH_2NH_2 groups have values of R_h 35% larger in neutral or acidic conditions than in a basic medium. These documented expansions are clearly due to electrostatic repulsion. Because measurements have been reported only at one concentration ($c \leq 10^{-3}$ mol/l) it is not clearly established that ionizable dendrimers act as polyelectrolytes [72]. The observation that ionic strength has little effect on dimension suggests that considerable screening occurs at the concentration of the measurements.

Dubin et al. measured [η] and D_o of seven generation of carboxylated PAMAM dendrimers in $NaNO_3$, NaH_2PO_4 aqueous buffer at pH 5.5 and 0.38 mol/l in order to minimize electrostatic interaction [85]. The values of $R_η$ of the 2.5 and 3.5 generation dendrimer agree closely with those of the corresponding methyl ester dendrimer in MeOH. Rather surprisingly, the quoted values of [η] do not go through a maximum expected for the higher generation dendrimers.

4
Dendrimer-Polymer Hybrids

4.1
Polymeric Dendrimers

The dendrimers discussed in the previous sections have short spacers containing usually between two and eight bonds between neighboring branch points. It is, however, also possible to work with polymeric chains rather than short spacers albeit with some some sacrifices to regularity and symmetry. Arborescent [86] and combburst [87] polymers contain linear polymer chains at the core and in each generation. These polymers have several trifunctional branch points along each polymer chain, rather than the end-standing branch points of the classical dendrimer. The branching functionality, considered per polymer chain is usually high (7–15) compared to 2 to 4 in a dendrimer. As a consequence the molecular weight and number of branch points increases very rapidly with each generation (see Scheme 8a and Table 1). The arborescent and combburst polymers are expected to have a high segment density and near spherical conformation. At this point there is no experimental evidence on the segment density distribution with a possible shell like structure or extensive backfolding.

Polymeric dendrimers have been synthesized by anionic, cationic and free radical polymerization. Gauthier and Möller and their collaborators used two alternating reactions of chloromethylation of polystyrene and grafting with polystyryllithium to create arborescent polystyrenes [86] (see Scheme 9a). Tomalia and co-workers used hydrolysis of poly(ethyloxazoline) to create secondary amines in poly(ethylene imine) onto which poly(ethyloxazoline) chains are grafted by means of the living oxazolinium end group [87] (see Scheme 9b). In the original work [86, 87] the individual polymer chains are kept small (DP=20

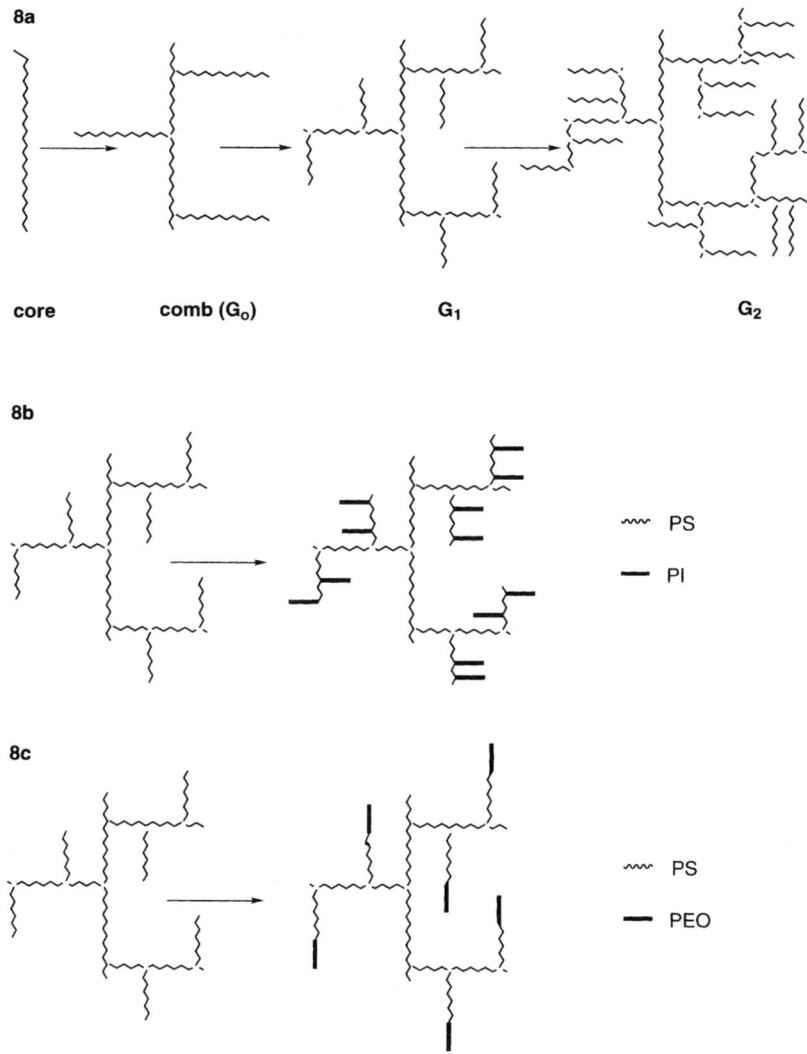

Scheme 8

or 50) and up to one third of the monomer units are converted into branch points. The spacers between two neighboring branch points are therefore small and stiff and do not have flexible polymer characteristics. Both laboratories noted that the grafting efficiency decreased with increasing generation. This is ascribed to steric congestion [86, 87]. It is also possible that there is strong thermodynamic repulsion between the highly branched polymer and linear polymer preventing a fast and complete grafting reaction. In later work the MW of

Table 1.A Characterization of arborescent polystyrene [88][a] (polymeric dendrimer)

Sample	$M_w(br) \times 10^{-3}$	M_w	M_w/M_n[b]	f_g	R_v (nm)
Core	–	9.6×10^3	1.07	–	–
Comb	10.2	1.56×10^5	1.15	14	9.8
G1	10.1	1.86×10^6	1.11	170	24.0
G2	9.5	1.57×10^7	1.80(?)	1460	45.6

[a] S10-series
[b] Apparent MWD by SEC

Table 1.B Characterization of Combburst poly(ethylene imine) [89][a] (polymeric dendrimer)

Sample	$M_w(br) \times 10^{-3}$	M_w	M_w/M_n[b]	f_g
Core	–	1×10^3	1.05	
G0	0.8	2.5×10^3	1.22	5
G1	7.7	1.38×10^5	1.34	26
G2	7.1	1.08×10^6	1.47	176
G3	(21.0)	1.04×10^7	1.20	745

[a] Mw of poly(ethylene imine) from MALLS
[b] Apparent MWD by SEC

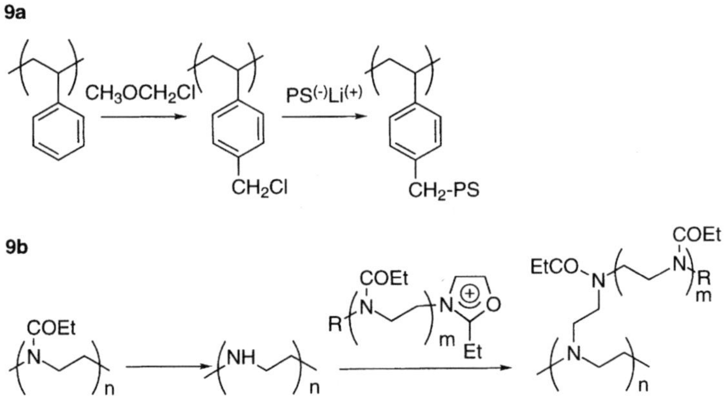

Scheme 9

the polymer chain has been increased to 30,000 D while keeping the branching functionality per chain between 5 and 15. In these new polymeric dendrimers, each spacer between neighboring branch points is a small polystyrene chain with several Kuhn steps [88–90]. It is worth noting that both research groups show that the apparent MWD remains reasonably narrow (Mw/Mn~1.2) through the preparation of several generations (see Table 1). Arborescent polybutadienes have also been prepared. A branch MW=10,000 is equivalent to about 200 monomers, 6% of which are converted to branch points [91].

The physical properties of these polymeric dendrimers have been studied to some extent. Intrinsic viscosity measurements combined with MW afford values of R_η according to Eq. (5). Alternatively, the translational diffusion coefficient leads to R_h according to Eq. (6). These equations may well be applicable, since it is observed that R_η and R_h scale with the 1/3 power of MW in support of the equal density hard-sphere assumption [88].

Comparison of R_η or R_h of two consecutive generations yields an apparent shell thickness. It is significant that the shell thickness of each generation increases with increasing generation at constant branch MW (see Table 1). The apparent thickness of the outer shell is always larger that the unperturbed end-to-end distance of the polymer chain. In some cases the value approaches the dimensions of the fully stretched chain. This appears to be a good indication that the addition of a new generation also enlarges the radius of the interior parent dendritic polymer. A full proof would require measurements of R_g on an inner/outer labeled polymeric dendrimer. Because the branching process is random, it is unlikely that the polymeric dendrimers have strongly segregated generational shells. Increasing crowding in each generation and the resulting tendency of *all chains* to stretch should, however, introduce some radial segregation of the material in consecutive shells.

The intrinsic viscosities of the dendritic polymers are extremely small compared with those of linear polymer of the same MW [86, 91]. Furthermore, the dendritic polymers expand very little in going from a θ solvent to a good solvent [92]. This is to be expected. When steric congestion forces the polymer chains to expand in a θ solvent, further expansion in a good solvent is limited. In this regard it is important to note that the θ condition must be carefully specified. It is known that branched polymers have different θ conditions to the linear counterpart [93].

Gauthier and coworkers have expanded the synthesis of dendritic polymers to dendritic graft copolymers in which the inner generations are polystyrene and the outer generation consists of polyisoprene [94] (Scheme 8b). They have also prepared amphiphilic graft copolymers in which the inner generations are polystyrene and the polymer chains in the peripheral shell are extended by poly(ethylene oxide) (PEO) [95] (Scheme 8c). In order to accomplish this, the last PS generation has been initiated with a lithium compound carrying a protected hydroxyl group. After deprotection, the hydroxyl group was activated with the potassium counterion for the polymerization of ethylene oxide. Comparison of the hydrodynamic radii before and after the extension with PEO indicated a rather small expansion due to the PEO chains, in spite of possible internal phase separation of the PS and PEO units in the polymer.

Monolayers of arborescent polystyrenes have been investigated by scanning force microscopy [90]. Dense polymers with a small branch spacing (M_b~500 D) are nearly spherical and become more spherical on annealing, thereby causing the break-up of the film. On the other hand less densely branched polymers

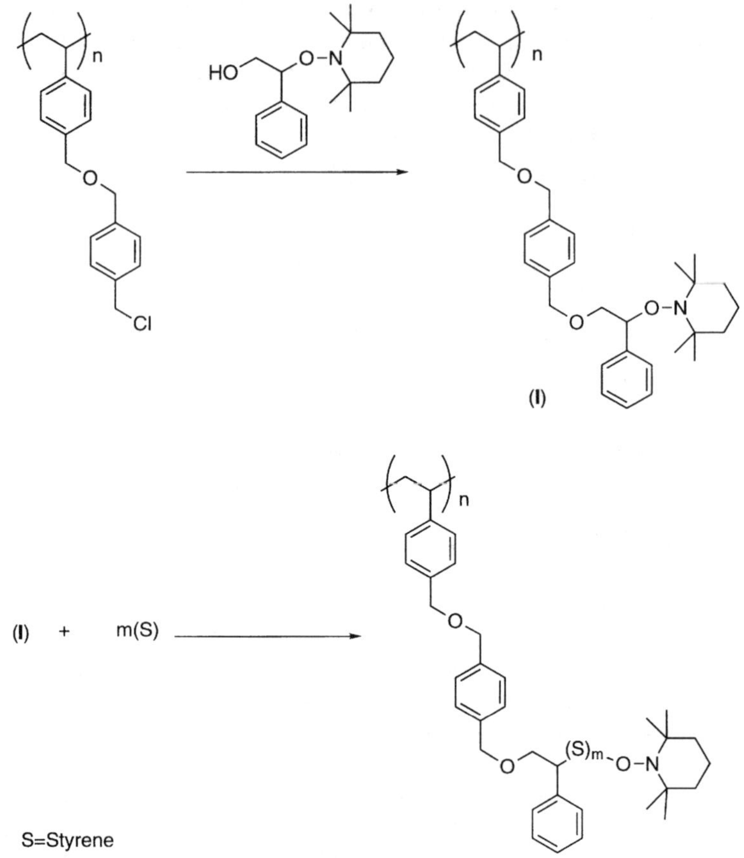

Scheme 10

(M_b~2000 D) are flattened on deposition and are unchanged on annealing, leaving a largely intact film.

Fréchet and coworkers recently described how living free radical polymerization can be used to make dendrigrafts. Either 2,2,6,6-tetramethylpiperidine oxide (TEMPO) modified polymerization or atom transfer radical polymerization (ATRP) can be used [96] (see Scheme 10). The method requires two alternating steps. In each polymerization step a copolymer is formed that contains some benzyl chloride functionality introduced by copolymerization with a small amount of *p*-(4-chloromethylbenzyloxymethyl) styrene. This unit is transformed into a TEMPO derivative. The TEMPO derivative initiates the polymerization of the next generation monomer or comonomer mixture. Alternatively, the chloromethyl groups on the polymer initiate an ATRP polymerization in the presence of $Cu^I Cl$ or $Cu^I Cl$-4,4' dipyridyl complex. This was shown to be the case for styrene and *n*-butylmethacrylate. SEC shows clearly the increase in molecu-

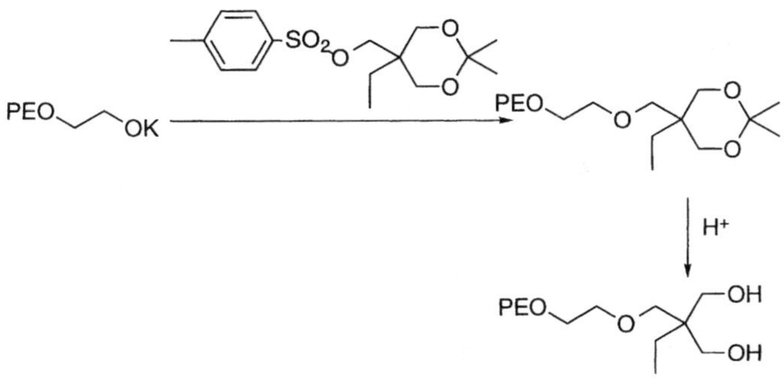

Scheme 11

lar weight of the polymer with each generation and comparison between the apparent MW from the universal calibration and absolute MW from SEC/MALLS proves the highly branched nature of the final product.

Truly polymeric dendrimers require an end-standing branch point with a specific functionality. This has been realized in the case of poly(ethylene oxide) [97]. Starting from a trifunctional anionic initiator a three-arm star is formed. The living potassium alcoholate end groups are reacted with 2,2-dimethyl-5-ethyl-5-tosyloxymethyl-1,3-dioxolane (see Scheme 11). Hydrolysis of the 1,3-dioxolane ring and activation of the hydroxyl groups with diphenylmethylpotassium initiates the second generation ethylene oxide. The resulting polymer has nine branches. The Gaussian chain shrinkage factor [98] $g=<s^2>/<s^2>_{lin}=0.605$ for this polymeric dendrimer. This value is intermediate between that of a regular three-arm star ($g=0.778$) and a regular nine-arm star ($g=0.309$). Experimentally, $g'=[\eta]/[\eta]_{lin}\approx0.5$ has been found from SEC, assuming universal calibration to be valid. This ratio seems low because in lightly branched regular stars $g'\approx g^{1/2}$ [81]. An alternative method has recently been used to make higher generation polymeric dendrimers [99]. A true copolymeric dendrimer consisting of a central six-arm polystyrene star and 12 peripheral PEO arms has also been reported [100].

Similar polymers, slightly less perfect, are the umbrella star copolymers [101]. These polymers are based on a central polystyrene star with 25 arms. An average of five polybutadiene or poly(2-vinylpyridine) branches are grafted onto the end of each arm. Since these polymers are models for block copolymer micelles their properties have been studied in selective solvents. In particular, the PBd-PS umbrella-star copolymers are monomolecularly dissolved in non-solvents for the core-forming polystyrene.

4.2
Dendrimers on Polymers

4.2.1
Dendrimers on Flexible Polymers

In general, the polymerization of macromonomers leads to low MW materials and the DP of the backbone is often similar to the DP of the macromonomer. Consequently, these polymers are nearly spherical and resemble star polymers in dilute solution. However, it has been shown recently that high MW poly(macromonomers) have a rodlike conformation in dilute solution [102, 103]. In the particular case studied, polystyrene macromonomers linked by an ethylene oxide spacer to an end-standing methacrylate group have been polymerized to DP≈1000. It is concluded that the central methacrylic backbone adopts a stretched conformation due to the constraining long and bulky polystyrene side chains. The polymer resembles a "bottle brush" [104]. X-ray scattering experiments in semi-dilute solution further revealed a lyotropic phase consistent with the high aspect ratio of the polymer [105].

The second impetus to consider monomers with dendritic side groups comes from the work of Percec et al. on monomers with tapered side chains, e.g., polymerization of 3,4,5-tris (4'-dodecyloxybenzyloxy) benzoic acid ethylene glycol (n=1,2,3,4) methacrylates

$$n=1,2,3,4 \tag{1}$$

yields polymers which self-assemble into a tubular supramolecular architecture with radii between 27 and 36 Å [106]. Dendrimers on polymers may therefore be expected to influence the backbone conformation and be a possible source of new polymeric materials. The first macromonomers carrying dendrimers have been prepared by Fréchet and Hawker who attached their poly(benzyl ether) dendrimer [G-3], [G-4], and [G-5] (see Scheme 3) to a styrenic functional group [107] thereby placing seven bonds between the backbone carbon and the focal point. Copolymerization with styrene produced high MW material. The copolymer composition is equal to the monomer composition, implying that the dendritic substituent does not affect the reactivity of the styrenic double bond. Styrenic monomers with poly(benzyl ether) dendrimer of general structure

$$[G-1], [G-2], [G-3] \tag{2}$$

have been homopolymerized radically in concentrated solution in toluene or in the bulk [108]. The higher generation monomer requires higher polymerization temperatures and a higher radical initiator concentration. Yields are over 50%. Analysis by SEC (polystyrene equivalent) indicates that the polymers with larger dendrons appear to have a lower degree of polymerization. This apparent result can be due in part to the bulkiness of the substituents. More troublesome are the observed wide MWDs (in bulk MWD>10) which suggest that extensive chain transfer may occur in the bulk polymerization. Absolute MW determinations of two fractions based on the [G-2] dendrimer indicate MWs of 2×10^5 and 2×10^6 with very small radii of gyrations and hydrodynamic radii. A second generation dendrimer has also been linked to a styrenic moiety by a urethane bond [109]

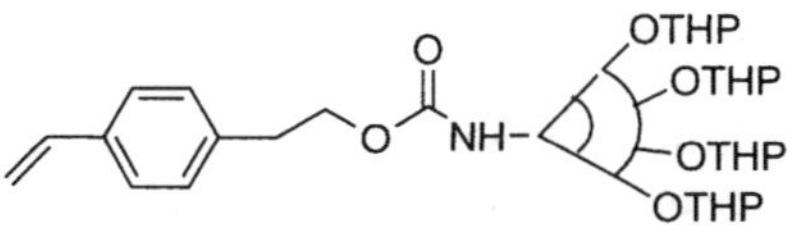

(3)

and polymerized to an apparent (SEC) DP=51. After hydrolysis of the peripheral tetrahydropyran groups a water soluble polymer is expected [109].

Schlüter and coworkers have made systematic studies of the polymerizability of methacrylic monomers modified by Fréchet type dendrons of generations 1–3 [109]. They observe that clean polymerization is limited to methacrylate monomer with generation 1 dendron when the dendron is directly linked to the monomer, i.e., when the focal branch point of the dendron is separated by four bonds from the nearest backbone carbon.

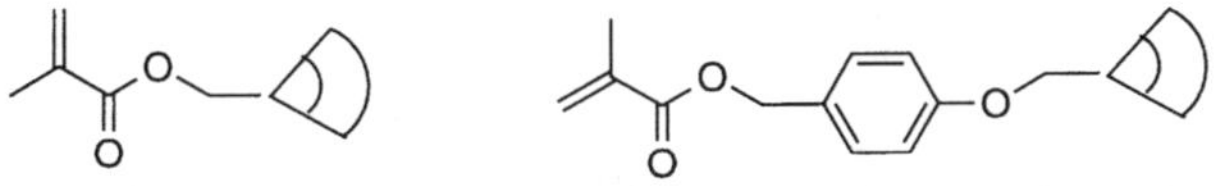

(4)

Insertion of a *p*-benzyloxy group causes a nine bond separation and affords homopolymerization of [G-1] and [G-2] substituted monomers. These observations on dendrimer carrying methacrylate monomers confirm earlier results [110, 111].

These observations are also consistent with results of Drahem and Ritter on the polymerization of methacrylamide monomers carrying four and eight peripheral ester functions dendritically bound to an L-aspartic acid chiral center [112].

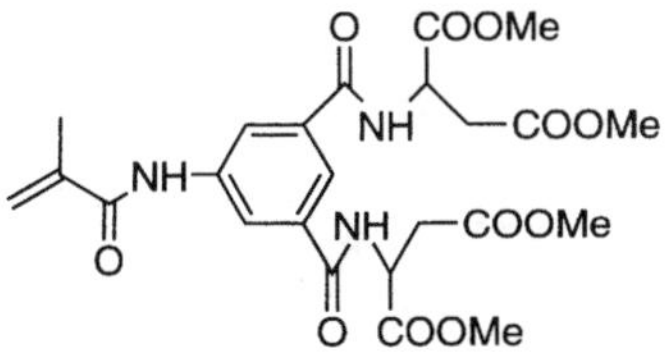

[G-1], [G-2] (5)

The number average DP of the first generation dendrimer monomer is estimated by SEC to be only 23, based on polystyrene equivalent SEC elution volume. Because of the branched nature of the substituent the true value is probably much larger. The optical rotations of the second generation monomer and its polymer are identical. This does not allow any conclusion about a possible secondary structure, e.g., helix formation in the polymer [112].

In conclusion it has turned out to be difficult to polymerize all but the first generation dendritic macromonomer. Higher generation dendrimers need to be attached through a long series of bonds to the polymer backbone. This in turn prevents any strong conformational influence on the backbone and no interesting new properties have been detected so far.

4.2.2
Dendrimers on Stiff Backbones

If dendrimers are introduced on stiff backbone polymers cylindrical molecules are created and an empty inner volume becomes available between the stiff backbone and the outer dendrimer shield. Furthermore the peripheral groups of the dendrimer can be manipulated to control the solubility characteristics of the resulting polymers.

Two routes to dendrimer modified stiff backbone polymers have been explored. The first involves the polycondensation of dendrimer substituted monomers. Monomers of the type

$$\text{[G-1], [G-2], [G-3]} \qquad (6)$$

in which the dendrimer is a Fréchet type poly(benzyl ether) have been cross-coupled with phenylene diboronic acid in the presence of Pd[PPh$_3$]$_4$ under Suzuki conditions. Good yields of polyphenylenes have been obtained in all cases [113]. SEC provides polystyrene equivalent MWs in the 20–60 K range. Absolute MW determination suggests that true MWs are about five times larger [113]. These results prove that the poly(paraphenylene) backbone can be covered with one third-generation dendrimer on every second phenylene ring.

Synthesis of more densely covered poly(paraphenylene) has also been attempted with two other monomers:

$$\text{[G-1]} \qquad (7)$$

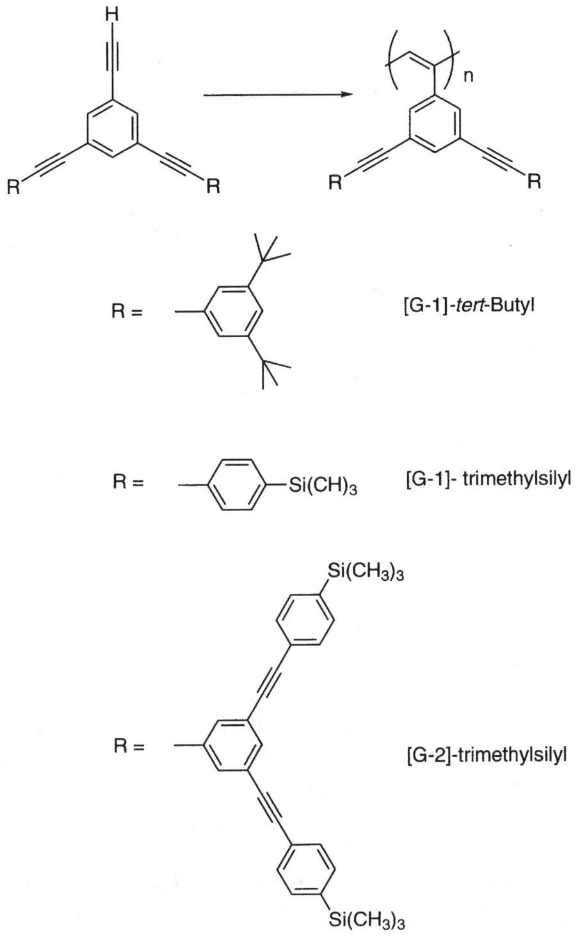

Scheme 12

In the first case the resulting polymer has two dendrons for every three phenylene rings and in the second case there are two dendrons on every second phenylene ring. The higher apparent MW of the polymers, measured by SEC, suggest that the first polymer is easier to prepare than the denser substituted second polymer [114].

Recently, Oikawa et al. succeeded in polymerizing a phenyl acetylene monomer carrying stiff dendrons of generation 1 and generation 2 [115] (see Scheme 12). The dendrimer is attached to every second carbon in the poly(acetylene) backbone. The apparent MW of the polymers is quite high ($>1\times10^6$) and the polymer solubility increases with increasing dendrimer generation. The replacement of *tert*-butyl groups by trimethylsilyl groups also improves solubilization. The maximum in the visible light absorption spectrum of the polymers

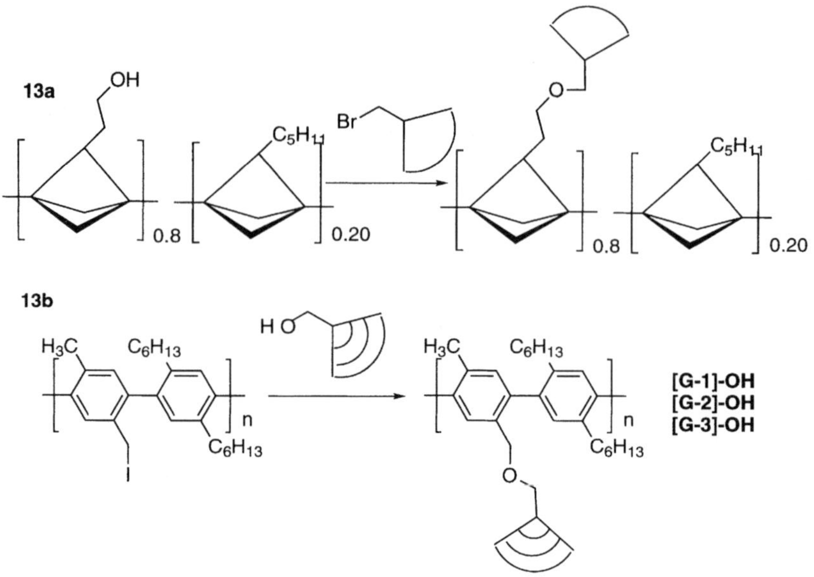

Scheme 13

is observed to move to higher wavelength with increased dendrimer size. In comparison with simple poly(phenyl acetylene) the π-conjugated system in the backbone is more extended and more uniform. This is evidence for the steric influence of the dendritic substituents on the conformation of the poly(phenyl acetylene) backbone.

The second method involves substitution reactions on a preformed polymer with a stiff backbone. The advantage of this method is that it provides better control over the MW of the starting polymer. However, the substitution may not be 100%. The first example involved the substitution on a poly(1,1,1-propellane) copolymer [116]. A copolymer with 80% functionalizable units has been successfully modified with a [G-1]-Br poly(benzyl ether) dendron (see Scheme 13a). Further studies exploring the feasibility and steric limitations of the substitution route have been performed on poly(p-phenylene) [113, 116]. The Williamson substitution reaction on

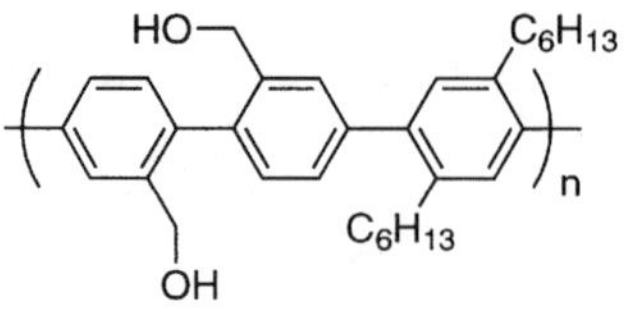

(8)

with [G-1]-Br and [G-2]-Br poly(benzyl ether) dendrons provided 100 and 50–60% functionalization, respectively. A [G-3]-Br substitution was therefore considered unlikely to be successful. The reverse Williamson substitution, using a iodo leaving group on the polymer, is more successful yielding 100, 100, and 70% substitution for [G-1]-OH, [G-2]-OH, and [G-3]-OH respectively (see Scheme 13b).

The most successful substitution reaction is via the isocyanate route. The poly(paraphenylene) backbone polymer chosen for this work

$$\tag{9}$$

has one extra carbon-carbon bond between the phenyl backbone and reactive hydroxyl group. Reaction with isocyanate focal group of a third generation Fréchet type poly(benzyl ether) dendrimer gave a 91–92% substitution yield, the highest attained with a third generation dendrimer [113, 117]. The status of this type of research at this point indicates that only one dendrimer, up to generation three, can be incorporated in every second phenyl ring along the polymer backbone. Moreover, the chemical route followed is of importance. The isocyanate route with a reaction center slightly further removed from the backbone gives the highest rate of substitution.

4.3
Linear Polymers on Dendrimers

4.3.1
Single Polymer-Dendrimer Hybrids

Hybrids of linear polymers and dendrimers are expected to be unusual block copolymers because they combine, in one molecule, a long flexible chain with a random coil conformation with a dense globular dendrimer. Gitsov et al. prepared the first polymer-dendrimer hybrids by reacting monofunctional narrow MWD PEO with their [G-3] and [G-4] poly(benzyl ether) dendrimers [118–121] (see Scheme 14a). This linear polymer-dendrimer hybrid is comparable with an AB type block copolymer. The ABA hybrids are similarly obtained from difunctional poly(ethylene glycols) (PEG) (see Scheme 14b). The Williamson reactions are performed in dry THF at room temperature with a small excess of dendrimer. Yields are in excess of 90%. A four-arm star PEO has similarly been modified with [G-2],[G-3], and [G-4] dendrimers by means of the focal benzyl bromide group in yields of 90, 85, and 80%, respectively [122]. The same hybrids can also be obtained by transesterification in the melt catalyzed by tin or cobalt salts (see Scheme 14c). Yields range between 50 and 80%. There is one report in which the linear polymer and dendrimer are joined by means of a peripheral group rather

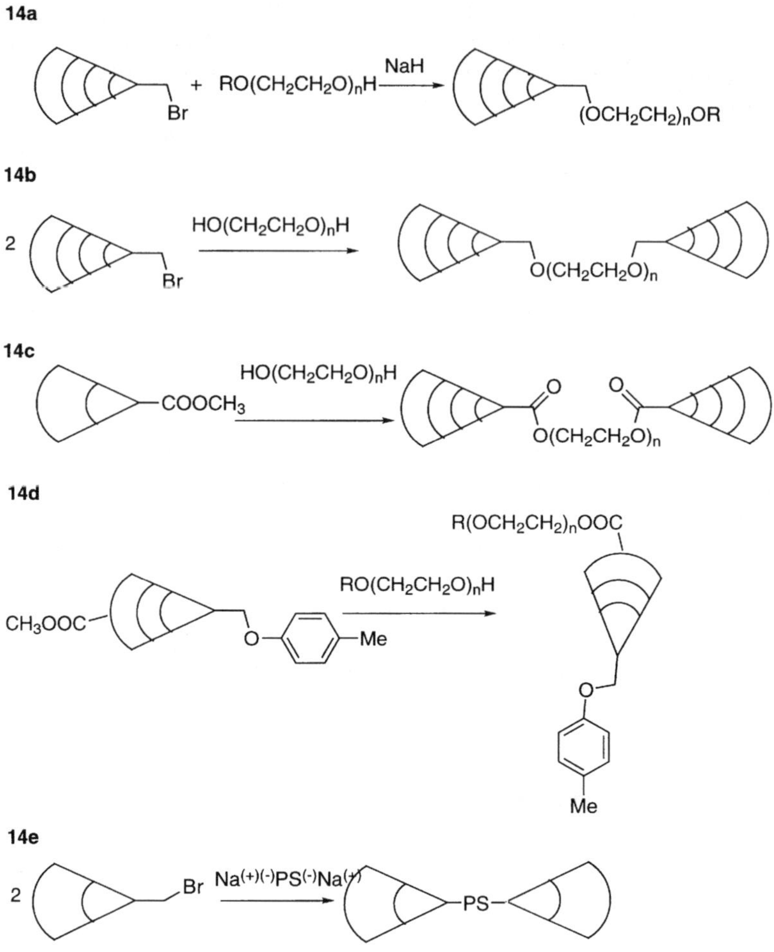

14a

14b

14c

14d

14e

Scheme 14

than the focal group (see Scheme 14d). This transesterification is performed in the melt at 225 °C and 0.03 Tor [123].

Living anionic polymerization has been used to place a central polystyrene chain between two dendrimers [124]. Prior to the coupling reaction at –78 °C the polystyrylpotassium reactivity is reduced by end-capping with diphenylethylene (Scheme 14e).

The single focal point of the dendrimer can be used to initiate the polymerization of monomer, e.g., of ε-caprolactone [125] (see Scheme 15a). The efficiency of the [G-4]-OK initiator is 100%. The polymerization is essentially complete in 6 min at 0 °C. Experimental and calculated MWs agree and the MWD is narrow. At longer times, M_n remains constant but the MWD increases due to trans-

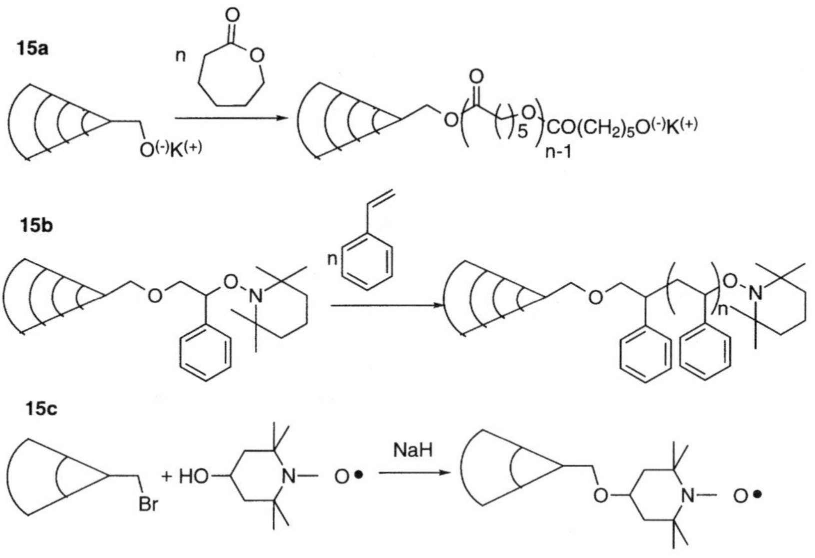

Scheme 15

esterification. These results are very different from those obtained with potassium *tert*-butoxide or the initiator derived from [G-1]-OH. It is therefore proposed that the large dendrimer provides temporary protection for the growing chain against transesterification and backbiting reactions.

The focal functional group of the dendrimer has also been used to initiate living free radical polymerization. For the TEMPO mediated free radical polymerization of styrene the focal point is converted to a TEMPO derivative [126] (see Scheme 15b). This species initiates the bulk living polymerization of styrene at 123 °C. Alternatively, the benzyl bromide focal group of poly(benzyl ether) dendrimers can be used in ATRP. In the presence of $Cu^{I}Cl/4,4'$ (di-*n*-heptyl)-2,2' dipyridyl complex styrene is polymerized at 130 °C [126]. The formation of block copolymers of polystyrene and dendrimer is confirmed by the absence of free dendrimer in SEC analysis of the crude product. Furthermore, the polystyrene peak in SEC has a superposable UV-absorption at 283 nm which is the characteristic wavelength of absorption of the dendrimer. In both systems the MWs vary between 10,000 and 100,000 D. Calculated and experimental MWs agree at low MW but the experimental MW tends progressively to lower values above 30,000 D with a concomitant widening of MWD from 1.19 to 1.5. Radical transfer reactions to the benzyl ether groups on the dendrimer seem not to be important. The TEMPO modified dendrimers [G-3] and [G-4] have been used to polymerize 4-acetoxystyrene. The resulting copolymer can be hydrolyzed to an amphilic block copolymer consisting of a hydrophilic poly(vinylphenol) and hydrophobic poly(benzyl ether) dendrimer.

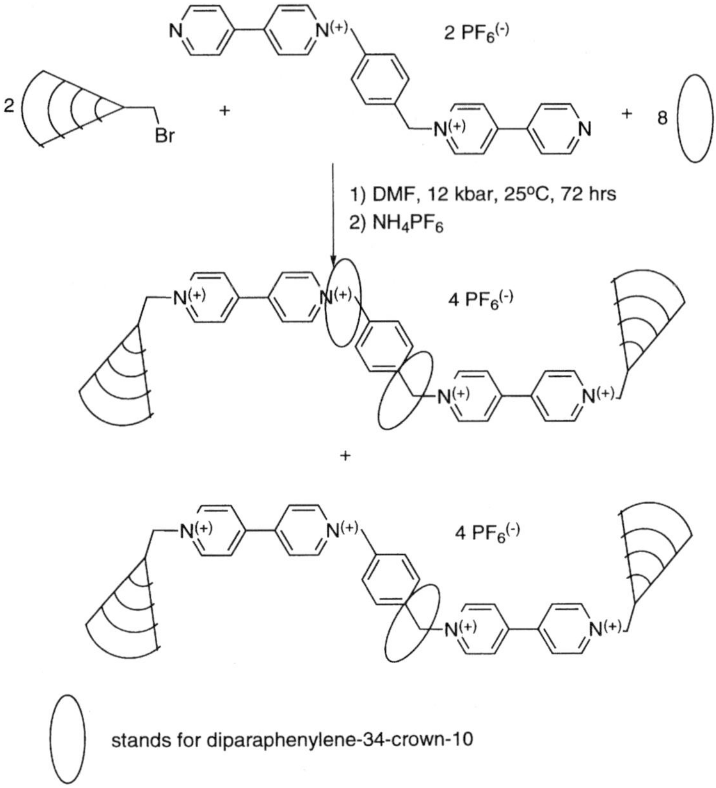

Scheme 16

Poly(benzyl ether) [G-2]-TEMPO, [G-3]-TEMPO, and [G-4]-TEMPO compounds have been synthesized and used as additives in the benzoyl peroxide initiated polymerization of styrene [127] (see Scheme 15c). After an induction period, chain growth is observed. However, the MWD is larger than in a dendrimer-free TEMPO modulated system (Mw/Mn≈2). The expectation that the dendrimer would isolate the growing chain end and prevent side reactions is not borne out. Polymerizations of methylmethacrylate, vinylacetate, and n-butylacrylate with the same initiator/TEMPO recipe are disappointing.

A special type of ABA block copolymer is formed when dendrimers are used as stoppers on rotaxanes. Rotaxanes are linear polymers onto which one or several large macrocycles are threaded. In order to prevent unthreading, sufficiently bulky groups must be attached to the two ends of the rotaxane. The [G-3] poly(benzyl ether) dendrimer proved to be highly suited for trapping [128] (see Scheme 16). The linear polymer is made of stiff bipyridinium groups that spontaneously assemble with diparaphenylene-34-crown-10 to form the rotaxane. The dendrimers are attached via quaternation of pyridine with benzyl bromide

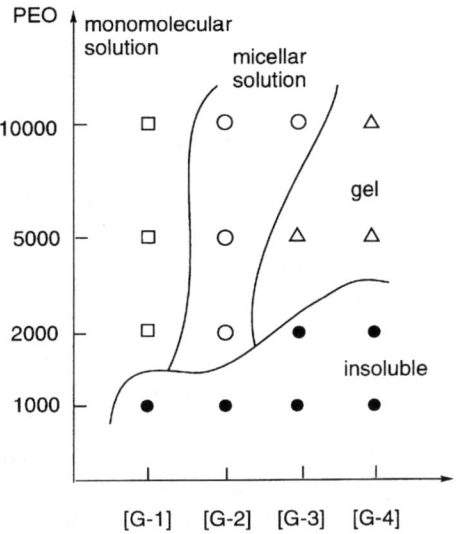

Fig. 6. Phase diagram of ABA poly(benzyl ether) dendrimer PEO in MeOH:H$_2$O (1:1)

at the focal point. The nonpolar dendrimer exerts a large solubilizing effect on the highly polar rotaxane which allows its study in a wide range of solvents.

Dendrimers that are prepared by the divergent method lack the focal group to attach them to a polymer chain. Such dendrimers must be grown from a preformed polymer having an end-standing functional group. The first synthesis by this method used glycine modified PEO and poly(α,ε-lysine) to grow a poly(α,ε-L-lysine) dendrimer [129] (See Fig. 3b for structure). Up to four generations of lysine have been added. The PEO-dendrimer is purified at each generation by precipitation in diethylether similar to the Merrifield procedure. At the fourth generation the MW of the poly(α,ε-L-lysine) dendrimer and PEO are comparable. These amphiphilic block copolymers decrease the surface tension of water and form micelles in solution.

Poly(propylene imine) dendrimers (see Scheme 1 for structure) have been constructed step by step onto an amine functionalized polystyrene [38, 130, 131]. The challenge in this synthesis is finding conditions for the poly(propylene imine) synthesis under which the low MW polystyrene (MW=3200) is soluble [38]. Similar poly(imine) dendrimers with carboxylic acid end groups have also been prepared [130]. The poly(propylene imine) dendrimer has also been synthesized on an amino-terminated poly(2-methyl-2-oxazoline) [132].

Because of the commercial interest in amphiphilic block copolymers the properties of the amphiphilic linear polymer-dendrimer have been most extensively studied. An illustration of the various phases that can be observed is shown in Fig. 6 for the case of an ABA type block copolymer, where A is the hydrophobic dendrimer and B the hydrophilic PEO [119]. The exact location of the phase boundaries is dependent on the polymer concentration. The boundary between the monomolecular and micellar phases depends on whether the experimental concentration is above or below the critical micelle concentration. The boundary between micellar solution and gel is defined by the equilibrium between loops and bridges formed by the soluble PEO blocks. Bridges are entropically favored. A similar, less complete, phase diagram is also obtained for AB type polymer-dendrimer block copolymers. However, in that case, no gel phase is produced for lack of bridging.

The structures formed by polystyrene-poly(propylene imine) dendrimers have also been analyzed. Block copolymers with 8, 16, and 32 end-standing amines are soluble in water. They have a critical micelle concentration of the order of 10^{-7} mol/l. At 3×10^{-4} mol/l they form different types of micelles. The dendrimer with eight amine groups (80% PS) form bilayers. The dendrimer with 16 amine groups (65% PS) forms cylinders and the dendrimer with 32 amine groups (50% PS) forms spherical micelles [38, 130, 131]. These are the classical lamellar, cylindrical, and spherical phases of block copolymers. However, the boundary between the phases occurs at very different volume fractions, due to the very different packing requirements of the linear polymer and spherical dendrimer at the interphase.

4.3.2
Multiple Polymer-Dendrimer Hybrids

Dendrimers with their multiple end-standing functional groups are ideally suited for the construction of star-shaped polymers. Indeed, the end-standing functional groups can be used as initiators for polymerization ("grafting from" method) or as functional groups for "grafting onto". They can also be used as redistribution centers in equilibrium polymerization.

"Grafting from" has not been a successful method in anionic polymerization because the required low molecular weight multifunctional organometallic initiators are almost always insoluble and this is also expected to be the case when dendrimers are modified. However, in cationic polymerizations the dormant species is less polar and more soluble. For example, the hexabenzyl bromide

$$\left(BrH_2C-\bigcirc-O\right)_2 P\underset{N\underset{P}{\sim}N}{\overset{N}{\sim}}P\left(O-\bigcirc-CH_2Br\right)_2 \quad (10)$$
$$\left(O-\bigcirc-CH_2Br\right)_2$$

has been used in the polymerization of 2-methyl-2-oxazoline [133] and it can be envisioned that larger dendritic initiators based on phosphorus can also be used [134].

The "grafting from" method is also successful in other ring opening polymerizations. For example, ε-caprolactone polymerization is initiated from poly(propylene imine) dendrimers with n=2–16 end-standing amine groups [135]. To be

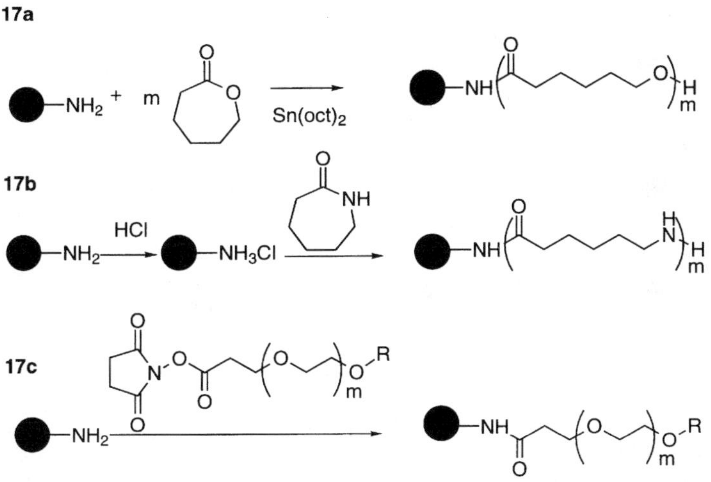

Scheme 17

successful this method requires stable dendrimers and preferably little or no reversibility leading to depolymerization and ring formation (see Scheme 17a). The polymerization in bulk at 230 °C of ε-caprolactam with poly(ethylene imine) dendrimers is also described [136] (Scheme 17b). The properties of the resulting semicrystalline six-arm star polymers are in all aspects similar to those of the linear nylon-6 except for a 40% lower melt viscosity [137] which is of significant industrial importance.

Recently, PEO has been grafted onto PAMAM dendrimers by means of N-succinimidyl propionic acid spacers [138] (see Scheme 17c). When n=16 and 32, experimental and expected functionalization agree satisfactorily. However, for n= 64, 128, and 256, progressively lower functionalization is observed. It is not clear whether this result is due to imperfections of the PAMAM dendrimer used or to steric limitations on the extent of the substitution reaction. At present this reaction scheme has been tested only on PEO chains with MW=5000.

Anionic polymerization is uniquely suited for the preparation of star polymers via grafting onto dendrimers. At least for a few monomers under well defined conditions narrow MWD chains having stable but reactive end groups are available for reaction with multifunctional dendrimers. It is required, however, that the dendrimers are essentially of a hydrocarbon nature except for the electophilic functional groups. In the original disclosure of the formation of 18-arm star polyisoprenes the required octadecachlorocarbosilane (see Fig. 3a for related structure) coupling agent was prepared by an embryonic recursive method which foreshadowed divergent dendrimer synthesis [139]. This octadecachlorocarbosilane compound is equivalent to a generation 1.5 dendrimer. The grafting-onto process consist of the nucleophilic displacement of Cl with the carbanionic end groups of living polymer chains. The coupling reaction between the

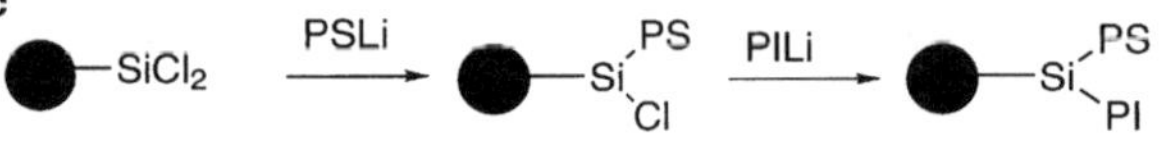

Scheme 18

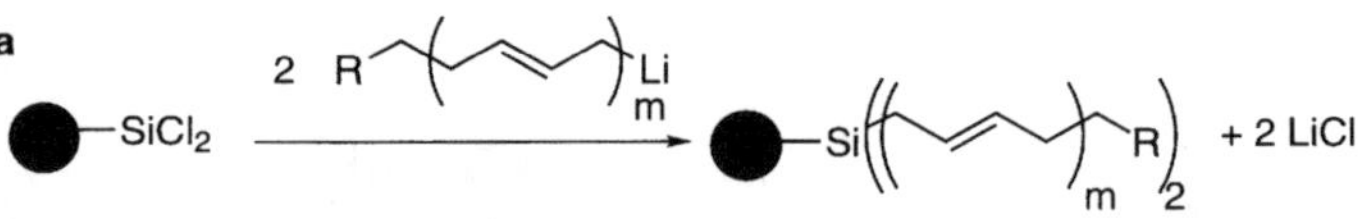

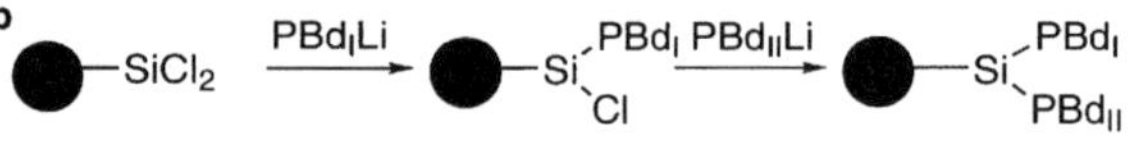

multifunctional chlorosilane compound and the living polymeric carbanion is represented in Scheme 18a. No side reactions have been observed when this reaction is performed at room temperature in hydrocarbon medium; although an excess of living polymer and relatively long reaction times are required [64, 140]. This "grafting onto" route has been extended to the preparation of star polymers with 32- [140], 64-, and 128-arm star polybutadienes [64] which are built onto carbosilane dendrimers of generations 2.5, 3.5, and 4.5, respectively [79]. It is worth noting that no steric limitations are observed when polybutadienyllithium is used. Steric limitations, however, are a problem with polyisoprenyllithium and polystyryllithium. The same coupling agents have been used to prepare star block copolymers in which each arm of the star consists of a diblock copolymer of styrene (30%) and butadiene (70%) [141].

The "grafting onto" reaction can be controlled to some extent. For example, substitution of the first half of the Si-Cl bonds with one polymer chain is known to be kinetically favored. The remaining Si-Cl is then substituted with another polymer. Two examples of this method have been explored. In the first, fuzzy star polymers have been prepared according to Scheme 18b. These polymers are called fuzzy because the two polybutadiences have different MWs, thereby presumably, creating a more diffuse surface [141]. In the second application, a mikto arm star polymer with eight polystyrene and eight polyisoprene arms has been synthesized [142] (Scheme 18c). Although the arms of such multiblock copolymers are forced to intermingle extensively, the polymers spontaneously form micro-separated two-phase systems.

The polymer redistribution reaction with functionalized dendrimers has been investigated by Van Aert et al. for the case of the transetherification of poly-(2,6-dimethyl-1,4-phenylene ether) (PPE) by means of phenols attached to dendrimers [143]. The number average molecular weight of the arms is controlled by the ratio of moles of PE units and the moles of added phenol. The phenols have been attached to poly(propylene imine) dendrimers by means of a *tert*-butyloxycarbonyl tyrosine(Scheme 19a). The redistribution rate is slow but can be increased by adding CuCl / 4-dimethylaminopyridine catalyst. Oxygen-free

Scheme 19

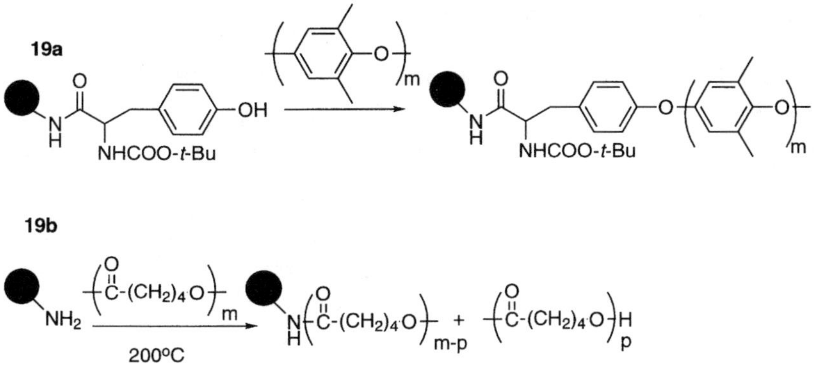

conditions are required during the redistribution in order to avoid oxidative polymerization and oxidative side reactions. The star polymers are different from the previously described model star polymers obtained by anionic polymerization because the arm MW has a most probable MWD. Interestingly, it was observed that the hydrodynamic volume of the star polymers as measured by SEC does not vary much with increasing functionality from f=4 to f=64 [143], although measured [η] and calculated MW suggest an important increase in R_η [144]. The hydrodynamic volume measured by dynamic light scattering is clearly an increasing function of the functionality of the stars at constant arm length, as well as under conditions of increasing arm length and constant dendrimer functionality. This apparent difference may be due to different averages probed by the two methods. The SEC maximum is close to a weight average while the dynamic light scattering measures a z-average value of the hydrodynamic volume.

The method of transetherification or the related transesterification should be applicable to other dendrimer polymer systems. Transamidation has been performed with poly(propylene imine) dendrimer and poly(ε-caprolactone). The linear chain is fragmented according to Scheme 19b, and a mixture of star and linear polymer is formed [135]. The average MW of the arms is equal to the MW of the linear fragments.

5
Hybrids of Dendrimers and Biological Polymers

5.1
Dendrimer-Peptide Hybrids

Tam has pioneered the use of dendrimers as templates for simultaneous growth of identical polypeptides [145]. Tam et al. were also the first to use the Merrifield approach to the synthesis of dendrimers [145–148]. Starting from phenylaceta-

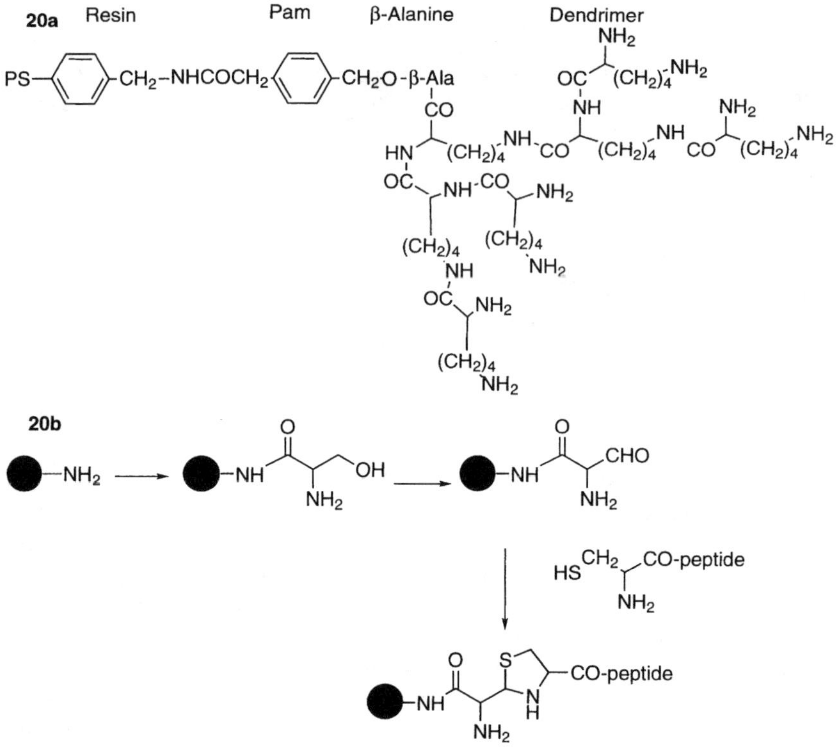

Scheme 20

midomethyl (Pam) modified poly(styrene-co-divinylbenzene) beads, a third generation of poly (α,ε-L-lysine) dendrimer with eight primary amines was obtained (see Scheme 20a). After capping each amine with a glycine spacer eight copies of a desired polypeptide are grown. A detailed critical description of the special conditions required in the synthesis of a dendrimer on a cross-linked polymer, including low loading of the resin beads, has been given [145]. The long hydrolyzable Pam spacer between the non-polar polymer and the more polar dendrimer is also important. The polypeptides chosen for the synthesis are specific antigens. The clusters of the antigens attached to the dendrimer elicit good antibody response without showing the problems associated with antigens bound to carrier proteins.

In recent work Tam and Rao have attached unprotected peptides directly to a third generation poly(α,ε-L-lysine) dendrimer [149]. As shown in Scheme 20b, the dendrimer is modified to contain eight aldehyde groups. These are then reacted with a cysteine terminated polypeptide with formation of a thiazolidine ring. The addition of the first five polypeptides occurs within 2 h at 18 °C but addition of subsequent peptide chains becomes progressively slower. After 67 h a mixture of hepta-and octapeptide is obtained. The method which allows the use of unprotected polypeptides was expanded to other reactions producing oxime and hydrazone dendrimer-peptide linkages [150, 151]. These are, however, less stable than the thiazolidine ring at elevated pH.

5.2
Dendrimer-DNA Complexes

Dendrimers with primary amines form complexes with nucleotides by electrostatic interaction between the cationic ammonium and anionic phosphate groups [152–157]. Such interactions are also known to occur with polylysine and other linear polycations. The first account of such interaction has been described for the case of PAMAM dendrimers with plasmid DNA encoding for either luciferase or bacterial β-galactosidase [156]. The complexation is monitored by DNA immobilization in electrophoresis [152, 155, 156] or by enhanced fluorescence anisotropy [153]. Complexes are completely immobile when the terminal amine and nucleotide base stoichiometry is around 1:1 to 1.5:1 [152, 155, 156]. Excess nucleotide shows no retardation of mobility, while excess dendrimer amine groups, i.e., 5:1 and higher, leads to positively charged complexes which migrate to the cathode [152, 155, 156]. It is observed that complexes are better prepared at low pH and low ionic strength [152]. Nevertheless, the dendrimer-DNA complexes are quite stable against variations in pH between 3 and 10 [152, 155] with some loss of complexation above pH 10 [153]. Ionic strengths between 50 mmol/l and 1.5 mol/l are tolerated. Disruption of the complex is only effected with strong ionic detergents [155, 157]. An important advantage of the complex is that it makes the DNA quite resistant to nuclease digestion [157].

The structure of the dendrimer-DNA complexes is not well characterized. It has been estimated that under conditions of maximum transfection, i.e., the process of bringing the DNA through the cell wall into the cell, the size of the luciferase plasmid is such that about 320 PAMAM dendrimers are present, although not all may be directly involved in the complexation [156]. Electron microscopy suggests that the dendrimer-DNA complexes form condensed aggregates [157]. Delong et al. have studied the complex formed between a third generation PAMAM dendrimer and a small 15 base oligonucleotide [152]. They found a broad distribution of MW by ultrafiltration with most material between 10 and 30 kD. Light scattering experiments, however, are obscured by the presence of a small amount of very high MW material.

The PAMAM dendrimer-DNA complexes have been shown to promote transfection [152–156]. The earlier studies used reporter luciferase which allows for measurement of luciferase activity in host cells [155, 156]. Although there is some dependence on the cell type being invaded [155], experiments have shown that generations 6–8 dendrimers are most efficient in transfection [155, 156]. There is little transfection with generations 2–4 PAMAM dendrimers and no further improved efficiency with generations 9 and 10 dendrimers. The preferred dendrimer to DNA stoichiometry is 5:1 or higher [155, 156]. It has since become clear that perfect generation 6 PAMAM-DNA complexes are not effectively transfected [158]. Experiments with solvolytically degraded PAMAM-DNA complexes suggest that partially degraded high MW complexes are the true transfection agents. This has been confirmed by the synthesis of defective PAMAM dendrimers which on complexation with DNA showed high transfection activity [158]. In the case of oligonucleotides, generations 4 and 5 are about equally efficient [153].

The uptake of the dendrimer-DNA is an active energy dependent endocytosis in living cells [155]. Flow cytometry has shown that the complex is located in the cell [152, 153] and microscopy shows cytoplasmic as well as nuclear uptake [152]. The cell uptake of DNA is also measured by its biological consequences. The biological activity of the DNA in the DNA-dendrimer complex seems to be reduced as less transcription to m-RNA is observed [157]. However, in cases where antisense oligonucleotides are transfected, the activity of a luciferase carrying cell is downgraded by the introduction of an antisense oligonucleotide specific to a portion of the luciferase RNA.

An important question in cell uptake of DNA is the potential toxicity of the dendrimer nucleotide complexes. In general, the toxicity has been found to be lower for dendrimer-nucleotides than for other transfection systems including polylysine [154–156].

5.3
Dendrimer-Antibody Conjugates

Antibodies are ideal for targeting specific cells due to their specific interaction with antigens on the cell surface. They can therefore be used as carriers of radioisotopes, toxins, and cytotoxins to the selected cells. This method requires the linkage of the agent to the antibody without affecting the immunoreactivity of the antibody. Direct linkage of the agent to the antibody is known to be limited and to decrease the immunoreactivity. Roberts et al. coupled porphyrin molecules with a third (24 NH_2 groups) or fourth (48 NH_2 groups) generation PAMAM dendrimer and then linked rabbit immunoglobin by means of remaining amino groups [159]. The porphyrin rings have then been used to complex with radioactive $^{67}CuCl_2$. They established that the dendrimer is attached to the carbohydrate region of the glycoprotein and that the immune response is retained.

Gansow and his coworkers developed conjugates of monoclonal antibodies and complexing agent modified PAMAM dendrimers [160] (Scheme 21). All reactions are performed in water. After purification and creation of the sulfhydryl group these molecules are linked to monoclonal antibodies that have been premodified with maleimide groups. The dendrimer-antibody ratio is always near unity. The uptake of ^{90}Y by the complexing sites in the conjugate takes about 30 min at 35 °C and about 3 h at room temperature. The amount of ^{90}Y labeling and the resulting radioactivity exceed by far that obtainable by direct linkage of the complexing agent to the antibodies. This result is obtained with loss of immunoreactivity.

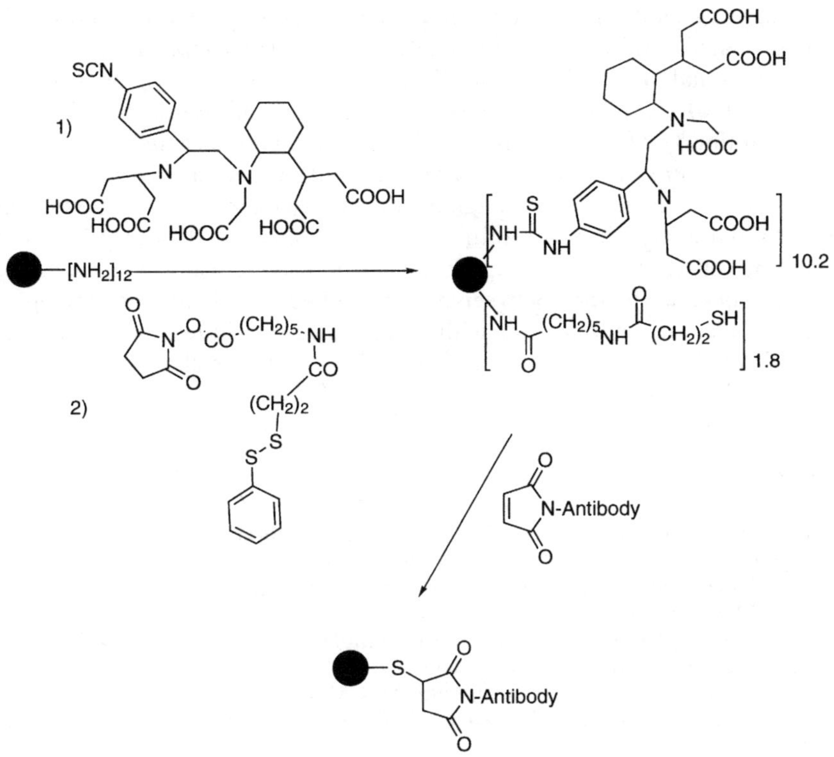

Dendrimers and Dendrimer-Polymer Hybrids

Scheme 21

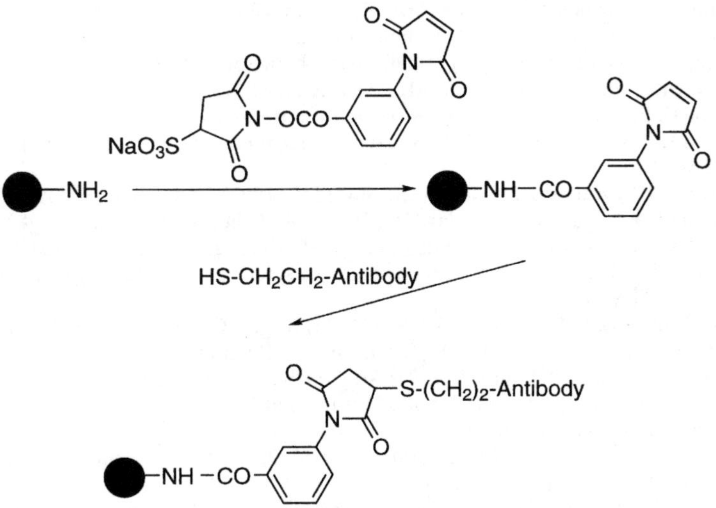

Scheme 22

A fourth generation PAMAM dendrimer containing 48 amino groups has been boronated with $Na(CH_3)_3NB_{10}H_8NCO$ followed by coupling with a thiolated mono-clonal antibody via a difunctional linking agent as shown in Scheme 22 [161]. This sequence allows for the incorporation of 1700 boron atoms per antibody. This level of boronation is potentially sufficient to deliver the required 10^9 atoms of ^{10}B to each tumor cell and to sustain a lethal reaction of ^{10}B with thermal neutrons. Although the boronated dendrimer-antibody conjugate retained a high in vitro affinity and specificity for the antigens on B16 melanoma cells, in vivo tumor uptake has been disappointing. A similar conjugation has been performed between boronated PAMAM dendrimer and epidermal growth factor, a polypeptide with 53 amino acid residues that can be transported through the brain barrier and target brain tumor cells [162].

6
References

1. Corey EJ, Cheng XM (1989) The logic of chemical synthesis. Wiley, New York
2. Buhleier E, Wehner W, Vögtle F (1978) Synthesis 155
3. Tomalia DA, Baker H, Dewald J, Hall M, Kallos G, Martin S, Roeck J, Ryder J, Smith P (1985) Polymer J 17:117
4. Wörner C, Mülhaupt R (1993) Angew Chem Int Ed Engl 32:1306
5. de Brabander-van den Berg EMM, Meijer EW (1993) Angew Chem Int Ed Engl 32:1308
6. de Brabander EMM, Brackman J, Mure-Mak M, de Man H, Hogeweg M, Keulen J, Scherrenberg R, Coussens B, Mengerink Y, van den Wal Sj (1996) Macromol Symp 102:9
7. Smith PB, Martin SJ, Hall MJ, Tomalia DA (1988) Appl Polym Analysis Charact 2:357
8. Newkome GR, Yao Z, Baker GR, Gupta VK (1985) J Org Chem 50:2003
9. Miller TM, Neenan TX (1990) Chem Mater 2:346
10. Hawker CJ, Fréchet JMJ (1990) J Am Chem Soc 112:7638
11. Wooley KL, Hawker CJ, Fréchet JMJ (1991) J Am Chem Soc 113:4252
12. Spindler R, Fréchet JMJ (1993) J Chem Soc, Perkin Trans I:913
13. Leon JW, Kawa M, Fréchet JMJ (1996) J Am Chem Soc 118:8847
14. Wooley KL, Hawker CJ, Fréchet JMJ (1994) Angew Chem Int Ed Engl 33:82
15. L'abbé G, Forier B, Dehaen W (1996) Chem Commun 2143
16. Chang H-T, Chen C-T, Kondo T, Siuzdal G, Sharpless KB (1996) Angew Chem Int Ed Engl 35:182
17. Kawachugi T, Walker KL, Wilkins CL, Moore JS (1995) J Am Chem Soc 117:2159
18. Tomalia DA, Naylor MA, Goddard WA III (1990) Angew Chem Int Ed Engl 29:138
19. Tomalia DA, Hedstrand DM, Wilson LR (1990). In: Mark HF, Bikales NM, Overberger CG, Mendes G (eds) Encyclopedia of polymer science and technology, 2nd edn, index volume. Wiley, New York
20. Mekelburger H-B, Jaworek W, Vögtle F (1992) Angew Chem Int Ed Engl 31:1571
21. Tomalia DA, Durst HD (1993) Topics in Current Chemistry 165:197
22. Issberner J, Moors R, Vögtle F (1994) Angew Chem Int Ed Engl 33:2413
23. Ardoin N, Astruc D (1995) Bull Soc Chim Fr 132:875
24. Voit BH (1995) Acta Polym 46:87
25. Newkome GR, Moorefield CN, Vögtle F (1996) Dendritic molecules: concept, synthesis, perspectives. VCH, Weinheim
26. Fréchet JMJ, Hawker CJ (1996). In: G Allen (ed) Comprehensive polymer science, 2nd suppl. Pergamon Elsevier Science, Oxford

27. Zeng F, Zimmerman SC (1997) Chem Rev 97:1681
28. Peerlings HWI, Meijer EW (1997) Chem Eur J 3:1563
29. Matthews OA, Shipway AN, Stoddart JF (1998) Progr Polym Sci 23:1
30. Newkome GR (ed) (1994-1996) Advances in dendritic macromolecules. vols 1-3. JAI Press, Greenwich, CT
31. Buyle Padias A, Hall HK Jr, Tomalia DA, McConnell JR (1987) J Org Chem 52:5305
32. Seyferth D, Son DY, Rheingold AL, Ostrander RL (1994) Organometallics 13:2682
33. van Genderen MHP, Baars MWPL, van Hest JCM, de Brabander-van den Berg EMM, Meijer EW (1994) Rec Trav Chim Pays-Bas 113:573
34. Kallos GJ, Tomalia DA, Hedstrand DM, Lewis S, Zhou J (1991) Rapid Commun Mass Spectrom 5:383
35. Schwartz BL, Rockwood AL, Smith RD, Tomalia DA, Spindler R (1995) Rapid Commun Mass Spectrom 9:1552
36. Hummelen JC, van Dongen JLJ, Meijer EW (1997) Chem Eur J 3:1489
37. Stöckigt D, Lohmer G, Belder D (1996) Rapid Commun Mass Spectrom 10:521
38. van Hest JCM, Delnoye DAP, Baars HWPL, Elissen-Romàn C, van Genderen MHP, Meijer EW (1996) Chem Eur J 2:1616
39. Sahota HS, Lloyd PM, Yeates SG, Derrick PJ, Taylor PC, Haddleton DM (1994) J Chem Soc, Chem Commun 2445
40. Leon JW, Fréchet JMJ (1995) Polym Bull 35:449
41. Szwarc M (1968) Carbanion, living polymer and electron transfer processes. Interscience Publ, New York
42. Frey H, Lorenz K, Mülhaupt R, Rapp U, Mayer-Posner FJ (1996) Macromol Symp 102:19
43. Martinez CA, Hay AS (1997) J Polym Sci: A: Polym Chem 35:1781
44. Xu Z, Kahr M, Walker KL, Wilkins CL, Moore JS (1994) J Am Chem Soc 116:4537
45. Walker KL, Kahr MS, Wilkins CL, Xu Z, Moore JS (1994) J Am Soc Mass Spectrom 5:731
46. Hayes W, Freeman AW, Fréchet JMJ (1997) PMSE Preprints 77:136
47. McCrackin FL, Wagner HL (1980) Macromolecules 13:685
48. Mourey TH, Turner SR, Rubinstein M, Fréchet JMJ, Hawker CJ, Wooley KL (1992) Macromolecules 25:2401
49. de Gennes P-G, Hervet H (1983) J Phys Lett 44:L351
50. Maciejewski M (1982) J Macromol Sci Chem Phys A17:689
51. Lescanec RL, Muthukumar M (1990) Macromolecules 23:2280
52. Mansfield ML, Klushin LI (1993) Macromolecules 26:4262
53. Boris D, Rubinstein M (1996) Macromolecules 29:7251
54. Murat M, Grest GS (1996) Macromolecules 29:1278
55. Zimm B, Stockmayer WH (1949) J Chem Phys 17:1301
56. Kurata M, Abe M, Iwana M, Matsushima M (1972) Polym J 3:729
57. Naylor AM, Goddard WA III, Kiefer GE, Tomalia DA (1989) J Am Chem Soc 111:2339
58. Mansfield ML (1994) Polymer 35:1827
59. Mansfield ML, Klushin LI (1992) J Phys Chem 96:3994
60. Chen ZY, Cui S-M (1996) Macromolecules 29:7943
61. Lue L, Prausnitz JM (1997) Macromolecules 30:6650
62. Huber K, Burchard W, Fetters LJ (1984) Macromolecules 17:541
63. Richter D, Farago B, Fetters LJ, Huang JS, Ewen B (1990) Macromolecules 23:1845
64. Bauer BJ, Fetters LJ, Graessley WW, Hadjichristidis N, Quack GE (1989) Macromolecules 22:2337
65. Roovers J (1999). In: Mishra MK, Kobayashi S (eds) Star and hyperbranched polymers. Marcel Dekker, New York
66. Roovers J, Zhou L-L, Toporowski PM, van der Zwan M, Iatrou H, Hadjichristidis N (1993) Macromolecules 26:4324
67. Burchard W, Kajiwara K, Nerger D (1982) J Polym Sci, Polym Phys Ed 20:157
68. Hammouda B (1992) J Polym Sci: Part B: Polym Phys 30:1387

69. Aharoni SM, Murthy NS (1983) Polym Commun 24:132
70. Bauer BJ, Topp A, Prosa TJ, Amis EJ, Yin R, Qin D, Tomalia DA (1997) PMSE Preprints 77:87
71. de Brabander E, Brackman J, Froehling P, Scherrenberg R, Put J (1997) PMSE Preprints 77:84
72. Aharoni SM, Crosby CR III, Walsh EK (1982) Macromolecules 15:1093
73. Tomalia DA, Hall M, Hedstrand DM (1987) J Am Chem Soc 109:1601
74. Einaga Y, Koyama H, Konishi T, Yamakawa H (1989) Macromolecules 22:3419
75. Hawker CJ, Malmström EE, Franck CW, Kampf JP (1997) J Am Chem Soc 119:9903
76. Stechemesser S, Eimer W (1997) Macromolecules 30:2204
77. Tomalia DA, Berry V, Hall M, Hedstrand DM (1987) Macromolecules 20:1164
78. de Brabander-van den Berg EMM, Nijenhuis A, Mure M, Keulen J, Reintjens R, Vandenbooren F, Bosman B, de Raat R, Frijns T, van de Wal S, Castelijns M, Put J, Meijer EW (1994) Macromol Symp 77:51
79. Zhou L-L, Roovers J (1993) Macromolecules 26:963
80. Morikawa A, Kakimoto M, Imai Y (1991) Macromolecules 24:3469
81. Roovers J (1996) In: Salamone JC (ed) Polymer materials encyclopedia, vol 1. CRC Press, Boca Raton, p 850
82. Kato Y, Itsubo A, Yamamoto Y, Fujimoto T, Nagasawa M (1975) Polymer J 7:123
83. Newkome GR, Young JK, Baker GR, Potter RL, Audoly L, Cooper D, Weiss CD, Morris K, Johnson CS Jr (1993) Macromolecules 26:2394
84. Young JK, Baker GR, Newkome GR, Morris KF, Johnson CS Jr (1994) Macromolecules 27:3463
85. Dubin PL, Edwards SL, Kaplan JI, Mehta MS, Tomalia DA, Xia J (1992) Anal Chem 64:2344
86. Gauthier M, Möller M (1991) Macromolecules 24:4548
87. Tomalia DA, Hedstrand DM, Ferritto MS (1991) Macromolecules 24:1435
88. Gauthier M, Möller M, Burchard W (1994) Macromol Symp 77:43
89. Yin R, Swanson DR, Tomalia DA (1995) PMSE Preprints 73:277
90. Sheiko SS, Gauthier M, Möller M (1997) Macromolecules 30:2343
91. Hempenius MA, Michelberger W, Möller M (1997) Macromolecules 30:5602
92. Gauthier M, Li W, Tichagwa L (1995) PMSE Preprints 73:232
93. Candau F, Rempp P, Benoit H (1972) Macromolecules 5:627
94. Kee RA, Gauthier M (1995) PMSE Preprints 73:335
95. Gauthier M, Tichagwa L, Downey JS, Gao S (1996) Macromolecules 29:519
96. Grubbs RB, Hawker CJ, Dao J, Fréchet JMJ (1997) Angew Chem Int Ed Engl 36:270
97. Six JL, Gnanou Y (1995) Macromol Symp 95:137
98. Zimm BH, Stockmayer WH (1949) J Chem Phys 17:1301
99. Bera TK, Taton D, Gnanou Y (1997) PMSE Preprints 77:126
100. Cloutet E, Six JL, Taton D, Gnanou Y (1995) PMSE 73:133
101. Wang F, Roovers J, Toporowski PM (1995) Macromol Reports A32(5&6):951
102. Wintermantel M, Schmidt M, Tsukahara Y, Kajiwara K, Kohiya S (1994) Macromol Rapid Commun 15:279
103. Wintermantel M, Gerle M, Fischer K, Schmidt M, Wataoka I, Urakawa H (1996) Macromolecules 29:978
104. Wintermantel M, Fischer K, Gerle M, Schmidt M, Kajiwara K, Urakawa H, Wataoka I (1995) Angew Chem Int Ed Engl 34:1472
105. Flory PJ (1956) Proc Roy Soc London Ser A 234:73
106. Percec V, Heck J, Tomazos D, Falkenberg F, Blackwell H, Ungar G (1993) J Chem Soc Perkin Trans I 2799
107. Hawker CJ, Fréchet JMJ (1992) Polymer 33:1507
108. Neubert I, Amoulong-Kirstein E, Schlüter A-D, Dautzenberg H (1996) Macromol Rapid Commun 17:517
109. Neubert I, Klopsch R, Claussen W, Schlüter A-D, (1996) Acta Polym 47:455

110. Fréchet JMJ, Gitsov I (1995) Macromol Symp 98:441
111. Chen YM, Chen CF, Liu WH, Li YF, Xi F (1996) Macromol Rapid Commun 17:401
112. Draheim G, Ritter H (1995) Macromol Chem Phys 196:2211
113. Karakaya B, Claussen W, Gessler K, Saenger W, Schlüter A-D (1997) J Am Chem Soc 119:3296
114. Claussen W, Schulte N, Schlüter A-D (1995) Macromol Rapid Commun 16:89
115. Kaneko T, Horie T, Asano M, Aoki T, Oikawa E (1997) Macromolecules 30:3118
116. Freudenberger R, Claussen W, Schlüter A-D, Wallmeier H (1994) Polymer 35:4495
117. Karakaya B, Claussen W, Schäfer A, Lehmann A, Schlüter A-D (1996) Acta Polym 47:49
118. Gitsov I, Wooley KL, Fréchet JMJ (1992) Angew Chem Int Ed Engl 31:1200
119. Gitsov I, Fréchet JMJ (1993) Macromolecules 26:6536
120. Gitsov I, Wooley KL, Hawker CJ, Ivanova PT, Fréchet JMJ (1993) Macromolecules 26:5621
121. Fréchet JMJ, Gitsov I (1995) Macromol Symp 98:441
122. Gitsov I, Fréchet JMJ (1996) J Am Chem Soc 118:3785
123. Wooley KL, Fréchet JMJ (1992) PMSE Preprints 67:90
124. Gitsov I, Fréchet JMJ (1994) Macromolecules 27:7309
125. Gitsov I, Ivanova PT, Fréchet JMJ (1994) Macromol Rapid Commun 15:387
126. Leduc MR, Hawker CJ, Dao J, Fréchet JMJ (1996) J Am Chem Soc 118:11,111
127. Matyjaszewski K, Shigemoto T, Fréchet JMJ, Leduc M (1996) Macromolecules 29:4167
128. Amabilino DB, Ashton PR, Balzani V, Brown CL, Credi A, Fréchet JMJ, Leon JW, Raymo FM, Spencer N, Stoddart JF, Venturi M (1996) J Am Chem Soc 118:12,012
129. Chapman TM, Hillyer GL, Mahan EJ, Shaffer KA (1994) J Am Chem Soc 116:11,195
130. van Hest JCM, Baars MWPL, Elissen-Román C, van Genderen MHP, Meijer EW (1995) Macromolecules 28:6689
131. van Hest JCM, Delnoye DAP, Baars MWPL, van Genderen MHP, Meier EW (1995) Science 268:1592
132. Aoi K, Motada A, Okada M, Imae T (1997) Macromol Rapid Commun 18:945
133. Chang JY, Ji HJ, Han MJ, Rhee SB, Cheong S, Yoon M (1994) Macromolecules 27:1376
134. Slany M, Bardaji M, Caminade A-M, Chaudret B, Majoral JP (1997) Inorg Chem 36:1939
135. Nijenhuis A (1997) private communication
136. Warakomski J (1992) Chem Mat 4:1000
137. Risch BG, Wilkes GL, Warakomski JM (1993) Polymer 34:2330
138. Yen DR, Merrill EW (1997) Polymer Preprints 38:531
139. Fetters LJ, Hadjichristidis N (1980) Macromolecules 13:191
140. Zhou L-L, Hadjichristidis N, Toporowski PM, Roovers J (1992) Rubber Chem Techn 65:303
141. Roovers J (1994) Macromolecules 27:5359
142. Avgeropoulos A, Poulos Y, Hadjichristidis N, Roovers J (1996) Macromolecules 29:6076
143. van Aert HAM, Burkard MEM, Jansen JFGA, van Genderen MHP, Meijer EW, Oevering H, Buning GHW (1995) Macromolecules 28:7967
144. van Aert HAM, van Genderen MHP, Meijer EW (1996) Polym Bull 37:273
145. Tam JP (1988) Proc Natl Acad Sci USA 85:5409
146. Posnett DN, McGrath H, Tam JP (1988) J Biol Chem 263:1719
147. Chang K-J, Pugh W, Blanchard SG, McDermed J, Tam JP (1988) Proc Natl Acad Sci USA 85:4929
148. Defoort J-P, Nardelli B, Huang W, Ho DD, Tam JP (1992) Proc Natl Acad Sci USA 89:3879
149. Rao C, Tam JP (1994) J Am Chem Soc 116:6975
150. Shao J, Tam JP (1995) J Am Chem Soc 117:3893
151. Spetzler J, Tam JP (1995) Int J Pept Protein Res 45:78
152. Delong R, Stephenson K, Loftus T, Fisher M, Alahari S, Nolting A, Juliano RL (1997) J Pharmac Sci 86:762

153. Poxon SW, Mitchell PM, Liang E, Hughes JA (1996) Drug Delivery 3:255
154. Bielinska A, Kukowska-Latallo JF, Johnson J, Tomalia DA, Baker JR Jr (1996) Nucleic Acids Res 24:2176
155. Kukowska-Latallo JF, Bielinska AU, Johnson J, Spindler R, Tomalia DA, Baker JR Jr (1996) Proc Natl Acad Sci 93:4897
156. Haensler J, Szoka FC Jr (1993) Bioconjugate Chem 4:372
157. Bielinska AU, Kukowska-Latallo JF, Baker JR Jr (1997) Biochim Biophys Acta 1353:180
158. Tang MX, Redemann CT, Szoka FC Jr (1996) Bioconjugate Chem 7:703
159. Roberts JC, Adams YE, Tomalia DA, Mercer-Smith JA, Lavallee DK (1990) Bioconjugate Chem 1:305
160. Wu C, Brechbiel MW, Kozak RW, Gansow OA (1994) Bioorg Med Chem Lett 4:449
161. Barth RF, Adams DM, Soloway AH, Alam F, Darby MV (1994) Bioconjugate Chem 5:58
162. Capala J, Barth RF, Bendayan M, Lauzon M, Adams DM, Soloway AH, Fenstermacher RA, Carlsson J (1996) Bioconjugate Chem 7:7

Received: March 1998

Author Index Volumes 101–142

Author Index Volumes 1-100 see Volume 100

de Abajo, J. and *de la Campa, J.G.*: Processable Aromatic Polyimides. Vol. 140, pp. 23-60.
Adolf, D. B. see Ediger, M. D.: Vol. 116, pp. 73-110.
Aharoni, S. M. and *Edwards, S. F.*: Rigid Polymer Networks. Vol. 118, pp. 1-231.
Améduri, B., Boutevin, B. and *Gramain, P.*: Synthesis of Block Copolymers by Radical Polymerization and Telomerization. Vol. 127, pp. 87-142.
Améduri, B. and *Boutevin, B.*: Synthesis and Properties of Fluorinated Telechelic Monodispersed Compounds. Vol. 102, pp. 133-170.
Amselem, S. see Domb, A. J.: Vol. 107, pp. 93-142.
Andrady, A. L.: Wavelenght Sensitivity in Polymer Photodegradation. Vol. 128, pp. 47-94.
Andreis, M. and *Koenig, J. L.*: Application of Nitrogen-15 NMR to Polymers. Vol. 124, pp. 191-238.
Angiolini, L. see Carlini, C.: Vol. 123, pp. 127-214.
Anseth, K. S., Newman, S. M. and *Bowman, C. N.*: Polymeric Dental Composites: Properties and Reaction Behavior of Multimethacrylate Dental Restorations. Vol. 122, pp. 177-218.
Armitage, B. A. see O'Brien, D. F.: Vol. 126, pp. 53-58.
Arndt, M. see Kaminski, W.: Vol. 127, pp. 143-187.
Arnold Jr., F. E. and *Arnold, F. E.*: Rigid-Rod Polymers and Molecular Composites. Vol. 117, pp. 257-296.
Arshady, R.: Polymer Synthesis via Activated Esters: A New Dimension of Creativity in Macromolecular Chemistry. Vol. 111, pp. 1-42.

Bahar, I., Erman, B. and *Monnerie, L.*: Effect of Molecular Structure on Local Chain Dynamics: Analytical Approaches and Computational Methods. Vol. 116, pp. 145-206.
Baltá-Calleja, F. J., González Arche, A., Ezquerra, T. A., Santa Cruz, C., Batallón, F., Frick, B. and *López Cabarcos, E.*: Structure and Properties of Ferroelectric Copolymers of Poly(vinylidene) Fluoride. Vol. 108, pp. 1-48.
Barshtein, G. R. and *Sabsai, O. Y.*: Compositions with Mineralorganic Fillers. Vol. 101, pp.1-28.
Batallán, F. see Baltá-Calleja, F. J.: Vol. 108, pp. 1-48.
Barton, J. see Hunkeler, D.: Vol. 112, pp. 115-134.
Bell, C. L. and *Peppas, N. A.*: Biomedical Membranes from Hydrogels and Interpolymer Complexes. Vol. 122, pp. 125-176.
Bellon-Maurel, A. see Calmon-Decriaud, A.: Vol. 135, pp. 207-226.
Bennett, D. E. see O'Brien, D. F.: Vol. 126, pp. 53-84.
Berry, G.C.: Static and Dynamic Light Scattering on Moderately Concentraded Solutions: Isotropic Solutions of Flexible and Rodlike Chains and Nematic Solutions of Rodlike Chains. Vol. 114, pp. 233-290.
Bershtein, V. A. and *Ryzhov, V. A.*: Far Infrared Spectroscopy of Polymers. Vol. 114, pp. 43-122.
Bigg, D. M.: Thermal Conductivity of Heterophase Polymer Compositions. Vol. 119, pp. 1-30.
Binder, K.: Phase Transitions in Polymer Blends and Block Copolymer Melts: Some Recent Developments. Vol. 112, pp. 115-134.

Binder, K.: Phase Transitions of Polymer Blends and Block Copolymer Melts in Thin Films. Vol. 138, pp. 1-90.
Bird, R. B. see Curtiss, C. F.: Vol. 125, pp. 1-102.
Biswas, M. and *Mukherjee, A.*: Synthesis and Evaluation of Metal-Containing Polymers. Vol. 115, pp. 89-124.
Boutevin, B. and *Robin, J. J.*: Synthesis and Properties of Fluorinated Diols. Vol. 102. pp. 105-132.
Boutevin, B. see Amédouri, B.: Vol. 102, pp. 133-170.
Boutevin, B. see Améduri, B.: Vol. 127, pp. 87-142.
Bowman, C. N. see Anseth, K. S.: Vol. 122, pp. 177-218.
Boyd, R. H.: Prediction of Polymer Crystal Structures and Properties. Vol. 116, pp. 1-26.
Briber, R. M. see Hedrick, J. L.: Vol. 141, pp. 1-44.
Bronnikov, S. V., Vettegren, V. I. and *Frenkel, S. Y.*: Kinetics of Deformation and Relaxation in Highly Oriented Polymers. Vol. 125, pp. 103-146.
Bruza, K. J. see Kirchhoff, R. A.: Vol. 117, pp. 1-66.
Burban, J. H. see Cussler, E. L.: Vol. 110, pp. 67-80.

Calmon-Decriaud, A. Bellon-Maurel, V., Silvestre, F.: Standard Methods for Testing the Aerobic Biodegradation of Polymeric Materials. Vol 135, pp. 207-226.
Cameron, N. R. and *Sherrington, D. C.*: High Internal Phase Emulsions (HIPEs)-Structure, Properties and Use in Polymer Preparation. Vol. 126, pp. 163-214.
de la Campa, J. G. see de Abajo, , J.: Vol. 140, pp. 23-60.
Candau, F. see Hunkeler, D.: Vol. 112, pp. 115-134.
Canelas, D. A. and *DeSimone, J. M.*: Polymerizations in Liquid and Supercritical Carbon Dioxide. Vol. 133, pp. 103-140.
Capek, I.: Kinetics of the Free-Radical Emulsion Polymerization of Vinyl Chloride. Vol. 120, pp. 135-206.
Carlini, C. and *Angiolini, L.*: Polymers as Free Radical Photoinitiators. Vol. 123, pp. 127-214.
Carter, K. R. see Hedrick, J. L.: Vol. 141, pp. 1-44.
Casas-Vazquez, J. see Jou, D.: Vol. 120, pp. 207-266.
Chandrasekhar, V.: Polymer Solid Electrolytes: Synthesis and Structure. Vol 135, pp. 139-206
Charleux, B., Faust R.: Synthesis of Branched Polymers by Cationic Polymerization. Vol. 142, pp. 1-70.
Chen, P. see Jaffe, M.: Vol. 117, pp. 297-328.
Choe, E.-W. see Jaffe, M.: Vol. 117, pp. 297-328.
Chow, T. S.: Glassy State Relaxation and Deformation in Polymers. Vol. 103, pp. 149-190.
Chung, T.-S. see Jaffe, M.: Vol. 117, pp. 297-328.
Comanita, B. see Roovers, J.: Vol. 142, pp. 179-228.
Connell, J. W. see Hergenrother, P. M.: Vol. 117, pp. 67-110.
Criado-Sancho, M. see Jou, D.: Vol. 120, pp. 207-266.
Curro, J.G. see Schweizer, K.S.: Vol. 116, pp. 319-378.
Curtiss, C. F. and *Bird, R. B.*: Statistical Mechanics of Transport Phenomena: Polymeric Liquid Mixtures. Vol. 125, pp. 1-102.
Cussler, E. L., Wang, K. L. and *Burban, J. H.*: Hydrogels as Separation Agents. Vol. 110, pp. 67-80.

DeSimone, J. M. see Canelas D. A.: Vol. 133, pp. 103-140.
DiMari, S. see Prokop, A.: Vol. 136, pp. 1-52.
Dimonie, M. V. see Hunkeler, D.: Vol. 112, pp. 115-134.
Dodd, L. R. and *Theodorou, D. N.*: Atomistic Monte Carlo Simulation and Continuum Mean Field Theory of the Structure and Equation of State Properties of Alkane and Polymer Melts. Vol. 116, pp. 249-282.
Doelker, E.: Cellulose Derivatives. Vol. 107, pp. 199-266.
Dolden, J. G.: Calculation of a Mesogenic Index with Emphasis Upon LC-Polyimides. Vol. 141, pp. 189-245.
Domb, A. J., Amselem, S., Shah, J. and *Maniar, M.*: Polyanhydrides: Synthesis and Characterization. Vol.107, pp. 93-142.
Dubrovskii, S. A. see Kazanskii, K. S.: Vol. 104, pp. 97-134.

Dunkin, I. R. see Steinke, J.: Vol. 123, pp. 81-126.
Dunson, D. L. see McGrath, J. E.: Vol. 140, pp. 61-106.

Economy, J. and *Goranov, K.*: Thermotropic Liquid Crystalline Polymers for High Performance Applications. Vol. 117, pp. 221-256.
Ediger, M. D. and *Adolf, D. B.*: Brownian Dynamics Simulations of Local Polymer Dynamics. Vol. 116, pp. 73-110.
Edwards, S. F. see Aharoni, S. M.: Vol. 118, pp. 1-231.
Endo, T. see Yagci, Y.: Vol. 127, pp. 59-86.
Erman, B. see Bahar, I.: Vol. 116, pp. 145-206.
Ewen, B, Richter, D.: Neutron Spin Echo Investigations on the Segmental Dynamics of Polymers in Melts, Networks and Solutions. Vol. 134, pp. 1-130.
Ezquerra, T. A. see Baltá-Calleja, F. J.: Vol. 108, pp. 1-48.

Faust, R. see Charleux, B: Vol. 142, pp. 1-70.
Fekete, E see Pukánszky, B: Vol. 139, pp. 109-154.
Fendler, J.H.: Membrane-Mimetic Approach to Advanced Materials. Vol. 113, pp. 1-209.
Fetters, L. J. see Xu, Z.: Vol. 120, pp. 1-50.
Förster, S. and *Schmidt, M.*: Polyelectrolytes in Solution. Vol. 120, pp. 51-134.
Frenkel, S. Y. see Bronnikov, S. V.: Vol. 125, pp. 103-146.
Frick, B. see Baltá-Calleja, F. J.: Vol. 108, pp. 1-48.
Fridman, M. L.: see Terent'eva, J. P.: Vol. 101, pp. 29-64.
Funke, W.: Microgels-Intramolecularly Crosslinked Macromolecules with a Globular Structure. Vol. 136, pp. 137-232.

Galina, H.: Mean-Field Kinetic Modeling of Polymerization: The Smoluchowski Coagulation Equation. Vol. 137, pp. 135-172.
Ganesh, K. see Kishore, K.: Vol. 121, pp. 81-122.
Gaw, K. O. and *Kakimoto, M.*: Polyimide-Epoxy Composites. Vol. 140, pp. 107-136.
Geckeler, K. E. see Rivas, B.: Vol. 102, pp. 171-188.
Geckeler, K. E.: Soluble Polymer Supports for Liquid-Phase Synthesis. Vol. 121, pp. 31-80.
Gehrke, S. H.: Synthesis, Equilibrium Swelling, Kinetics Permeability and Applications of Environmentally Responsive Gels. Vol. 110, pp. 81-144.
de Gennes, P.-G.: Flexible Polymers in Nanopores. Vol. 138, pp. 91-106.
Giannelis, E.P., Krishnamoorti, R., Manias, E.: Polymer-Silicate Nanocomposites: Model Systems for Confined Polymers and Polymer Brushes. Vol. 138, pp. 107-148.
Godovsky, D. Y.: Electron Behavior and Magnetic Properties Polymer-Nanocomposites. Vol. 119, pp. 79-122.
González Arche, A. see Baltá-Calleja, F. J.: Vol. 108, pp. 1-48.
Goranov, K. see Economy, J.: Vol. 117, pp. 221-256.
Gramain, P. see Améduri, B.: Vol. 127, pp. 87-142.
Grest, G.S.: Normal and Shear Forces Between Polymer Brushes. Vol. 138, pp. 149-184
Grosberg, A. and *Nechaev, S.*: Polymer Topology. Vol. 106, pp. 1-30.
Grubbs, R., Risse, W. and *Novac, B.*: The Development of Well-defined Catalysts for Ring-Opening Olefin Metathesis. Vol. 102, pp. 47-72.
van Gunsteren, W. F. see Gusev, A. A.: Vol. 116, pp. 207-248.
Gusev, A. A., Müller-Plathe, F., van Gunsteren, W. F. and *Suter, U. W.*: Dynamics of Small Molecules in Bulk Polymers. Vol. 116, pp. 207-248.
Guillot, J. see Hunkeler, D.: Vol. 112, pp. 115-134.
Guyot, A. and *Tauer, K.*: Reactive Surfactants in Emulsion Polymerization. Vol. 111, pp. 43-66.

Hadjichristidis, N., Pispas, S., Pitsikalis, M., Iatrou, H., Vlahos, C.: Asymmetric Star Polymers Synthesis and Properties. Vol. 142, pp. 71-128.
Hadjichristidis, N. see Xu, Z.: Vol. 120, pp. 1-50.
Hadjichristidis, N. see Pitsikalis, M.: Vol. 135, pp. 1-138.

Hall, H. K. see *Penelle, J.*: Vol. 102, pp. 73-104.
Hammouda, B.: SANS from Homogeneous Polymer Mixtures: A Unified Overview. Vol. 106, pp. 87-134.
Harada, A.: Design and Construction of Supramolecular Architectures Consisting of Cyclodextrins and Polymers. Vol. 133, pp. 141-192.
Haralson, M. A. see *Prokop, A.*: Vol. 136, pp. 1-52.
Hawker, C. J. see *Hedrick, J. L.*: Vol. 141, pp. 1-44.
Hedrick, J. L., Carter, K. R., Labadie, J. W., Miller, R. D., Volksen, W., Hawker, C. J., Yoon, D. Y., Russell, T. P., McGrath, J. E., Briber, R. M.: Nanoporous Polyimides. Vol. 141, pp. 1-44.
Hedrick, J. L. see *Hergenrother, P. M.*: Vol. 117, pp. 67-110.
Hedrick, J.L. see *McGrath, J. E.*: Vol. 140, pp. 61-106.
Heller, J.: Poly (Ortho Esters). Vol. 107, pp. 41-92.
Hemielec, A. A. see *Hunkeler, D.*: Vol. 112, pp. 115-134.
Hergenrother, P. M., Connell, J. W., Labadie, J. W. and *Hedrick, J. L.*: Poly(arylene ether)s Containing Heterocyclic Units. Vol. 117, pp. 67-110.
Hervet, H. see *Léger, L.*: Vol. 138, pp. 185-226.
Hiramatsu, N. see *Matsushige, M.*: Vol. 125, pp. 147-186.
Hirasa, O. see *Suzuki, M.*: Vol. 110, pp. 241-262.
Hirotsu, S.: Coexistence of Phases and the Nature of First-Order Transition in Poly-N-isopropylacrylamide Gels. Vol. 110, pp. 1-26.
Hornsby, P.: Rheology, Compoundind and Processing of Filled Thermoplastics. Vol. 139, pp. 155-216.
Hunkeler, D., Candau, F., Pichot, C., Hemielec, A. E., Xie, T. Y., Barton, J., Vaskova, V., Guillot, J., Dimonie, M. V., Reichert, K. H.: Heterophase Polymerization: A Physical and Kinetic Comparision and Categorization. Vol. 112, pp. 115-134.
Hunkeler, D. see *Prokop, A.*: Vol. 136, pp. 1-52; 53-74.

Iatrou, H. see *Hadjichristidis, N.*: Vol. 142, pp. 71-128.
Ichikawa, T. see *Yoshida, H.*: Vol. 105, pp. 3-36.
Ihara, E. see *Yasuda, H.*: Vol. 133, pp. 53-102.
Ikada, Y. see *Uyama, Y.*: Vol. 137, pp. 1-40.
Ilavsky, M.: Effect on Phase Transition on Swelling and Mechanical Behavior of Synthetic Hydrogels. Vol. 109, pp. 173-206.
Imai, Y.: Rapid Synthesis of Polyimides from Nylon-Salt Monomers. Vol. 140, pp. 1-23.
Inomata, H. see *Saito, S.*: Vol. 106, pp. 207-232.
Irie, M.: Stimuli-Responsive Poly(N-isopropylacrylamide), Photo- and Chemical-Induced Phase Transitions. Vol. 110, pp. 49-66.
Ise, N. see *Matsuoka, H.*: Vol. 114, pp. 187-232.
Ito, K., Kawaguchi, S,: Poly(macronomers), Homo- and Copolymerization. Vol. 142, pp. 129-178.
Ivanov, A. E. see *Zubov, V. P.*: Vol. 104, pp. 135-176.

Jaffe, M., Chen, P., Choe, E.-W., Chung, T.-S. and *Makhija, S.*: High Performance Polymer Blends. Vol. 117, pp. 297-328.
Jancar, J.: Structure-Property Relationships in Thermoplastic Matrices. Vol. 139, pp. 1-66.
Joos-Müller, B. see *Funke, W.*: Vol. 136, pp. 137-232.
Jou, D., Casas-Vazquez, J. and *Criado-Sancho, M.*: Thermodynamics of Polymer Solutions under Flow: Phase Separation and Polymer Degradation. Vol. 120, pp. 207-266.

Kaetsu, I.: Radiation Synthesis of Polymeric Materials for Biomedical and Biochemical Applications. Vol. 105, pp. 81-98.
Kakimoto, M. see *Gaw, K. O.*: Vol. 140, pp. 107-136.
Kaminski, W. and *Arndt, M.*: Metallocenes for Polymer Catalysis. Vol. 127, pp. 143-187.
Kammer, H. W., Kressler, H. and *Kummerloewe, C.*: Phase Behavior of Polymer Blends - Effects of Thermodynamics and Rheology. Vol. 106, pp. 31-86.

Kandyrin, L. B. and *Kuleznev, V. N.*: The Dependence of Viscosity on the Composition of Concentrated Dispersions and the Free Volume Concept of Disperse Systems. Vol. 103, pp. 103-148.
Kaneko, M. see Ramaraj, R.: Vol. 123, pp. 215-242.
Kang, E. T., Neoh, K. G. and *Tan, K. L.*: X-Ray Photoelectron Spectroscopic Studies of Electroactive Polymers. Vol. 106, pp. 135-190.
Kato, K. see Uyama, Y.: Vol. 137, pp. 1-40.
Kawaguchi, S. see Ito, K.: Vol. 142, p 129-178.
Kazanskii, K. S. and *Dubrovskii, S. A.*: Chemistry and Physics of „Agricultural" Hydrogels. Vol. 104, pp. 97-134.
Kennedy, J. P. see Majoros, I.: Vol. 112, pp. 1-113.
Khokhlov, A., Starodybtzev, S. and *Vasilevskaya, V.*: Conformational Transitions of Polymer Gels: Theory and Experiment. Vol. 109, pp. 121-172.
Kilian, H. G. and *Pieper, T.*: Packing of Chain Segments. A Method for Describing X-Ray Patterns of Crystalline, Liquid Crystalline and Non-Crystalline Polymers. Vol. 108, pp. 49-90.
Kishore, K. and *Ganesh, K.*: Polymers Containing Disulfide, Tetrasulfide, Diselenide and Ditelluride Linkages in the Main Chain. Vol. 121, pp. 81-122.
Kitamaru, R.: Phase Structure of Polyethylene and Other Crystalline Polymers by Solid-State ^{13}C/MNR. Vol. 137, pp 41-102.
Klier, J. see Scranton, A. B.: Vol. 122, pp. 1-54.
Kobayashi, S., Shoda, S. and *Uyama, H.*: Enzymatic Polymerization and Oligomerization. Vol. 121, pp. 1-30.
Koenig, J. L. see Andreis, M.: Vol. 124, pp. 191-238.
Kokufuta, E.: Novel Applications for Stimulus-Sensitive Polymer Gels in the Preparation of Functional Immobilized Biocatalysts. Vol. 110, pp. 157-178.
Konno, M. see Saito, S.: Vol. 109, pp. 207-232.
Kopecek, J. see Putnam, D.: Vol. 122, pp. 55-124.
Koßmehl, G. see Schopf, G.: Vol. 129, pp. 1-145.
Kressler, J. see Kammer, H. W.: Vol. 106, pp. 31-86.
Kricheldorf, H. R.: Liquid-Cristalline Polyimides. Vol. 141, pp. 83-188.
Krishnamoorti, R. see Giannelis, E.P.: Vol. 138, pp. 107-148.
Kirchhoff, R. A. and *Bruza, K. J.*: Polymers from Benzocyclobutenes. Vol. 117, pp. 1-66.
Kuchanov, S. I.: Modern Aspects of Quantitative Theory of Free-Radical Copolymerization. Vol. 103, pp. 1-102.
Kuleznev, V. N. see Kandyrin, L. B.: Vol. 103, pp. 103-148.
Kulichkhin, S. G. see Malkin, A. Y.: Vol. 101, pp. 217-258.
Kummerloewe, C. see Kammer, H. W.: Vol. 106, pp. 31-86.
Kuznetsova, N. P. see Samsonov, G. V.: Vol. 104, pp. 1-50. Labadie, J. W. see Hergenrother, P. M.: Vol. 117, pp. 67-110.

Labadie, J. W. see Hedrick, J. L.: Vol. 141, pp. 1-44.
Lamparski, H. G. see O´Brien, D. F.: Vol. 126, pp. 53-84.
Laschewsky, A.: Molecular Concepts, Self-Organisation and Properties of Polysoaps. Vol. 124, pp. 1-86.
Laso, M. see Leontidis, E.: Vol. 116, pp. 283-318.
Lazár, M. and *RychlΩ, R.*: Oxidation of Hydrocarbon Polymers. Vol. 102, pp. 189-222.
Lechowicz, J. see Galina, H.: Vol. 137, pp. 135-172.
Léger, L., Raphaël, E., Hervet, H.: Surface-Anchored Polymer Chains: Their Role in Adhesion and Friction. Vol. 138, pp. 185-226.
Lenz, R. W.: Biodegradable Polymers. Vol. 107, pp. 1-40.
Leontidis, E., de Pablo, J. J., Laso, M. and *Suter, U. W.*: A Critical Evaluation of Novel Algorithms for the Off-Lattice Monte Carlo Simulation of Condensed Polymer Phases. Vol. 116, pp. 283-318.
Lesec, J. see Viovy, J.-L.: Vol. 114, pp. 1-42.
Liang, G. L. see Sumpter, B. G.: Vol. 116, pp. 27-72.
Lienert, K.-W.: Poly(ester-imide)s for Industrial Use. Vol. 141, pp. 45-82.

Lin, J. and *Sherrington, D. C.*: Recent Developments in the Synthesis, Thermostability and Liquid Crystal Properties of Aromatic Polyamides. Vol. 111, pp. 177-220.
López Cabarcos, E. see Baltá-Calleja, F. J.: Vol. 108, pp. 1-48.

Majoros, I., Nagy, A. and *Kennedy, J. P.*: Conventional and Living Carbocationic Polymerizations United. I. A Comprehensive Model and New Diagnostic Method to Probe the Mechanism of Homopolymerizations. Vol. 112, pp. 1-113.
Makhija, S. see Jaffe, M.: Vol. 117, pp. 297-328.
Malkin, A. Y. and *Kulichkhin, S. G.*: Rheokinetics of Curing. Vol. 101, pp. 217-258.
Maniar, M. see Domb, A. J.: Vol. 107, pp. 93-142.
Manias, E., see Giannelis, E.P.: Vol. 138, pp. 107-148.
Mashima, K., Nakayama, Y. and *Nakamura, A.*: Recent Trends in Polymerization of a-Olefins Catalyzed by Organometallic Complexes of Early Transition Metals. Vol. 133, pp. 1-52.
Matsumoto, A.: Free-Radical Crosslinking Polymerization and Copolymerization of Multivinyl Compounds. Vol. 123, pp. 41-80.
Matsumoto, A. see Otsu, T.: Vol. 136, pp. 75-138.
Matsuoka, H. and *Ise, N.*: Small-Angle and Ultra-Small Angle Scattering Study of the Ordered Structure in Polyelectrolyte Solutions and Colloidal Dispersions. Vol. 114, pp. 187-232.
Matsushige, K., Hiramatsu, N. and *Okabe, H.*: Ultrasonic Spectroscopy for Polymeric Materials. Vol. 125, pp. 147-186.
Mattice, W. L. see Rehahn, M.: Vol. 131/132, pp. 1-475.
Mays, W. see Xu, Z.: Vol. 120, pp. 1-50.
Mays, J.W. see Pitsikalis, M.: Vol.135, pp. 1-138.
McGrath, J. E. see Hedrick, J. L.: Vol. 141, pp. 1-44.
McGrath, J. E., Dunson, D. L., Hedrick, J. L.: Synthesis and Characterization of Segmented Polyimide-Polyorganosiloxane Copolymers. Vol. 140, pp. 61-106.
Mecham, S. J. see McGrath, J. E.: Vol. 140, pp. 61-106.
Mikos, A. G. see Thomson, R. C.: Vol. 122, pp. 245-274.
Mison, P. and Sillion, B.: Thermosetting Oligomers Containing Maleimides and Nadiimides End-Groups. Vol. 140, pp. 137-180.
Miyasaka, K.: PVA-Iodine Complexes: Formation, Structure and Properties. Vol. 108. pp. 91-130.
Miller, R. D. see Hedrick, J. L.: Vol. 141, pp. 1-44.
Monnerie, L. see Bahar, I.: Vol. 116, pp. 145-206.
Morishima, Y.: Photoinduced Electron Transfer in Amphiphilic Polyelectrolyte Systems. Vol. 104, pp. 51-96.
Mours, M. see Winter, H. H.: Vol. 134, pp. 165-234.
Müllen, K. see Scherf, U.: Vol. 123, pp. 1-40.
Müller-Plathe, F. see Gusev, A. A.: Vol. 116, pp. 207-248.
Mukerherjee, A. see Biswas, M.: Vol. 115, pp. 89-124.
Mylnikov, V.: Photoconducting Polymers. Vol. 115, pp. 1-88.

Nagy, A. see Majoros, I.: Vol. 112, pp. 1-11.
Nakamura, A. see Mashima, K.: Vol. 133, pp. 1-52.
Nakayama, Y. see Mashima, K.: Vol. 133, pp. 1-52.
Narasinham, B., Peppas, N. A.: The Physics of Polymer Dissolution: Modeling Approaches and Experimental Behavior. Vol. 128, pp. 157-208.
Nechaev, S. see Grosberg, A.: Vol. 106, pp. 1-30.
Neoh, K. G. see Kang, E. T.: Vol. 106, pp. 135-190.
Newman, S. M. see Anseth, K. S.: Vol. 122, pp. 177-218.
Nijenhuis, K. te: Thermoreversible Networks. Vol. 130, pp. 1-252.
Noid, D. W. see Sumpter, B. G.: Vol. 116, pp. 27-72.
Novac, B. see Grubbs, R.: Vol. 102, pp. 47-72.
Novikov, V. V. see Privalko, V. P.: Vol. 119, pp. 31-78.

O'Brien, D. F., Armitage, B. A., Bennett, D. E. and *Lamparski, H. G.*: Polymerization and Domain Formation in Lipid Assemblies. Vol. 126, pp. 53-84.

Ogasawara, M.: Application of Pulse Radiolysis to the Study of Polymers and Polymerizations. Vol.105, pp. 37-80.
Okabe, H. see Matsushige, K.: Vol. 125, pp. 147-186.
Okada, M.: Ring-Opening Polymerization of Bicyclic and Spiro Compounds. Reactivities and Polymerization Mechanisms. Vol. 102, pp. 1-46.
Okano, T.: Molecular Design of Temperature-Responsive Polymers as Intelligent Materials. Vol. 110, pp. 179-198.
Okay, O. see Funke, W.: Vol. 136, pp. 137-232.
Onuki, A.: Theory of Phase Transition in Polymer Gels. Vol. 109, pp. 63-120.
Osad'ko, I.S.: Selective Spectroscopy of Chromophore Doped Polymers and Glasses. Vol. 114, pp. 123-186.
Otsu, T., Matsumoto, A.: Controlled Synthesis of Polymers Using the Iniferter Technique: Developments in Living Radical Polymerization. Vol. 136, pp. 75-138.

de Pablo, J. J. see Leontidis, E.: Vol. 116, pp. 283-318.
Padias, A. B. see Penelle, J.: Vol. 102, pp. 73-104.
Pascault, J.-P. see Williams, R. J. J.: Vol. 128, pp. 95-156.
Pasch, H.: Analysis of Complex Polymers by Interaction Chromatography. Vol. 128, pp. 1-46.
Penelle, J., Hall, H. K., Padias, A. B. and *Tanaka, H.*: Captodative Olefins in Polymer Chemistry. Vol. 102, pp. 73-104.
Peppas, N. A. see Bell, C. L.: Vol. 122, pp. 125-176.
Peppas, N. A. see Narasimhan, B.: Vol. 128, pp. 157-208.
Pichot, C. see Hunkeler, D.: Vol. 112, pp. 115-134.
Pieper, T. see Kilian, H. G.: Vol. 108, pp. 49-90.
Pispas, S. see Pitsikalis, M.: Vol. 135, pp. 1-138.
Pispas, S. see Hadjichristidis: Vol. 142, pp. 71-128.
Pitsikalis, M., Pispas, S., Mays, J. W., Hadjichristidis, N.: Nonlinear Block Copolymer Architectures. Vol. 135, pp. 1-138.
Pitsikalis, M. see Hadjichristidis: Vol. 142, pp. 71-128.
Pospíšil, J.: Functionalized Oligomers and Polymers as Stabilizers for Conventional Polymers. Vol. 101, pp. 65-168.
Pospíšil, J.: Aromatic and Heterocyclic Amines in Polymer Stabilization. Vol. 124, pp. 87-190.
Powers, A. C. see Prokop, A.: Vol. 136, pp. 53-74.
Priddy, D. B.: Recent Advances in Styrene Polymerization. Vol. 111, pp. 67-114.
Priddy, D. B.: Thermal Discoloration Chemistry of Styrene-co-Acrylonitrile. Vol. 121, pp. 123-154.
Privalko, V. P. and *Novikov, V. V.*: Model Treatments of the Heat Conductivity of Heterogeneous Polymers. Vol. 119, pp 31-78.
Prokop, A., Hunkeler, D., Powers, A. C., Whitesell, R. R., Wang, T. G.: Water Soluble Polymers for Immunoisolation II: Evaluation of Multicomponent Microencapsulation Systems. Vol. 136, pp. 53-74.
Prokop, A., Hunkeler, D., DiMari, S., Haralson, M. A., Wang, T. G.: Water Soluble Polymers for Immunoisolation I: Complex Coacervation and Cytotoxicity. Vol. 136, pp. 1-52.
Pukánszky, B. and *Fekete, E.*: Adhesion and Surface Modification. Vol. 139, pp. 109 -154.
Putnam, D. and *Kopecek, J.*: Polymer Conjugates with Anticancer Acitivity. Vol. 122, pp. 55- 124.

Ramaraj, R. and *Kaneko, M.*: Metal Complex in Polymer Membrane as a Model for Photosynthetic Oxygen Evolving Center. Vol. 123, pp. 215-242.
Rangarajan, B. see Scranton, A. B.: Vol. 122, pp. 1-54.
Raphaël, E. see Léger, L.: Vol. 138, pp. 185-226.
Reichert, K. H. see Hunkeler, D.: Vol. 112, pp. 115-134.
Rehahn, M., Mattice, W. L., Suter, U. W.: Rotational Isomeric State Models in Macromolecular Systems. Vol. 131/132, pp. 1-475.
Richter, D. see Ewen, B.: Vol. 134, pp.1-130.
Risse, W. see Grubbs, R.: Vol. 102, pp. 47-72.

Rivas, B. L. and Geckeler, K. E.: Synthesis and Metal Complexation of Poly(ethyleneimine) and Derivatives. Vol. 102, pp. 171-188.
Robin, J. J. see Boutevin, B.: Vol. 102, pp. 105-132.
Roe, R.-J.: MD Simulation Study of Glass Transition and Short Time Dynamics in Polymer Liquids. Vol. 116, pp. 111-114.
Roovers, J., Comanita, B.: Dendrimers and Dendrimer-Polymer Hybrids. Vol. 142, pp xxx-xxx.
Rothon, R. N.: Mineral Fillers in Thermoplastics: Filler Manufacture and Characterisation. Vol. 139, pp. 67-108.
Rozenberg, B. A. see Williams, R. J. J.: Vol. 128, pp. 95-156.
Ruckenstein, E.: Concentrated Emulsion Polymerization. Vol. 127, pp. 1-58.
Rusanov, A. L.: Novel Bis (Naphtalic Anhydrides) and Their Polyheteroarylenes with Improved Processability. Vol. 111, pp. 115-176.
Russel, T. P. see Hedrick, J. L.: Vol. 141, pp. 1-44.
Rychlý, J. see Lazár, M.: Vol. 102, pp. 189-222.
Ryzhov, V. A. see Bershtein, V. A.: Vol. 114, pp. 43-122.

Sabsai, O. Y. see Barshtein, G. R.: Vol. 101, pp. 1-28.
Saburov, V. V. see Zubov, V. P.: Vol. 104, pp. 135-176.
Saito, S., Konno, M. and Inomata, H.: Volume Phase Transition of N-Alkylacrylamide Gels. Vol. 109, pp. 207-232.
Samsonov, G. V. and Kuznetsova, N. P.: Crosslinked Polyelectrolytes in Biology. Vol. 104, pp. 1-50.
Santa Cruz, C. see Baltá-Calleja, F. J.: Vol. 108, pp. 1-48.
Sato, T. and Teramoto, A.: Concentrated Solutions of Liquid-Christalline Polymers. Vol. 126, pp. 85-162.
Scherf, U. and Müllen, K.: The Synthesis of Ladder Polymers. Vol. 123, pp. 1-40.
Schmidt, M. see Förster, S.: Vol. 120, pp. 51-134.
Schopf, G. and Koßmehl, G.: Polythiophenes - Electrically Conductive Polymers. Vol. 129, pp. 1-145.
Schweizer, K. S.: Prism Theory of the Structure, Thermodynamics, and Phase Transitions of Polymer Liquids and Alloys. Vol. 116, pp. 319-378.
Scranton, A. B., Rangarajan, B. and Klier, J.: Biomedical Applications of Polyelectrolytes. Vol. 122, pp. 1-54.
Sefton, M. V. and Stevenson, W. T. K.: Microencapsulation of Live Animal Cells Using Polycrylates. Vol.107, pp. 143-198.
Shamanin, V. V.: Bases of the Axiomatic Theory of Addition Polymerization. Vol. 112, pp. 135-180.
Sherrington, D. C. see Cameron, N. R., Vol. 126, pp. 163-214.
Sherrington, D. C. see Lin, J.: Vol. 111, pp. 177-220.
Sherrington, D. C. see Steinke, J.: Vol. 123, pp. 81-126.
Shibayama, M. see Tanaka, T.: Vol. 109, pp. 1-62.
Shiga, T.: Deformation and Viscoelastic Behavior of Polymer Gels in Electric Fields. Vol. 134, pp. 131-164.
Shoda, S. see Kobayashi, S.: Vol. 121, pp. 1-30.
Siegel, R. A.: Hydrophobic Weak Polyelectrolyte Gels: Studies of Swelling Equilibria and Kinetics. Vol. 109, pp. 233-268.
Silvestre, F. see Calmon-Decriaud, A.: Vol. 207, pp. 207-226.
Sillion, B. see Mison, P.: Vol. 140, pp. 137-180.
Singh, R. P. see Sivaram, S.: Vol. 101, pp. 169-216.
Sivaram, S. and Singh, R. P.: Degradation and Stabilization of Ethylene-Propylene Copolymers and Their Blends: A Critical Review. Vol. 101, pp. 169-216.
Starodybtzev, S. see Khokhlov, A.: Vol. 109, pp. 121-172.
Steinke, J., Sherrington, D. C. and Dunkin, I. R.: Imprinting of Synthetic Polymers Using Molecular Templates. Vol. 123, pp. 81-126.
Stenzenberger, H. D.: Addition Polyimides. Vol. 117, pp. 165-220.
Stevenson, W. T. K. see Sefton, M. V.: Vol. 107, pp. 143-198.
Sumpter, B. G., Noid, D. W., Liang, G. L. and Wunderlich, B.: Atomistic Dynamics of Macromolecular Crystals. Vol. 116, pp. 27-72.

Suter, U. W. see *Gusev, A. A.*: Vol. 116, pp. 207-248.
Suter, U. W. see *Leontidis, E.*: Vol. 116, pp. 283-318.
Suter, U. W. see *Rehahn, M.*: Vol. 131/132, pp. 1-475.
Suzuki, A.: Phase Transition in Gels of Sub-Millimeter Size Induced by Interaction with Stimuli. Vol. 110, pp. 199-240.
Suzuki, A. and *Hirasa, O.*: An Approach to Artifical Muscle by Polymer Gels due to Micro-Phase Separation. Vol. 110, pp. 241-262.

Tagawa, S.: Radiation Effects on Ion Beams on Polymers. Vol. 105, pp. 99-116.
Tan, K. L. see *Kang, E. T.*: Vol. 106, pp. 135-190.
Tanaka, T. see *Penelle, J.*: Vol. 102, pp. 73-104.
Tanaka, H. and *Shibayama, M.*: Phase Transition and Related Phenomena of Polymer Gels. Vol. 109, pp. 1-62.
Tauer, K. see *Guyot, A.*: Vol. 111, pp. 43-66.
Teramoto, A. see *Sato, T.*: Vol. 126, pp. 85-162.
Terent´eva, J. P. and *Fridman, M. L.*: Compositions Based on Aminoresins. Vol. 101, pp. 29-64.
Theodorou, D. N. see *Dodd, L. R.*: Vol. 116, pp. 249-282.
Thomson, R. C., Wake, M. C., Yaszemski, M. J. and *Mikos, A. G.*: Biodegradable Polymer Scaffolds to Regenerate Organs. Vol. 122, pp. 245-274.
Tokita, M.: Friction Between Polymer Networks of Gels and Solvent. Vol. 110, pp. 27-48.
Tsuruta, T.: Contemporary Topics in Polymeric Materials for Biomedical Applications. Vol. 126, pp. 1-52.

Uyama, H. see *Kobayashi, S.*: Vol. 121, pp. 1-30.
Uyama, Y: Surface Modification of Polymers by Grafting. Vol. 137, pp. 1-40.

Vasilevskaya, V. see *Khokhlov, A.*: Vol. 109, pp. 121-172.
Vaskova, V. see *Hunkeler, D.*: Vol.:112, pp. 115-134.
Verdugo, P.: Polymer Gel Phase Transition in Condensation-Decondensation of Secretory Products. Vol. 110, pp. 145-156.
Vettegren, V. I.: see *Bronnikov, S. V.*: Vol. 125, pp. 103-146.
Viovy, J.-L. and *Lesec, J.*: Separation of Macromolecules in Gels: Permeation Chromatography and Electrophoresis. Vol. 114, pp. 1-42.
Vlahos, C. see *Hadjichristidis, N.*: Vol. 142, pp. 71-128.
Volksen, W.: Condensation Polyimides: Synthesis, Solution Behavior, and Imidization Characteristics. Vol. 117, pp. 111-164.
Volksen, W. see *Hedrick, J. L.*: Vol. 141, pp. 1-44.

Wake, M. C. see *Thomson, R. C.*: Vol. 122, pp. 245-274.
Wang, K. L. see *Cussler, E. L.*: Vol. 110, pp. 67-80.
Wang, S.-Q.: Molecular Transitions and Dynamics at Polymer/Wall Interfaces: Origins of Flow Instabilities and Wall Slip. Vol. 138, pp. 227-276.
Wang, T. G. see *Prokop, A.*: Vol. 136, pp.1-52; 53-74.
Whitesell, R. R. see *Prokop, A.*: Vol. 136, pp. 53-74.
Williams, R. J. J., Rozenberg, B. A., Pascault, J.-P.: Reaction Induced Phase Separation in Modified Thermosetting Polymers. Vol. 128, pp. 95-156.
Winter, H. H., Mours, M.: Rheology of Polymers Near Liquid-Solid Transitions. Vol. 134, pp. 165-234.
Wu, C.: Laser Light Scattering Characterization of Special Intractable Macromolecules in Solution. Vol 137, pp. 103-134.
Wunderlich, B. see *Sumpter, B. G.*: Vol. 116, pp. 27-72.

Xie, T. Y. see *Hunkeler, D.*: Vol. 112, pp. 115-134.
Xu, Z., Hadjichristidis, N., Fetters, L. J. and *Mays, J. W.*: Structure/Chain-Flexibility Relationships of Polymers. Vol. 120, pp. 1-50.

Yagci, Y. and *Endo, T.*: N-Benzyl and N-Alkoxy Pyridium Salts as Thermal and Photochemical Initiators for Cationic Polymerization. Vol. 127, pp. 59-86.

Yannas, I. V.: Tissue Regeneration Templates Based on Collagen-Glycosaminoglycan Copolymers. Vol. 122, pp. 219-244.

Yamaoka, H.: Polymer Materials for Fusion Reactors. Vol. 105, pp. 117-144.

Yasuda, H. and *Ihara, E.*: Rare Earth Metal-Initiated Living Polymerizations of Polar and Nonpolar Monomers. Vol. 133, pp. 53-102.

Yaszemski, M. J. see Thomson, R. C.: Vol. 122, pp. 245-274.

Yoon, D. Y. see Hedrick, J. L.: Vol. 141, pp. 1-44.

Yoshida, H. and *Ichikawa, T.*: Electron Spin Studies of Free Radicals in Irradiated Polymers. Vol. 105, pp. 3-36.

Zubov, V. P., Ivanov, A. E. and *Saburov, V. V.*: Polymer-Coated Adsorbents for the Separation of Biopolymers and Particles. Vol. 104, pp. 135-176.

Subject Index

Addition-fragmentation chain transfer polymerization 139
Aggregation number 108
-Aldehyde- -methacryloyl macromonomer 140
Aldol group transfer polymerization 43, 49
Aminolysis 184
Amphiphilic 13-16, 25, 28, 38-42, 50, 59, 65
Amphiphilic block copolymer 215
Amphiphilic microarm star 91, 106
Anionic polymerization 5, 29, 30, 45, 48, 63, 200, 216
Antisense oligonucleotide 222
Arborescent polymer 200
Arborols 181
ASTRAMOL 184
Atom transfer radical polymerization (ATRP) 204
ATRP 26

Backbone 41, 42, 45
Backfolding 194
Bacterial -galactosidase 221
Binary cluster integral 150
Biodegradable macromonomer 141
Biodegradable particle 161
1,3-Bis (1-phenylethyl)benzene 79
1,4-Bis (1-phenylethyl)benzene 72, 78
1,2-Bis (trichlorosilyl) ethane 85
Block copolymers 75, 106, 121, 122
Bootstrap effect 147
Boronation 223
Bottlebrushes, molecular 150, 151, 154, 206
Branching factor g 105
Branching functionality 181, 202
Brush adsorption 171
Butadienyl-ended macromonomer 141

Calixarene 24
Cascade 181

Cationic polymerization 200, 216
Chain-end modification 57-64
Chemical asymmetry 75, 78
Chloromethylation 200
Chlorosilane 77-81, 84, 85, 88, 89, 93, 96
CO_2-philic macromonomer 161
Comb-shaped polymer 42, 45, 48, 134, 149
Combburst polymer 200
Common good solvent 100, 102, 105, 106, 108
Common theta solvent 100, 102
Compatibility 144, 156
Composition distribution 146
Conformation 194
Conformational averages 100
Contour length 153
Convergent method 184
Copolymerization of macromonomer 145
Core destruction 10, 20, 25, 43
Core-shell structure 162
Core-shell type star polymer 139
Critical micelle concentration 39
Cross interactions 100, 102
Crosslinked core 4-17
Cyclophosphazine 91
Cyclosiloxane 34-37
Cytotoxin 222

Dendrimer 181
Dendrimer-antibody conjugate 222
Dendrimer-DNA complex 221
Dendrimer-peptide hybrid 219
Dendritic graft copolymer 203
Dichloromethylsilane 83
Diffusion coefficient 72, 106, 115, 116
Diffusion-controlled termination 143
Difunctional (divinylic) monomer 4-17
Dihydroxy-ended macromonomer 137
3-Dimethylaminopropyllithium 97
1,1-Diphenyl ethylene 79, 80, 95, 96, 98
Dispersion, non-aqueous 158
Dispersion polymerization 157

Divergent method 183
Divinylbenzene 72, 77-81, 89, 91, 104, 107, 123
Domb-Barret function 154
Double-comb polymer 135
Double-haired star polymer 135
Dynamic light scattering 151

Electron spray ionization mass spectrometry (ESI-MS) 188
Electrostatic repulsion 200
Emulsion polymerization 50, 52, 167
Epoxidized macromonomer 136
ESCA 171
ESR 142
Excluded volume 100
Excluded-volume effects 153
Excluded-volume parameter 150
Expansion factor 101, 153

Fast ion bombardment mass spectrometry (FAB-MS) 187
Flower-like conformation 134
Free radical polymerization 200
Free volume 155
Free-radical polymerization 43, 48-52, 55, 59, 64, 65
FTIR-ATR 171
Functionalized core 15

Glycopeptide macromonomer 141
Good solvent 104
Graft copolymer 41-65, 134
- -, micellar aggregate 156
Grafting from 41-44, 216
Grafting onto 44-47, 216
Grass transition temperature 155
Group transfer polymerization 3, 53
Growth, accelerated 186

Heteroarm star polymers 75
Hexamethylcyclotrisiloxane 93
Homogeneous nucleation theory 163
Homopolymerization of macromonomer 142
Hydrodynamic properties 108
Hydrodynamic radius 73, 109, 197
Hydrogenation 184
Hydrolysis 200
Hydrosilylation 34-37, 83, 88, 136
Hyperbranched polymer 65

Inifer technique 21
Initiator, multifunctional 17-30
-, orthogonal 138
Interface 114
Interfacial tension 112
Intrinsic viscosity 72, 104, 105, 108, 109, 197, 203
Inverse star block copolymers 98, 99

Janus-type star polymer 140

Kinetic chain length 142
Kratky-Porod chain 154
Kuhn segment length 151
Kuhn segment number 153

Light scattering, dynamic 151
Liquid crystal, lyotropic 155
Liquid crystalline comb polymer 140
Living cationic isomerization polymerization 51
Living cationic polymerization 81
Living coupling agent 39
Living free radical polymerization 138, 147
Luciferase 221
Lyotropic liquid crystal 155

Macromolecular coupling agent 76
Macromonomers 48-65, 81, 87, 91, 133, 134, 206
Mark-Houwink-Sakurada equation 151
Matrix-assisted laser desorption/ionization time of flight mass spectrometry (MALDI-TOF) 189
Mayo-Lewis equation 145
Mean-square radius of gyration 150
Megamolecules 74, 125
Melt viscosity 15
Mesogen-substituted macromonomer 140
Metal-free anionic polymerization 139
Methacrylate end group 136
Methyltrichlorosilane (trichloromethylsilane) 75, 76, 93, 97, 98
Micellar aggregate of graft copolymer 156
Micellar polymerization 144, 148
Micelles 107, 108, 117
Michael addition 184
Microemulsion polymerization 170
Microgel core 4-17
Microphase-separated copolymer 135

Subject Index

Microphases 112, 116, 119
Microsphere 135, 157
Molecular bottlebrushes 150, 151, 154
Molecular weight asymmetry 75
Monoclonal antibody 222
Monolayer 203
Monte Carlo simulation 102, 103, 105, 154
Morphology 112, 116-118
Multiarm star polymer 4-41
Multibin kinetic model 165
Multibranched radical 142
Multifunctional coupling agent 30-41
Multifunctional initiator 17-30
Multifunctional macromonomer 135
Mushroom-brush transition 171

NMR spectroscopy 156, 172, 187
Non-aqueous dispersion 158
-Norbornenyl macromonomer 138, 139
Nucleation mechanism 163
Nucleotide 221

Order-disorder transition 72, 110, 116, 120-122
Orthogonal initiator 138

1,3-Pentadienyl-terminated macromonomer 139
Persistence length 153
Phase boundaries 216
Phase diagram 112
Phenyl acetylene 209
Photobleaching, recovery after 197
Poly(amidoamine)s (PAMAM) 184
Poly(benzyl ether) 197
- dendrimer 184, 189, 211
Poly(carbosilane) dendrimer 191
Poly(ethylene oxide) macromonomer 139
Poly (L-lysine) dendrimer 195, 197, 215, 221
Poly(macromonomer) 134, 149, 155
Poly(paraphenylene) 208
Poly(phenylacetylene) 186
- dendrimer 191
Poly(propylene imine) 6
- dendrimer 215
Polyacrylate macromonomer 139
Polybutadienes, asymmetric 76
Polycondensation 137
Polydispersity 193
Polyisoprenes, asymmetric 77
Polymacromonomers, bulk properties 154
Polymer redistribution reaction 218

Polymer-dendrimer hybrid 211
Polymer-polymer reaction 146, 147
Polymers, asymmetric -functionalized 97
Polyolefin macromonomer 136
Polysiloxane macromonomer 141
Polystyrene macromonomer 137
Polystyryllithium 76, 78
Pour point 11
PS, end-capped 76, 85

Quaterpolymers 81, 96

Radical polymerization kinetics 142
Radioisotope 222
Radius of gyration 73, 105, 195
Reactivity ratio 145, 165
Recovery after photobleaching 197
Renomalization group theory 73, 105
Retrosynthetic analysis 182
Ring-opening methathesis polymerization 137
Rotaxanes 214

SAXS 151
Scaled excluded-volume parameter 153
Scanning force microscopy 203
SEC-DV 149
SEC-LALLS 149, 150
SEC-MALLS 149
Sedimentation experiment 151
Segment density 194, 200
Segmental volume fraction 199
Selective gas permeable membrane 156
Shear stability 12, 34
Shell thickness 203
Shrinkage factor 150, 205
Simulation 194
Size exclusion chromatography (SEC) 193
Smith-Ewart theory 170
Soap-free emulsion polymerization 169
Solvent, common good 100-108
-, theta 100, 102
Sphericity 194
Star homopolymers 77, 81
Star polymers, asymmetric 71, 74, 77, 100, 123, 124
- -, core-shell type 139
- -, double-haired 135
- -, heteroarm 75
- -, Janus-type 140
- -, multiarm 4-41
Star-block copolymers 79, 123

- - -, inverse 98, 99
Star-shaped polymer 134, 149
Starburst 181
Stars, amphiphilic microarm 91, 106
-, end-functionalized 79
-, symmetric 74, 103, 104
Static structure factor 110
Stereoregular macromonomer 139
Steric saturation 194
Sterically stabilized particle 164
Stiff backbone polymers 208
STM 155
Stockmayer-Fixman plot 149
Stretching free energy 112, 114
p-Styrylalkyl end group 137
Sugar-graft 141
Surface area per terminal group 199
Surface control 156
Suzuki conditions 208

Tapping scanning force microscopy 155, 171
Telechelic macromonomer 135
TEMPO 204
Tensile strength 15, 28, 29, 44, 59
Terminally-attached adsorption 171
Terpolymers 81, 93, 96
Tetrachlorosilane 85, 88, 89, 96
2,2,6,6-Tetramethylpiperidine oxide (TEMPO) 204

Thermoplastic elastomer 28, 45
Thermosensitive microsphere 160
Theta solvent 104
Topological asymmetry 75, 98
Toxicity 222
Toxin 222
Transamidation 219
Transcription 222
Transesterification 212
Transfection 221
Translational diffusion coefficient 197, 203
Tricontinuous cubic structure 116, 120

Umbrella copolymers 87, 88, 108
Umbrella star copolymer 88, 205
Uniform macromonomer 139, 151

Vergina star copolymer 19, 48
Vinyl ether-functionalized macromonomer 139
Viscoelastic behavior 115
Viscosity 11, 12, 42, 47
Williamson reaction 184, 211
Williamson substitution 210
Wormlike chain 153

Yamakawa-Stockmayer-Shimada theory 153

Springer and the environment

At Springer we firmly believe that an international science publisher has a special obligation to the environment, and our corporate policies consistently reflect this conviction.

We also expect our business partners – paper mills, printers, packaging manufacturers, etc. – to commit themselves to using materials and production processes that do not harm the environment. The paper in this book is made from low- or no-chlorine pulp and is acid free, in conformance with international standards for paper permanency.

Printing: Saladruck, Berlin
Binding: Buchbinderei Lüderitz & Bauer, Berlin